RB40.
Y0-BCF-267
Contemporary topics in analytica
3 3073 00289013 3

Contemporary Topics in

Analytical and Clinical Chemistry

Volume 2

A Continuation Order Plan is available for this series. A continuation order will bring delivery of each new volume immediately upon publication. Volumes are billed only upon actual shipment. For further information please contact the publisher.

Contemporary Topics in Analytical and Clinical Chemistry

Volume 2

Edited by
David M. Hercules
University of Pittsburgh
Pittsburgh, Pennsylvania

Gary M. Hieftje
Indiana University
Bloomington, Indiana

Lloyd R. Snyder
Technicon Instruments Corporation
Tarrytown, New York

and

Merle A. Evenson
University of Wisconsin
Madison, Wisconsin

PLENUM PRESS • NEW YORK AND LONDON

Library of Congress Cataloging in Publication Data

Main entry under title:

Contemporary topics in analytical and clinical chemistry.

Includes bibliographical references and index.
1. Chemistry, Clinical. 2. Chemistry, Analytical. I. Hercules, David M. [DNLM: 1. Chemistry, Analytical–Periodicals. 2. Chemistry, Analytical–Instrumentation–Periodicals. 3. Chemistry, Clinical–Periodicals. 4. Chemistry, Clinical–Instrumentation–Periodicals. W1 CO769ZE]
RB40.C66 616.07'56 77-8099
ISBN 0-306-33522-0 (v. 2)

A Division of Plenum Publishing Corporation
227 West 17th Street, New York, N.Y. 10011

Printed in the United States of America

Contributors

J. A. Carney Technicon Methods and Standards Laboratory, London, England

Lawrence J. Felice Department of Chemistry, Purdue University, West Lafayette, Indiana

R. W. Giese Northeastern University, Boston, Massachusetts

B. L. Karger Northeastern University, Boston, Massachusetts

Peter T. Kissinger Department of Chemistry, Purdue University, West Lafayette, Indiana

J. Landon Department of Chemical Pathology, St. Bartholomew's Hospital, London, England

D. J. Langley Technicon Methods and Standards Laboratory, London, England

David J. Miner Department of Chemistry, Purdue University, West Lafayette, Indiana

T. C. O'Haver Department of Chemistry, University of Maryland, College Park, Maryland

Carl R. Preddy Department of Chemistry, Purdue University, West Lafayette, Indiana

Ronald E. Shoup Department of Chemistry, Purdue University, West Lafayette, Indiana

Barton A. Smith Department of Chemistry, Harvard University, Cambridge, Massachusetts

L. R. Snyder Technicon Instruments Corporation, Tarrytown, New York

B. R. Ware Department of Chemistry, Harvard University, Cambridge, Massachusetts

Contents

1

Wavelength Modulation Spectroscopy

T. C. O'Haver

1. Introduction

The term *wavelength modulation* refers to the kind of spectroscopy experiment in which the wavelength to which the monochromator (or other spectral selection device) is tuned is repetitively scanned more-or-less rapidly over a discrete spectral interval called the modulation interval, $\Delta\lambda$. It is useful to distinguish this idea from *derivative spectroscopy,* which refers to the practice of recording or computing the first or higher wavelength derivatives of spectral intensity or absorbance with respect to wavelength. These two concepts are related in that wavelength modulation is a commonly used method of obtaining derivative spectra; it is widely known that if the modulation interval is small compared to the width of a spectral band, then the resulting ac component of the photosignal at the modulation frequency or its harmonics will be closely proportional to the wavelength derivatives. However, derivative spectra may be obtained in other ways not involving wavelength modulation, and there are other applications of wavelength modulation which do not involve recording derivative spectra. For example, as we will see in this chapter, wavelength modulation can be used to measure the intensity of an atomic spectral line superimposed on an intense background. In such applications the modulation intervals commonly used are far too large to produce a true wavelength derivative of the spectral line. In another application, wavelength modulation can be used in fluorescence spectrometry as a component-resolution technique to record the normal (not derivative) spectra of individual components in a mixture of overlapping fluorescence spectra. Since these latter applications

---

T. C. O'Haver • Department of Chemistry, University of Maryland, College Park, Maryland 20742

of wavelength modulation normally involve large modulation intervals and since the application to derivative spectroscopy necessarily involves small modulation intervals (compared to the width of the spectral lines or bands being studied), we will treat the large modulation case and the small modulation case individually. For completeness, we will also consider derivative techniques not involving wavelength modulation.

2. Historical Background

Britton Chance was evidently first to use the wavelength modulation principle in spectrophotometry. As part of a long-term research effort to develop better spectrophotometric instrumentation to measure small, rapid changes in the absorbance of solutions of enzymes, Chance had described in 1942 a dual-wavelength filter photometer which measured the transmission of a solution at two wavelengths *simultaneously*, using a single light source with two filters and two phototubes.[1] But this system was not suitable for investigation of enzymes which could only be obtained as a turbid suspension because the light loss due to scattering was excessive. Thus, in 1951, Chance described a dual-wavelength system in which a single phototube (by then, a photomultiplier was used) was placed as close to the sample cell as practical.[2] By itself, this approach minimized scattered light losses, but it no longer allowed simultaneous measurements at two wavelengths. Rather, the system employed an oscillating mirror to direct the exit beams from two monochromators *alternately* through the sample cell. Appropriate electronics demodulated the resulting time-shared photosignal. This was essentially a wavelength modulation spectrometer using a more-or-less square-wave modulation waveform. Chance clearly demonstrated the idea of "balancing out" the absorption of an interfering band by setting the monochromators to two different wavelengths at which the absorption of the interfering component is the same.

Chance's two-monochromator system was a bit bulky and expensive, so the idea of modifying a single monochromator so its wavelength could be modulated rapidly was an attractive one. This seems to have been done first by Hammond and Price,[3] who used a vibrating Littrow mirror to effect the modulation and a tuned amplifier to selectively detect the resulting modulation in the photocurrent. The system was used to reduce the effects of stray light in the spectrometer.

At this point the further development of the wavelength modulation principle split into two more-or-less separate paths: the dual-wavelength approach based directly on Chance's work, and the modulated-monochromator approach represented by the instrument of Hammond and Price. The difference is essentially one of modulation waveform, the former approach resulting in a square waveform and the latter generally resembling a sinusoidal waveform due to the natural oscillatory properties of vibrating masses. The dual-wavelength approach has remained a widely used instrumental technique to this day; several commercial instruments are now and have been available for years. The modulated-monochromator approach is less well represented commercially, but conventional monochromators can be modified for wavelength modulation by the experimenter if desired. The dual-wavelength approach does have the advantage of greater versatility, since it is usually possible to set either of the two wavelengths independently to any value within the range of the instrument. On the other hand most schemes for modulation of a single monochromator restrict the modulation interval, $\Delta\lambda$, to a comparatively small value, although of course this interval can be positioned anywhere within the instrument's wavelength range. This limitation is of no consequence when it is desired to obtain wavelength modulation derivative spectra; in fact if a dual-wavelength instrument is to be used for this purpose, some means of synchronizing the wavelengths of the two monochromators to a fixed displacement while they are being scanned must be provided. This is a standard feature of the modern commercial models.

The idea of obtaining derivative spectra by wavelength modulation dates from about 1954. Morrison[(4)] used an analogous idea in studying electron impact ionization efficiency curves, i.e., modulation of the ionizing electron voltage. He explained how to obtain second derivatives by utilizing second-harmonic detection and included a discussion of the expected signal-to-noise advantages of the modulation technique. In the same year, Stacy French and co-workers[(5)] described an optical absorption spectrophotometer in which wavelength modulation was produced by vibrating one of the slits. There had been at that time some interest among applied spectroscopists in the problems of interpreting systems of overlapping absorption bands.[(6)] Giese and French proposed in 1955 that measurements of the first derivative of transmission with respect to wavelength would be useful in studying complex overlapping bands.[(7)] The possible use of the derivative techniques for *quantitative* analysis, however, was not discussed. This idea was investigated in 1956 by Collier and Singleton[(8)] in a study of the quantitative infrared analysis of mixtures. These workers preferred to use electronic differentiation rather than wavelength modulation, in part because of the mistaken notion that it would not be possible to obtain second derivatives by the latter approach. However, they explained clearly for the first time two important properties of the derivative technique: the idea of making measurements for one component at the "cross-over" point of an interfering component (subsequently called the *zero-crossing* technique), and the idea that in mixtures where the nature of the interference is unknown, measurements based on the second-derivative central maximum generally provide improved accuracy compared to measurements of total absorption at the analyte maximum. This latter idea turns out to be quite correct for spectral bands of Lorentzian shape because of the comparatively small ratio of the heights of the negative-side lobes to the positive central maximum in the second derivative.[9] However, the band shapes measured by Collier and Singleton were dominated by a distorted triangular slit function for which no systematic predictions could be made. Nevertheless, they did point out that for very complex mixtures, one might expect some degree of cancellation of negative lobes of one interference with positive lobes of another, and thus might generally obtain somewhat better accuracy second-derivative measurement. This they did obtain for very complex (40–50 components) mixtures of crude tar phenols, for which it proved possible to determine several major components with some accuracy (1–20%).

Although Collier and Singleton,[8] and subsequently other workers,[10–15] found electronic differentiation useful, the wavelength modulation technique has received more attention.[16] The most popular and, in most cases, the most convenient method of wavelength modulation is the oscillating transparent refractor plate technique introduced in 1959 by McWilliam.[17] This is an especially useful technique for obtaining small, well-controlled wavelength modulation intervals. Larger modulation intervals are more easily obtained by means of an oscillating mirror, prism, or grating.

3. Theoretical Aspects

3.1. The Small Modulation Case: Derivative Spectroscopy

The important questions about derivative spectroscopy to be addressed theoretically are: how are derivatives produced by wavelength modulation; how does differentiation affect signal-to-noise ratio; and what selectivity advantages are to be expected?

The first of these questions has received considerable attention in the literature[18–25]; the most detailed and general analyses have been given by Hager and Anderson.[24] Essentially the process of differentiation by wavelength modulation may be considered as a form of spatial frequency filtering of the high-pass, low-cut kind. That is, in a derivative spectrum, the high-spatial-frequency components (fine structure) are emphasized and the low-spatial-frequency components (broad features) are attenuated. In the limit of small wavelength modulation interval, the recorded derivative approaches the true mathematical derivative. All of this is, of course, intuitively obvious, but the theoretical work does allow a more rigorous understanding, permits calculation of the "distortion" caused by finite modulation intervals, and provides a comparison of square and sinusoidal modulation waveforms (sinusoidal is marginally better in the sense that the true derivative is more closely approached at a given modulation interval). Hager and Anderson[24] showed that the spatial frequency filters are Bessel functions of the first kind, of order equal to the order of the derivative. They also derived a generalized instrument function for the *m*th derivative.

The question of the effect of differentiation on signal-to-noise ratio (SNR) is an important one which unfortunately has been largely neglected. There are two parts to this question: first, how does the SNR of a derivative spectrum compare to the normal spectrum; and second, how do different methods of obtaining derivatives compare in SNR, i.e., wavelength modulation vs. electronic differentiation. The second aspect has been addressed in a strictly qualitative way by Bonfiglioli and Brovetto,[18] who argued that wavelength modulation would in general provide better SNR than electronic differentiation. On the other hand Gunders and Kaplan[19] concluded that there was no significant difference between these techniques, but their analysis was based on some rather drastic simplifying assumptions, not the least of which was that only photon shot noise was considered. It now seems clear that wavelength modulation can be expected to reduce the effect of low-frequency noises which are additive with respect to the photosignal modulation. Such noises are common in absorption measurements; some examples are source fluctuation noise and background absorption fluctuation noise. In measurements dominated by noises of this type, one may expect that the SNR of the derivatives obtained by wavelength modulation will be superior to those obtained by electronic differentiation and may even be superior to the SNR of the normal spectrum. This is due essentially to the independence of the modulation frequency and the cutoff frequency of the low-pass filter following the phase detector; the former can be made relatively high to reduce the effect of low-frequency fluctuation noises while the latter can be made much lower to attenuate high-frequency noises.[22,26,27] In contrast, there is no way electronic differentiation can distinguish between a change in signal caused by scanning through a genuine spectral feature and one caused by a strictly temporal intensity fluctuation. This disadvantage may no longer be significant if the predominant noise is a white noise such as photomultiplier shot noise. Thus, O'Haver and Green[15] have found that in condensed-phase fluorescence spectrometry, wavelength modulation and electronic differentiation give equivalent results. They also found, both experimentally[15] and on the basis of a computer simulation[28] that the SNRs of the normal, first, and second derivative spectra decrease by about twofold in each successive order of differentiation. This result is not necessarily of general applicability, however, and in fact one would expect this to be very much a function of the extent of data smoothing, i.e., low-pass filtering, etc., employed.

The question of selectivity advantage is a central one which has received disproportionately little theoretical attention. Bonfiglioli and Brovetto[18] gave a quantitative treatment for the case of a Gaussian peak superimposed on an exponential background. They predicted a 4- to 20-fold improvement in peak detectability in the first-derivative mode, depending on the particular case. O'Haver and Green[28] gave a more extensive analysis of the case of a Gaussian peak superimposed on a Gaussian background. The selectivity advantage was found to be mostly a function of the peak-to-background *width ratio*; if the background is broader than the peak, there is usually some selectivity advantage in the derivative mode, depending on the position of the peak relative to the background maximum.

One might reasonably expect that, since the magnitude of the derivative of a peak is inversely proportional to its width, the selectivity advantage would be given simply by the width ratio (or, for the second derivative, the square of the width ratio). In fact, the realizable gain is often greater because it is possible to optimize the selection of one of several possible ways to measure the derivative "intensity." Some of these derivative measures are shown in Figure 1. In this figure the band to be measured (the "analyte" band) is the narrow one located at $X = 5$, and the larger band to the left is the interfering background band. In this example the analyte band has a height and width half that of the interfering band. In the first and second derivatives, we see the features of the analyte spectrum emphasized twofold and fourfold, respectively, just as we would expect on the basis of the width ratio. But when we attempt to utilize these spectra for quantitative analysis purposes, we find that there are several different ways we can derive quantitative measures from these curves. For example, in the normal spectrum, the simplest approach would be just to measure the total intensity (or absorbance, as appropriate) at the analyte maximum (TI) and ignore the interfering band. This measure will clearly be subject in this case to a large *systematic error*, which we define here as the relative percent difference between the measure taken with and without the interfering band. A popular alternative is the *tangential baseline* method, labeled OB, although in this and some other cases the baseline cannot be drawn uniquely because of the lack of the required points of tangency. The baseline method can also be applied to the first and second derivatives (1B and 2B, respectively). In addition, we can measure the "peak-to-peak" distance between adjacent maxima and minima in the derivatives (1P, 2PN, and 2PF). All of these measures have different systematic errors and only the best for each derivative need be retained. The selectivity advantage of a derivative measurement can then be expressed as the ratio of the systematic error of the derivative measure to that of the TI measure. This ratio

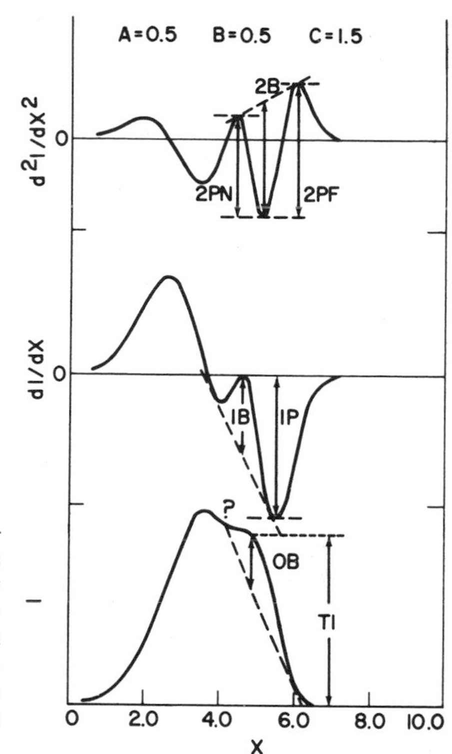

Figure 1. The zeroth, first, and second derivatives of a pair of overlapping Gaussian bands. The "analyte" band, whose intensity we seek to measure here, is located at x = 5 and has a height and width one half that of the "interfering" band located at x = 3.5. The arrows indicate various ways one might graphically obtain a measure of the intensity of the analyte band.

is plotted against the background/analyte width ratio in Figure 2 for the 1P and 2B measures of an analyte band which is positioned near the maximum of a background band of equal height. The dotted and dashed lines represent error ratios equal to the width ratio and the square of the width ratio, respectively. The observed error ratios are always less than that predicted from the width ratios alone. This is generally the case for the best measures if the width ratio is greater than about 1.5. For width ratios less than this, the derivative measures of Figure 1 do not work well. An alternative is to use the zero-crossing technique of Collier and Singleton.[8]

The decision as to whether a derivative technique will be beneficial depends on the trade-off between random and systematic errors; the selectivity advantage will result in a reduction of systematic errors due to overlapping interfering bands, but the reduced SNR will increase random measurement errors. We would not, therefore, expect an improvement in over-all error unless the systematic errors are reduced more than the random errors are increased. If we make, for the sake of argument, the assumption that the random errors will increase by a factor of 2 for each

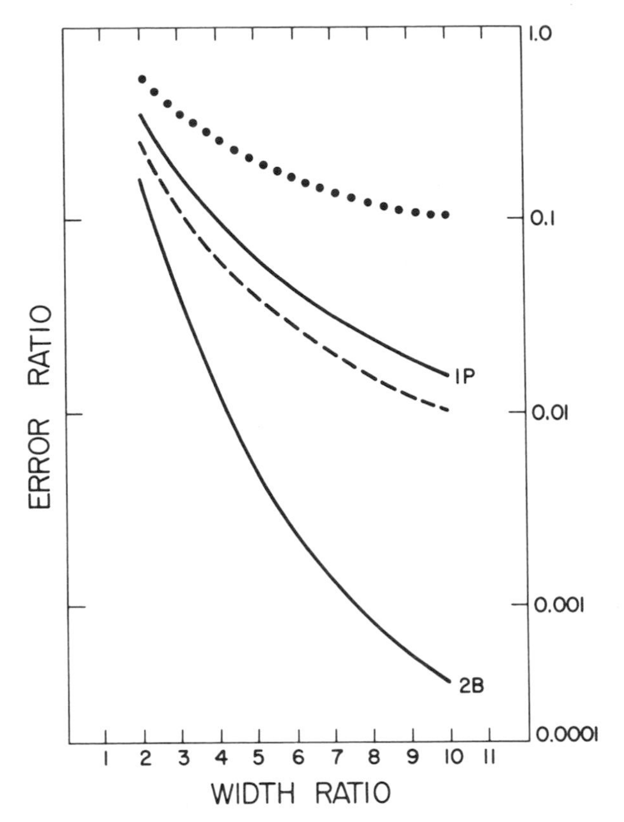

Figure 2. The ratio of the systematic errors of graphical derivative measures to those of the total intensity (TI) measures, as a function of the ratio of the width of the interfering ("background") band to that of the analyte band. The two solid lines represent the error ratios of the 1P (first derivative peak-to-peak) and 2B (second derivative baseline) measures shown in Figure 1. For comparison, the dotted and dashed lines illustrate the expected behavior for an error ratio which is inversely proportional to (dotted line) and inversely proportional to the square of (dashed line) the width ratio. In every case the error ratio is smaller (i.e., derivative measures are better) at larger width ratios (i.e., when the analyte band is narrower). The 1P error ratio is always lower than reciprocal of the width ratio and the 2B error ratio is lower than the reciprocal of the square of the width ratio.

order of differentiation,[28] and if we generalize from Figure 2 that the reduction in systematic error will be *at least* equal to the width ratio (or its square, for the second derivative), then it is clear that we can generally expect some over-all improvement if the width ratio is at least two and if the random noise error in the normal (nonderivative) measurement is no greater than the systematic error. This was found to be the case for several overlapping Gaussian band pairs for which the signal-to-noise ratio of the normal spectra were very good (100–200, based on the rms noise). For a width ratio of two, the total error (a weighted average of systematic error and random error) was found to be improved by 3- to 60-fold, depending on the height ratio and band separation, whereas for a width ratio of 10, the improvement was 30- to 130-fold. Naturally, if the initial SNR is poorer, the improvement will be less dramatic.

For width ratios near unity, the zero-crossing technique[8] may be the only useful derivative method. Although theoretically this method suffers no systematic error if the wavelength of measurement is properly chosen, random errors can occasionally be very large.[28]

3.2. The Large Modulation Case

In the large modulation case the modulation conditions are generally chosen to optimize signal-to-noise ratio. The important variables are the shape of the measured spectral peak; the frequency, waveform and interval of modulation; the nature of the dominant noise sources. The optimum value of the interval of modulation, $\Delta\lambda$, is related to the full width of the peak, $\delta\lambda$, being that defined by the spectral bandpass of the measuring monochromator or by the true spectral profile width. Balslev[29] reasoned that $\Delta\lambda$ should be made equal to $\delta\lambda$, while Snelleman[26] preferred to use a $\Delta\lambda$ "a few times larger" than $\delta\lambda$. A more detailed analysis[30] shows that the optimum $\Delta\lambda$ depends on the modulation waveform, the detection mode (i.e. first or second harmonic), and to some extent on the nature of the dominant noise source. The second-harmonic ("2F") mode is generally favored, because of its ability to reject a sloping linear background. For sinusoidal modulation waveforms and 2F mode detection, the maximum signal is obtained at $\Delta\lambda = 2\,\delta\lambda$ for either Gaussian, Lorentzian, or triangular peak shapes. If signal-carried photon shot noise is dominant, such as in atomic fluorescence measurements in the absence of background, then the maximum signal-to-noise ratio is obtained at $\Delta\lambda = 2.5\,\delta\lambda$. In the 1F mode slightly smaller optimum $\Delta\lambda$'s are predicted. For square modulation there is no optimum $\Delta\lambda$ in the usual sense, since the SNR increases asymptotically with $\Delta\lambda$. The possibility of spectral interferences with nearby lines limits the maximum $\Delta\lambda$ employed in practice.

As in the small modulation case, the effect of modulation on the amplitude of the signal and on the SNR depends on the nature of the major noise sources. One expects an improvement in SNR if low-frequency additive noises are dominant. On the other hand, if the dominant noise is photon shot noise caused by a continuum background emission upon which the desired spectral line or band is superimposed, then the noise will not be influenced by wavelength modulation. However, the signal, and thus the SNR, will be reduced two- to fourfold, depending on the modulation waveform, profile shape, and detection mode. The advantage of square-wave modulation is twofold in the 2F mode. If signal-carried photon shot noise is dominant, as in the measurement of an isolated atomic emission or fluorescence line or band without background, the reduction in the average dc level of the photosignal caused by modulation will result in a corresponding reduction in shot noise and the SNR will be reduced by roughly the square root of the factors in Table 1. In either of these latter two cases, wavelength modulation will actually degrade SNR to some extent.

4. Experimental Techniques

4.1. Wavelength Modulation

Most schemes for producing wavelength modulation are based on appropriately modifying a dispersive spectrometer, for example by the addition of a rotating or oscillating mirror. This approach has been used to obtain square wave[2,21] and sinusoidal[3,18,20,23,25,26,31] modulation. Oscillation of one of the slits is also a possibility.[5,29,32–37] This normally results in sinusoidal modulation; however, square-wave modulation can be achieved by using two adjacent stationary exit (or entrance) slits and a rotating mask which allows light to pass through the slits alternately.[38]

One of the most popular techniques for wavelength modulation has been the use of a vibrating or rotating refractor plate placed inside the monochromator or polychromator.[17,22,39–50] The technique requires a minimum of modification of the spectrometer. In the most common arrangement, a thin plane-parallel quartz plate is mounted just inside the entrance slit and is either rotationally oscillated about an axis parallel to the slit or is moved into and out of the beam by means of a vibrating vane or tuning fork.[49] Refraction of the entrance beam by the plate causes a lateral displacement of the beam and a corresponding displacement of the spectral image at the focal plane in the direction of wavelength dispersion. For small angles, the displacement, d, is given by

$$d = \alpha t \left(1 - \frac{1}{n}\right)$$

where d is the displacement in mm, t is the thickness of the plate in mm, α is the angle of incidence in radians, and n is the refractive index of the plate material. The resulting wavelength modulation interval can then be calculated knowing the linear dispersion of the spectrometer. The presence of the plate also increases the effective focal length of the spectrometer to an extent which depends on the plate thickness and the speed of the

Table 1. Applications of Wavelength Modulation to Flame and Graphite Furnace Atomic Emission Spectrometry

Element	Line (nm)	Detection limit (μg/ml)	Sources of spectral interference other than atomizer background	Atomizer	Reference
Al	396.2	0.01	1000 μg/ml K	N_2O/C_2H_2	74
	396.2	0.3		N_2O/C_2H_2	73
	396.2	0.00004		Graphite furnace	75
Ba	553.6	5[a]	1000 μg/ml Ca	H_2/O_2	42
		0.2[a]	2000 μg/ml Ca	N_2O/C_2H_2	50
		0.00008		Graphite furnace	75
Ca	422.7	0.5–3	1000 μg/ml Li, Na, K, Rb, Mg, Fe	H_2/O_2	42
	422.7	0.009	1000 μg/ml La	N_2O/C_2H_2	52
Cr	425.4	0.0004		Graphite furnace	75
	425.4	0.1		N_2O/C_2H_2	73
	425.4	0.004	1% Na	N_2O/C_2H_2	50
	425.4	1.3[a]	1% fly ash matrix, 0.1% K	N_2O/C_2H_2	47
Cu	324.7	0.0004		Graphite furnace	75
	324.7	1.2[a]	1% fly ash matrix	N_2O/C_2H_2	47
Ga	417.2	0.003			50
K	766.5	0.03–2	1000 μg/ml Li, Na, Rb, Mg, Ca, Fe	H_2/O_2	42
Li	670.8	0.02	1000 μg/ml Na, K, Rb, Mg, Ca, Fe	H_2/O_2	42
Mg	285.2	0.002		Graphite furnace	75
Mn	279.4	2.2		N_2O/C_2H_2	73
	403.09	5[a]	1% fly ash + 0.1% K	N_2O/C_2H_2	47
Na	589.0	0.000004		Grahpite furnace	75
Ni	341.48	1[a]	1% fly ash	N_2O/C_2H_2	47
	352.4	0.004		Graphite furnace	75
Pb	405.78	0.7[a]	1% fly ash	N_2O/C_2H_2	47
	405.78	0.06		Graphite furnace	75
Sr	460.73	13[a]	1% fly ash + 0.1% K	N_2O/C_2H_2	47
Ti	396.4	4.4		N_2O/C_2H_2	73
	399.9	0.01		Graphite furnace	75

[a] Detection limits not reported. This is the concentration in solution actually measured with an SNR much greater than 2. Thus, detection limits are predicted to be much lower than these concentrations.

spectrometer. This effect may in some instances be negligible,[48] but if not, the spectrometer may be refocused to compensate.[50]

A low-cost alternative to a modulated dispersive spectrometer is a tilting interference filter photometer.[51,52] The large throughput of filters is an advantage of this approach, particularly for low-light-level applications.[51]

One of the potential sources of error with any of the modulation schemes discussed is that in the process of obtaining wavelength modulation, they may inadvertently produce some direct intensity modulation as well. Thus, even a completely flat structureless background spectrum would give rise to an AC photosignal component which could not be distinguished from a signal resulting from genuine spectral slope or curvature in the modulation interval. This is a particular problem with the vibrating slit devices because of uneven slit illumination in the direction of motion at the slit (or, in the case of a vibrating *exit* slit, uneven sensitivity of the photocathode). Further, in systems utilizing alternately masked fixed slits, the slit areas must be accurately controlled. In the vibrating-mirror approach, intensity modulation may be expected for large modulation intervals if the camera lens or mirror collects only a portion of the dispersed beam from the grating or prism; this would generally be the case for monochromators, which typically have collimator and camera elements of equal width, but not in spectrographs or polychromators. Masking the collimator can reduce this effect.[53] In the case of the refractor-plate modulator, intensity modulation results from the reflective properties of the air/quartz interface. This is especially serious in the systems which produce square-wave modulation by moving a tilted plate into and out of the beam.[40] In the more common arrangement, however, the plate stays in the beam all the time, and one must contend only with the *change* in reflection with the angle of incidence. This change is fortunately small at small angles,[48] and in any case it is possible[52] to reduce the effect even further by adjusting the rest angle of the plate so that the intensity modulation occurs predominantly at the first harmonic of the modulation frequency and the spectral signal occurs at the second harmonic (or vice versa). The lock-in amplifier will then reject the intensity modulation almost completely.

Wavelength modulation may also be achieved by direct oscillation of the dispersing element.[15,54] This rather straightforward approach has been applied to grating monochromators when very large modulation intervals (up to 100 nm) are required.[55] Modification of the drive mechanism is obviously required and the frequency of modulation is restricted ($\leq$15 Hz) because of the inertia of the grating. However, the technique is free from adventitious intensity modulation, except insofar as the transmission factor of the monochromator is a function of wavelength.

At the other extreme of wavelength modulation interval are those techniques used when extremely high resolution and small modulation intervals are required, such as in continuum-source atomic-absorption spectrometry. For this purpose, it is possible to wavelength-modulate a Fabry-Perot interferometer by attaching the etalon plates to piezoelectric transducers.[56] Wavelength modulation can also be produced without resorting to mechanical moving parts by electronic modulation of an image detector. In this way derivative spectra have been obtained with a vidicon tube[57] and with an image-dissector photomultiplier tube.[58]

We should also mention here the dual-wavelength approach to wavelength modulation, which normally results in square-wave modulation and ordinarily is restricted to the first-harmonic mode.[2,21,59–62] Dual-wavelength absorption spectrometers are commercially available, but tend to be rather costly.

The selection of the optimum waveform and frequency of wavelength modulation often involves a trade-off between theoretical advantages and practical advantages. On the basis of signal amplitude and, more important, signal-to-noise ratio, square-wave modulation clearly has the advantage if lock-in signal processing is utilized. The SNR advantage compared to sinusoidal modulation can be as much as a factor of 1.8.[30] However, it is more difficult to generate square-wave signals at high frequencies by wavelength modulation, particularly in the 2F mode. The vibrating-mirror, refractor-plate, or slit systems are not normally fast enough to permit square-wave modulation except at low frequencies.[50] Systems using tilted refractor plates which moved into and out of the beam[49] can produce square-wave modulation in the 1F mode at sufficiently high frequencies, but are unsuitable for 2F mode operation.

Sinusoidal (including sine, triangular, and trapezoidal) modulation, on the other hand, has the following advantages: (1) sinusoidal oscillation is the "natural" waveform of vibrating bodies and is therefore mechanically easier to generate than square or other waveforms, particularly at high modulation frequencies; (2) the sinusoidal waveform permits a real-time $x-y$ oscilloscope display of the entire spectral profile vs. wavelength (not possible with square-wave modulation); (3) the same waveform, at a slightly different amplitude, can be used in both first (1F) and second (2F) harmonic detection modes; (4) sinusoidal modulation is easily performed in a manner which does not modulate the background intensity to a detectable extent; and (5) the modulation interval $\Delta\lambda$ is continuously adjustable.

With respect to the *frequency* of wavelength modulation, the highest practical frequency is generally desirable to reduce the effects of low-frequency noise and background signal transients.[22,26] In analytical flames, low-frequency noise has been observed at frequencies below about 50 Hz. In "nonflame" electrically heated furnace atomizers, very large background signals are often observed, due to thermal emission from the atomizer walls and/or background absorption and scatter from matrix constituents in complex samples. Steady background signals of this type are easily corrected by wavelength modulation at any frequency. However, in electrically heated atomizers, these background signals occur as transients, often of quite short duration. Transient (pulse-type) signals have frequency spectra which are weighted toward low frequencies, like low frequency noise; but the shorter the transient duration, the more the frequency spectrum spreads out toward higher frequencies. Thus, higher modulation frequencies may be necessary in some cases to avoid spurious response due to these transients.

4.2. Derivative Techniques Not Involving Wavelength Modulation

The use of an electronic differentiator to obtain derivative spectra[8,10–15] is based on the proportionality between time derivatives and wavelength derivatives at a constant scan rate:

$$\frac{\delta A}{\delta t} = \frac{\delta A}{\delta \lambda} \frac{\delta \lambda}{\delta t}$$

where $\delta A/\delta t$ is the time rate of change of absorbance (or intensity), $\delta A/\delta \lambda$ is the slope of the normal spectrum, and $\delta \lambda/\delta t$ is the scan rate. A significant disadvantage of the electronic differentiation technique is that the scan rate must be constant and it is not possible to monitor the derivative at one point in a spectrum as a function of time. Also, as mentioned before, the signal-to-noise ratio of an electronic derivative is in many cases inferior to a wavelength modulation derivative.[18,22,26] However, the electronic approach is simple and inexpensive and requires no modification of the spectrometer. Modern differentiators are usually designed around operational amplifiers.[15]

In some types of recording spectrometers, a derivative signal may be obtained from the recorder-pen tachometer generator used to critically damp the pen motion. The output of the tachometer is an ac signal whose amplitude is proportional to the pen velocity and thus to the derivative of the spectrum recorded. Appropriate synchronous (phase) detector and filter circuitry can be used to convert this signal into a recordable derivative. This technique has received relatively little attention.[63–65]

Numerical differentiation techniques combined with some sort of data smoothing have been widely used by spectroscopists[66–71] and will no doubt see increased use as microprocessors become a standard part of the electronics of spectrometric instruments.

5. Applications

5.1. Atomic Spectroscopy

In this section we consider applications of derivative and wavelength modulation to the spectroscopy of atoms in the gas phase. In these applications the object is to measure the intensity of an isolated atomic emission, absorption, or fluorescence line, usually superimposed on an intense, probably unstable, background. In general the most useful technique will involve wavelength modulation across the entire line profile, which we have called the large modulation case.

5.1.1. Emission

Flame Atomizer. In flame atomic-emission spectrometry (FAES), wavelength modulation can reduce the effect of flame background and matrix emission[47,48,52,72–74] and can often provide a modest (typically twofold) improvement in precision[47,52,73] and reduction in detection limits.[52,72] Useful secondary benefits include a reduction in the solution volume required to obtain a background-corrected emission reading and a considerable increase in the speed and convenience with which instrumental conditions such as burner position and flow rates can be optimized for maximum analyte line emission intensity. Published applications of wavelength modulation to FAES, organized by element and type of interfering background, are summarized in Table 1. In a few cases standard reference materials have been analyzed with the aid of FAES wavelength modulation techniques; these include aluminum in steel, iron, and phosphate rock,[74] calcium in blood serum and in limestone,[52] and several elements in fly ash.[47] Pertinent details are given in Table 2.

A unique capability of wavelength modulation is the cancellation of line-overlap spectral interference using the zero-crossing technique.[47] This idea is an extension of a technique described by Collier and Singleton.[8] The requirements for its use are that the wavelength of the interfering line must be known; however, its intensity need not be. Measurements are taken at the zero-crossing of the second-harmonic response function of the interfering band. The technique has been used in several modifications of atomic spectroscopy; line pairs which have been dealt with in this way are summarized in Table 3. Note that the technique is applicable even when the separation of the two lines is much less than their width.

Furnace Atomizer. Wavelength modulation is an ideal way to compensate for the large, transient blackbody wall emission observed in graphite furnace atomic-emission spectrometry. Epstein *et al.*[75] have shown that detection limits can be improved 10- to 300-fold when wavelength

Table 2. Analysis of Reference Samples by Wavelength Modulation Spectrometry

Element	Line (nm)	Method[a]	Sample	Found	Comparison value	Reference
Al	396.2	FAES	SRM 20 g steel	397 ± 6 μg/g	400[b] μg/g	74
			SRM 59a ferrosilicon	3420 ± 10 μg/g	3500[b] μg/g	74
			SRM 1134 high-Si steel	3322 ± 40 μg/g	3290[b] μg/g	74
			SRM 1142 ductile iron	883 ± 2 μg/g	890[b] μg/g	74
Ca	422.7	FAES	SRM 1a limestone	322.3 ± 1.5 μg/ml	324[b] ± 0.9 μg/ml	52
			Moni-Trol blood serum	93.5 ± 0.9 μg/ml	94 ± 2 μg/ml	52
Cd	228.8	FAAC	Ni alloy	1.1 ± 0.3 μg/g	1.2 μg/g	85
Cu	324.7	GFAES	Spinach	11.5 ± 0.4 μg/g	12.1 ± 0.1 μg/g	75
		GFAES	Pine needles	2.8 ± 0.3 μg/g	2.9 ± 0.1 μg/g	75
		FAAC	Pine needles	3.0 ± 0.3 μg/g	2.9 ± 0.1 μg/g	83
		FAAC	Tomato leaves	10.9 ± 0.1 μg/g	10.8 ± 0.1 μg/g	83
		FAES	SRM-1633 fly ash	118 ± 2[c] μg/g	120 μg/g	47
Cr	425.43	FAES	SRM-1633 fly ash	129 ± 1.8 μg/g	132 ± 5[b] μg/g	47
Mg	285.2	FAAC	Ni alloy	2.6 μg/g	3 μg/g	85
				8.1 μg/g	10 μg/g	85
Mn	403.08	FAES	SRM-1633 fly ash	519 ± 4 μg/g	495 ± 30[b] μg/g	47
Na	589.0	GFAES	Water	0.1 ± 0.03 ng/ml	0.1 ng/ml	75
Ni	341.5	FAES	SRM-1633 fly ash	95 ± 5 μg/g	98 ± 3[b] μg/g	47
	361.9	FAES	SRM-1633 fly ash	102 ± 18 μg/g	98 ± 3[b] μg/g	47
Pb	216.9	FAAC	Ni alloy	4.4 μg/g	4.6 μg/g	85
			SRM-1633 fly ash	10.9 μg/g	9 μg/g	85
	405.78	FAES	SRM-1633 fly ash	76 ± 9 μg/g	70 ± 2[b] μg/g	47
Sr	460.73	FAES	Ni alloy	1340 ± 13 μg/g	1380[c] μg/g	47
Zn	213.8	FAAC		11.5 μg/g	12 μg/g	85
			Copper	33.5 μg/g	36 μg/g	85
				411 ± 34 μg/g	409[d] ± 10 μg/g	85
				10.3 ± 0.8 μg/g	10.9[d] ± 0.5 μg/g	85

[a] FAES = flame atomic emission spectrometry; FAAC = flame atomic absorption spectrometry; GFAES = graphite furnace atomic emission spectrometry.
[b] NBS certified value.
[c] Information value—not certified.
[d] Line-source AA value after electrolytic separation of Cu.

Table 3. Atomic Line-Overlap Interferences Which Have Been Compensated by Wavelength Modulation

Analyte		Interference		Separation (nm)	Spectrometer bandpass (nm)	Reference
^6Li	670.8	^7Li	670.8	0.015	0.02	79
Cr	359.35	Co	359.49	0.14	0.4	47
Fe	361.88	Ni	361.94	0.062	0.08	47
Co	360.208	Ni	360.228	0.02	0.08	47
Co	388.19	CN	388.34[a]	0.15	0.4	47
Ni	352.45	Co	352.34	0.11	0.13	47
		Sr^+	352.68	0.23		
Ca	422.7	Fe	421.6	1.1	13	52
Zn	213.856		213.850	0.003	0.002[b]	85
Zn	213.856	Cu	213.853	0.003	0.002[b]	85

[a] Molecular band.
[b] Total line FWHM (spectrometer bandpass and true line profile width) ≅ 0.003 nm.

modulation is utilized. Representative data obtained by this technique are included in Table 1.

Plasma Atomization. Background emission problems can be as bad in plasma and arc sources as in any flame, and it is therefore no surprise that wavelength modulation has found an application there.[38,77–81] Leys[38] reported a 3- to 10-fold improvement in detection limit of copper in the presence of excess $AgNO_3$. Lichte and Skogerboe[77] used wavelength modulation to reduce spectral interferences caused by carbon monoxide, nitrogen oxides, and hydrocarbons in the microwave plasma emission spectroscopy of mercury vapor. Skogerboe and co-workers[77–79,81] have described single-element and multielement[81] microwave plasma emission systems utilizing wavelength modulation background correction. In thermal volatilization[89] and chromatography detector[82] applications, where transient background emission often results, the dynamic character of wavelength modulation background correction is particularly valuable. Thus, in a gas-chromatography microwave-excited plasma-emission detector system used for studies of selenium and lead alkyls in the environment, wavelength modulation was found to be essential.[82] (The selectivity ratios for dimethyl selenide and for lead alkyls with respect to methanol were improved 100- and 10-fold, respectively.)

Elements which have been measured in plasmas with wavelength modulation are listed in Table 4.

5.1.2. *Absorption*

As conventionally practiced, atomic absorption spectrophotometry utilizes separate narrow-line primary sources for each element, usually

Table 4. Applications of Wavelength Modulation in Plasma Emission Spectrometry

Element	Line (nm)	Detection limit (μg/ml)	Reference
As	193.7	0.03	78
		0.01[a]	89
B	249.8	0.01	78
Cd	228.8	0.0004	78
	228.8	0.005	81
	326.1	0.0002[a]	89
Co	240.7	0.06	78
Cu	324.7	0.3[b]	38
	327.4	0.009	81
	324.7	0.001[a]	89
Fe	372.0	0.01	81
Hg	253.7	0.003	77, 78
	253.7	0.0006[a]	89
Mn	403.5	0.0001[a]	89
Pb	405.8	0.005	78
	405.8	0.008	81
	283.7	55 pg[c]	82
	405.8	0.0001[a]	89
Se	196.0	0.04	78
	196.0	15 pg[c]	82
V	437.9	0.08	78
Zn	213.9	0.0006	78
	213.9	0.0005	81
	213.9	0.0002[a]	89

[a] Thermal volatilization of 5μl sample into plasma.
[b] DC arc; all others are microwave plasmas.
[c] Absolute detection limit in microwave plasma used as GC detector.

either hollow cathode lamps or electrodeless discharge lamps. Although such sources provide excellent sensitivity and analytical curve linearity, the separate lamps are an obvious inconvenience if several elements must be determined in one sample, and they preclude the possibility of a simple compact simultaneous multielement analysis system. Moreover, the correction for nonanalyte absorption is not straightforward, and in fact the commercial background correctors commonly used can result in subtle analytical errors in some cases.[82] Thus, ever since the earliest days of atomic absorption, investigators have sought ways to utilize continuum primary sources effectively. This work has been reviewed by Zander *et al.*[9,83] Snelleman[20] was the first to point out that the use of wavelength modulation in conjunction with a continuum primary source would correct for nonspecific absorption and would reduce low-frequency source fluctuation noise. This idea has been developed by other workers,[56,83–85] and it now appears that, largely because of technological advances in the design of high-intensity continuum sources and high-resolution spectrometers, continuum-source atomic absorption has much to offer. The particular advantages of wavelength modulation in this context are illustrated in Figure 3, which shows the continuum-source atomic-absorption signals at the Pb 283.3 nm line of a series of lead standard solutions and two solutions

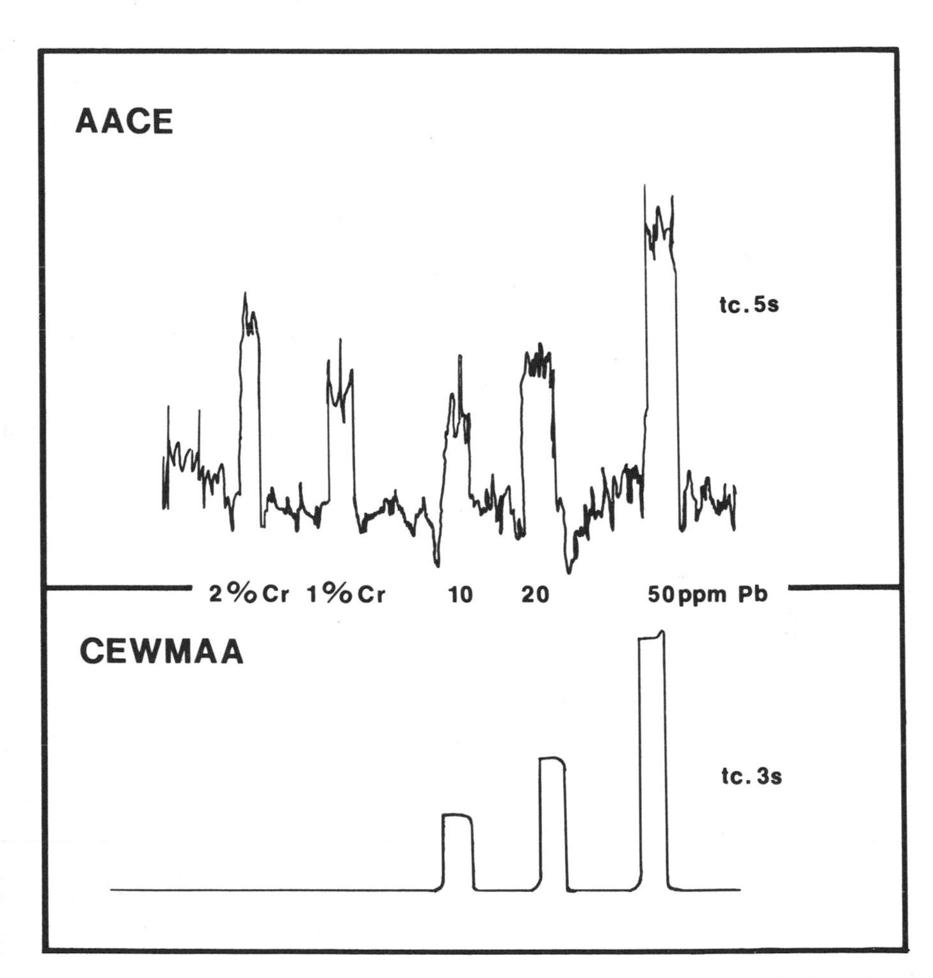

Figure 3. The effect of wavelength modulation in continuum-source atomic absorption. Recorder traces are shown of the flame analysis for lead of five solutions containing, from left to right, 2% by weight Cr but no Pb, 1% by weight Cr but no Pb, and 10, 20, and 50 μg/ml Pb in water. The upper trace is without wavelength modulation and the lower trace is with wavelength modulation. Note the dramatic improvement in signal-to-noise ratio, reduction of baseline drift, and correction for nonspecific absorption.

containing 1% and 2% chromium but no lead. The top trace, without wavelength modulation, exhibits low SNR, baseline drift, and spurious signals due to the light scattering and molecular absorption of the (lead-free) chromium solutions. All of these defects are corrected by wavelength modulation (bottom trace). It is also possible to use the zero-crossing technique to compensate for several of the well-known line-overlap interferences which can occasionally be a problem in the atomic-absorption analysis of alloys.[85] Some of these cases are also included in Table 3.

Detection limits obtained by various workers using wavelength modulation techniques in flame atomic absorption are listed in Table 5.

The very large transient background absorption signals often observed in graphite furnace atomic absorption of complex samples is easily compensated by wavelength modulation.[86,87] Table 6 lists some detection limits

Table 5. Applications of Wavelength Modulation in Flame Atomic Absorption Spectrometry

Element	Line (nm)	Detection limit (μg/ml)	Reference
Zn	213.8	0.03	83
Cd	216.9	0.07[a]	83
		0.02[b]	9
		10	84
		0.2	56
Ni	232.0	2	84
		0.1	83
Co	240.7	0.2	83
Fe	248.3	0.4	84
		0.08	83
Mn	279.5	0.03	83
Pb	283.3	0.15	83
Mg	285.2	0.1	84
		0.03	56
		0.002	83
Cu	324.7	0.5	84
		0.1	56
		0.018	83
Cr	357.9	0.4	84
		0.4	56
		0.3	83
Ca	422.7	0.1	84
		0.07	56
		0.018	83

[a] 1 s time constant.
[b] 10 s time constant.

Table 6. Detection Limits in Complex Matrices Obtained by Graphite Furnace[a] Wavelength Modulation Continuum Source Atomic Absorption

Element	Matrix	Detection limit (μg/ml)	Reference
Ba	Sea water	0.001[b]	86
Cr	Blood serum	0.0003[b]	86
Cu	3% NaCl	0.0024[c]	86
Pb	3% NaCl	0.001[c]	86
Cd	3% NaCl	0.002[c]	87

[a] Perkin-Elmer HGA-2000.
[b] 25-μl sample volume.
[c] 20-μl sample volume.

measured in complex matrices by the technique of wavelength modulation continuum-source atomic absorption.

Thus it seems possible that, for very difficult samples plagued by serious spectral interference, wavelength modulation continuum-source atomic absorption may well be the method of choice. Applications of the technique to various reference samples are included in Table 2.

5.1.3. Fluorescence

In continuum-source atomic-fluorescence spectrometry, wavelength modulation can reduce the effects of light scattering from particles in the flame.[(88)] The technique has not been widely applied.

5.1.4. Simultaneous Multielement Systems

The idea of using a refractor plate in the entrance beam of a polychromator (e.g., direct reader) to provide simultaneous wavelength modulation of all channels has been utilized by Skogerboe[(80)] and by Kirkbright[(89,90)] to provide background correction in multielement microwave plasma emission systems. Particularly in the case of Kirkbright's system, which utilizes thermal volatilization from electrically heated substrates into the plasma, the dynamic characteristic of wavelength modulation background correction is a considerable advantage over the older "spectrum shifter" idea, in which separate readings are taken off the lines and subtracted from the readings on the lines. On the other hand, a multichannel wavelength modulation system involves more complicated signal-processing electronics. This problem can be handled either by using separate synchronous ac detectors for each channel[89] or by using a computer-based data system.[80]

5.2. Molecular Spectroscopy

In this section we will consider applications of derivative and wavelength modulation techniques to molecular and/or condensed-phase spectroscopy. In these applications there is often useful information contained in the detailed shape and structure of the spectral bands, so one finds a greater application of derivative techniques, for example by small-amplitude wavelength modulation, electronic differentiation, or s[illegible]ther technique aimed at obtaining a wavelength derivative of t[illegible]al band.

5.2.1. *Spectroscopy of Solids*

Interestingly, the most extensive applications of derivative spectroscopy have *not* been in mainstream quantitative analytical chemistry but in solid-state physics, where derivative techniques have been widely used since the mid-1960s in studies of the optical properties of solids, particularly semiconductors.[23,25,91–128] The topic has been the subject of several reviews.[91–93] The optical spectra of many semiconductors exhibit a strong absorption edge on which are superimposed very subtle features which are related to the electronic structure of the material. The observation of these subtle features is greatly facilitated by generating the derivative spectra, which is almost always done by wavelength modulation. This approach has been applied to both absorption and reflectance spectra and has been used to study electronic band structure,[103,106,110,113,119,126,127] phonon[115,120,123] and exciton[108,114,116,121,122] phenomena, dielectric functions,[106,124] and phase transitions.[120] The materials which have been studied are listed in Table 7.

5.2.2. *Spectroscopy of Molecules in Solution*

Absorption Spectrophotometry. Many applications of derivative and wavelength modulation techniques in absorption spectrophotometry have been in biochemical analysis.[2,13,14,63,65,129–132] These techniques have been used in the study of enzymes,[2,14] steroids,[65] chlorophyll,[13] and bacteriochlorophyll–protein complexes.[63] These applications have mostly been of a qualitative nature, in which the characteristic enhancement of spectral detail in the derivative spectra are used to distinguish between similar compounds or to follow subtle changes in spectral features. Thus, for example, in a study of testosterone and hydroxylated progesterones, Olson

Table 7. Solid Materials Studied by Derivative Spectroscopy

Material	Reference
Antimony sulfide iodide	119
Cadmium indium gallium sulfide	125
Copper metal	103, 105
Cuprous oxide	108, 117, 121, 122
Gallium arsenide	25, 111
Gallium phosphade	106, 107, 113
Germanium metal	107, 112
Gold metal crystals	103, 105
[illegible]lmium oxide	111
[illegible] gallium phosphide	123, 126
[illegible]osphide	106
Ir[illegible]	116
Lead sulfide, selenide, and telluride	118
Lead tin telluride	103, 109
Manganese oxide	128
Molybdenum sulfide	23
Potassium tantalate	115
Silver metal crystals	103, 105
Silicon	107
Strontium titanate	115, 120
Tin diselenide	118, 127
Tin disulfide	127

and Alway[65] were able to distinguish α and β epimers and to locate the position of substitution in some cases.

Applications to quantitative analysis have also been fruitful. An early paper by Collier and Singleton[8] describes an extensive investigation into the application of derivative techniques to the IR analysis of mixtures of the various isomers of cresol, xylenol, and mesitol. More recent applications include the determination of acetone in chloroform,[40] mixtures of di- and trichlorophenols,[61] *l*-methylpyridinium chloride absorbed on clay,[62] mixtures of isophthalic and terephthalic acids,[62] mixtures of methyl violet and eosin *y*,[60,62] bilirubin in the presence of albumin,[132] and mixtures of various polycyclic aromatic hydrocarbons such as anthracene, phenanthrene, chrysene, and pyrene.[133] Quantitative applications to inorganic species in solution include the determination of nickel in the presence of cobalt,[134] Pr and Nd nitrates in the presence of Cu nitrate,[135] and mixtures of various rare-earth salts.[61,62]

Fluorescence Spectrometry. In recent years derivative techniques have been profitably applied to fluorescence spectrometry for both qualitative structure-enhancement applications[13,15,67,136–140] and for the quantitative analysis of mixtures.[15,55,136–138] In addition to the more-or-less straightforward application of derivative concepts to fluorescence spectrometry, there has also been the development of a wavelength-modulation component resolution technique called *selective modulation*[15,137,138] which is capable of extracting the spectrum of one component in a mixture of two fluorescent species whose spectra overlap heavily. Both this and conventional derivative techniques have recently been applied to problems in environmental analysis. Fox and Staley,[139] in a study of polycyclic aromatic hydrocarbons in atmospheric particulate matter, utilized derivative and selective modulation techniques to detect and identify traces of benzo[*k*]fluoranthene coeluting with benzo[*e*]pyrene and of benz[*a*]anthracene coeluting with chrysene. Kolb and Shearin[140] applied derivative techniques to petroleum oil fingerprinting based on low-temperature fluorescence spectra.

It seems likely that derivative and wavelength modulation techniques will find increasing use in fluorescence spectrophotometry.

5.2.3. *Spectroscopy of Molecules in the Gas Phase*

Derivative spectroscopy has proved to be a very useful approach to the analysis of small molecules in the gas phase. Hager and Anderson[32–36] have developed a second-derivative method and have applied it to the trace analysis of various pollutant gases in air, including NO, NO_2, O_3, SO_2, NH_3, etc. A commerical second-derivative spectrometer based on their work is manufactured by Lear-Siegler Instruments. Hunt and Williams[142] have used this instrument to measure NH_3 in human breath at sub-ppm levels. Strojek, Yates, and Kuwana[143] have designed an elaborate rapid-scanning derivative spectrometer which they have applied to the analysis of SO_2 in air. These applications have demonstrated excellent sensitivity and selectivity.

6. References

1. B. Chance, *Rev. Sci. Instrum., 13,* 158 (1942).
2. B. Chance, *Rev. Sci. Instrum., 22,* 634 (1951).
3. V. J. Hammond and W. C. Price, *J. Opt. Soc. Am., 43,* 924 (1953).
4. J. D. Morrison, *J. Chem. Phys., 22,* 1219 (1954).
5. C. S. French, A. B. Church, and R. W. Eppley, *Carnegie Inst. Wash. Year Book, 53,* 182 (1954).
6. J. M. Vandenbelt and C. Heinrich, *Appl. Spectrosc., 7,* 171 (1953).
7. A. T. Giese and C. S. French, *Appl. Spectrosc., 9,* 78 (1955).
8. G. L. Collier and F. Singleton, *J. Appl. Chem., 6,* 495 (1956).
9. A. T. Zander, PhD dissertation, University of Maryland (1976).
10. G. L. Collier and A. C. M. Panting, *Spectrochim. Acta, 14,* 104 (1959).
11. A. E. Martin, *Nature, 180,* 231 (1957).

12. J. P. Walters and H. V. Malmstadt, *Appl. Spectrosc., 20*(3), 193 (1966).
13. W. L. Butler, *Arch. Biochem. Biophys., 93,* 413 (1961).
14. T. Shiga, K. Shiga, and M. Kuroda, *Anal. Biochem., 44*(1), 291 (1971).
15. G. L. Green and T. C. O'Haver, *Anal. Chem., 46,* 2191 (1974).
16. T. C. O'Haver and G. L. Green, *Am. Lab.,* 15 (1975).
17. I. G. McWilliam, *J. Sci. Instrum., 36,* 51 (1959).
18. G. Bonfiglioli and P. Brovetto, *Appl. Opt., 3*(12), 1417 (1964).
19. E. Gunders and B. Kaplan, *J. Opt. Soc. Am., 55*(9), 1094 (1965).
20. F. Aramu and A. Rucci, *Rev. Sci. Instrum., 37*(12), 1696 (1966).
21. F. R. Stauffer and H. Sakai, *Appl. Opt., 7,* 61 (1968).
22. A. Perregaux and G. Ascarelli, *Appl. Opt., 7*(10), 2031 (1968).
23. B. L. Evans and K. T. Thompson, *J. Sci. Instrum. (J. Phys. Eng.), 2,* 327 (1969).
24. R. Hager, Jr. and R. Anderson, *J. Opt. Soc. Am., 60*(11), 1444 (1970).
25. R. Zucca and Y. R. Shen, *Appl. Opt., 12,* 1293 (1973).
26. W. Snelleman, *Spectrochim. Acta, 23B,* 403 (1968).
27. T. C. O'Haver, General analytical considerations, in: *Trace Analysis: Spectroscopic Methods,* Chapter 2 (J. D. Winefordner, ed.), Wiley, New York (1976).
28. T. C. O'Haver and G. L. Green, *Anal. Chem., 48,* 312 (1976).
29. I. Balslev, *Phys. Rev., 143,* 636 (1966).
30. T. C. O'Haver, M. S. Epstein, and A. T. Zander, *Anal. Chem., 49,* 458 (1977).
31. J. Overend, A. Gilby, J. Russell, C. Brown, J. Beutel, C. Bjork, and H. Paulat, *Appl. Opt., 6*(3), 457 (1967).
32. D. T. Williams and R. N. Hager, Jr., *Appl. Opt., 9*(7), 1597 (1970).
33. R. N. Hager, Jr., Paper No. 71-1045, Joint Conference on Sensing Environmental Pollutants (1971).
34. R. N. Hager, Jr., *Anal. Chem., 45,* 1131A (1973).
35. R. N. Hager, Jr., Paper 20E, 15th Eastern Analytical Symposium (1973).
36. R. N. Hager, Jr., and A. M. Garcia, Paper No. 240, 1974 Pittsburgh Conference (1974).
37. K. Visser, F. M. Hamm, and P. B. Zeeman, *Appl. Spectrosc., 39,* 72 (1976).
38. J. A. Leys, *Anal. Chem., 41,* 396 (1969).
39. A. Gilgore, P. Stoller, and A. Fowler, *Rev. Sci. Instrum., 38,* 1535 (1967).
40. I. G. McWilliam, *Anal. Chem., 41,* 674 (1969).
41. R. Roldan, *Rev. Sci. Instrum., 40,* 1388 (1969).
42. W. Snelleman, T. Rains, K. Yee, H. Cook, and O. Menis, *Anal. Chem., 42,* 394 (1970).
43. K. L. Shaklee and J. E. Rowe, *Appl. Opt., 9,* 627 (1970).
44. M. Nordmeyer, *Spectrochim. Acta, 2713,* 377 (1972).
45. D. G. Mitchell, J. M. Rankin, and B. W. Bailey, *Spectrosc. Lett., 5,* 87 (1972).
46. W. Fowler, D. Knapp, and J. D. Winefordner, *Anal. Chem., 46,* 4 (1974).
47. M. S. Epstein and T. C. O'Haver, *Spectrochim. Acta, 30B,* 135 (1975).
48. R. W. Spillman and H. V. Malmstadt, *Anal. Chem., 48,* 303 (1976).
49. R. K. Skogerboe, P. J. Lamothe, G. J. Bastiaans, S. J. Freeland, and G. N. Coleman, *Appl. Spectrosc., 30,* 495 (1976).
50. S. R. Koirtyohann, E. D. Glass, D. A. Yates, E. Hinderberger, and F. F. Lichte, *Anal. Chem., 49,* 1121 (1977).
51. R. H. Eather and D. L. Reasoner, *Appl. Opt., 8,* 227 (1969).
52. R. J. Sydor and G. M. Hieftje, *Anal. Chem., 48,* 535 (1976).
53. M. S. Epstein, National Bureau of Standards, private communication.
54. T. C. O'Haver, G. L. Green, and B. R. Keppler, *Chem. Instrum., 4,* 197 (1973).
55. T. C. O'Haver and W. M. Parks, *Anal. Chem., 46,* 1886 (1974).
56. G. J. Nitis, V. Svoboda, and J. D. Winefordner, *Spectrochim. Acta, 27B,* 345 (1972).

57. T. E. Cook, H. L. Pardue, and R. E. Santini, *Anal. Chem., 48,* 451 (1976).
58. A. Danielsson and P. Lindblom, *Appl. Spectrosc., 30,* 151 (1976).
59. J. P. Pemsler, *Rev. Sci. Instrum., 28,* 274 (1957).
60. Perkin-Elmer Corp., UV/FL Prod. Dept., Tech. Memo. No. 4,5,34,44 (1970).
61. S. Shibata, M. Furukawa, and K. Goto, *Anal. Chim. Acta, 65,* 49 (1973).
62. T. J. Porro, *Anal. Chem., 44,* 93A (1972).
63. M. P. Klein and E. A. Dratz, *Rev. Sci. Instrum., 39,* 397 (1968).
64. R. Cook, Cary Instrument Tech. Memo. UV-70-8 (1970).
65. E. C. Olson and D. Alway, *Anal. Chem., 32,* 370 (1960).
66. A. Savitzky and M. Golay, *Anal. Chem., 36,* 1627 (1964).
67. F. Grum, D. Paine, and L. Zoeller, *Appl. Opt., 11,* 93 (1970).
68. D. Lewis, P. Merkel, and W. Hamill, *J. Chem. Phys., 53,* 2750 (1970).
69. W. L. Baun and M. Chamberlain, *Rev. Sci. Instrum., 44,* 1421 (1973).
70. J. F. Brandts and L. J. Kaplan, *Biochemistry, 12,* 2011 (1973).
71. J. Piereaux, *Appl. Spectrosc., 30,* 219 (1976).
72. W. Snelleman, T. C. Rains, K. W. Yee, H. D. Cook, and O. Menis, *Anal. Chem., 42,* 394 (1970).
73. I. S. Maines, D. G. Mitchell, J. M. Rankin, and B. W. Bailey, *Spectrosc. Lett., 5,* 251 (1972).
74. T. C. Rains and O. Menis, *Anal. Lett., 7,* 715 (1974).
75. M. S. Epstein, T. C. Rains, and T. C. O'Haver, *Appl. Spectrosc., 30,* 324 (1976).
76. F. E. Lichte and R. K. Skogerboe, *Anal. Chem., 44,* 1321 (1972).
77. F. E. Lichte and R. K. Skogerboe, *Anal. Chem., 45,* 399 (1973).
78. F. E. Lichte and R. K. Skogerboe, *Appl. Spectrosc., 28,* 354 (1974).
79. W. A. Gordon, K. M. Hambidge, and M. L. Franklin, in: *Applications of Newer Techniques of Analysis* (I. L. Simmons and G. W. Ewing, eds.), p. 23, Plenum, New York (1973).
80. R. K. Skogerboe, P. J. Lamothe, G. J. Bastiaans, S. J. Freeland, and G. N. Coleman, *Appl. Spectrosc., 30,* 495 (1976).
81. D. Reamer, T. O'Haver, and W. Zoller, Determination of Alkyl Lead and Selenium Compounds in the Atmosphere Using a GC/MPD with Wavelength Modulation, 8th Materials Research Symposium, National Bureau of Standards, Gaithersburg, Maryland (Sept. 23, 1976).
82. A. T. Zander, *Am. Lab., 8*(11), 11 (1976).
83. A. T. Zander, T. C. O'Haver, and P. N. Keliher, *Anal. Chem., 48,* 1166 (1976).
84. R. C. Elser and J. D. Winefordner, *Anal. Chem., 44,* 698 (1972).
85. A. T. Zander, T. C. O'Haver, and P. N. Keliher, *Anal. Chem., 49,* 838 (1977).
86. T. C. O'Haver, G. C. Turk, and J. M. Harnly, The Analysis of High-Solids Samples by Wavelength Modulation Atomic Absorption in a Graphite Furnace Atomizer, Paper No. 150, Pittsburgh Conference on Analytical Chemistry and Applied Spectroscopy, Cleveland, Ohio (March 1, 1977).
87. J. M. Harnly and T. C. O'Haver, *Anal. Chem., 49,* 2187 (1977).
88. W. K. Fowler, D. O. Knapp, and J. D. Winefordner, *Anal. Chem., 46,* 601 (1974).
89. MPS-600 Preliminary Specifications, EDT Research, London.
90. G. Kirkbright, D. Alger, and D. J. Johnson, A New Microwave-Excited Atmospheric Argon Plasma Spectrometer for Simultaneous Multi-Element Analysis, Paper 454, 28th Pittsburgh Conference, Cleveland, Ohio (March, 1977).
91. B. Batz, *Semicond. Semimetals, 9,* 315 (1972).
92. Y. R. Shen, *Surf. Sci., 37,* 522 (1973).
93. M. Cardona, *Modulation Spectroscopy,* Academic Press, New York (1969).
94. I. Balslev, *Solid State Commun., 3,* 213 (1965).

95. E. Matatagui, A. Thompson, and M. Cardona, *Phys. Rev., 176*(3), 950 (1968).
96. K. Shaklee, R. Rowe, and M. Cardona, *Phys. Rev., 174,* 828 (1968).
97. R. Braunstein, P. Schrieber, and M. Welkowsky, *Solid State Commun., 6,* 627 (1968).
98. J. Rowe, M. Cardona, and K. Shaklee, *Solid State Commun., 7,* 441 (1969).
99. J. Rowe, F. Pollak, and M. Cardona, *Phys. Rev. Lett., 22*(18), 933 (1969).
100. R. Zucca and Y. Shen, *Phys. Rev., B1,* 2668 (1970).
101. C. Fong, M. Cohen, R. Zucca, J. Stokes, and Y. Shen, *Phys. Rev. Lett., 25*(21), 1486 (1970).
102. D. M. Kom, M. Welkowsky, and R. Braunstein, *Solid State Commun., 9,* 2001 (1971).
103. M. Welkowsky and R. Braunstein, *Solid State Commun., 9,* 2139 (1971).
104. M. Welkowsky and R. Braunstein, *Rev. Sci. Instrum., 43*(3), 399 (1972).
105. J. Stokes, Y. Shen, Y. Tsang, M. Cohen, and C. Fong, *Phys. Lett., 38A*(5), 347 (1972).
106. C. de Alverez, J. Walter, M. Cohen, J. Stokes, and Y. Shen, *Phys. Rev., 6*(4), 1412 (1972).
107. T. Nishino and Y. Hamakawa, *Phys. Status Solidi, 50,* 345 (1972).
108. A. Daunois, J. C. Merle, J. L. Diess, and S. Nikitine, *Phys. Status Solidi, 50,* 691 (1972).
109. D. M. Korn and R. Braunstein, *Phys. Rev., 5,* 4837 (1972).
110. P. J. Stiles, *U.S. Natl. Tech. Inform. Serv.,* AD Rep. No. 746271, *Gov. Rep. Announce, 72,* 220 (1972).
111. G. P. Hart, J. A. Neely, and R. J. Kearney, *Rev. Sci. Instrum., 44,* 37 (1973).
112. E. Schmidt, *Scr. Fac. Sci. Nat. Univ. Purkynianae Brun., 2,* 43 (1972).
113. T. Nishino, M. Takeka, and Y. Hamakawa, *Surf. Sci., 37,* 404 (1973).
114. M. Iliev and M. Balera, *Surf. Sci., 37,* 585 (1973).
115. K. W. Blazey, *Surf. Sci., 37,* 251 (1973).
116. S. H. Wemple, *Surf. Sci., 37,* 297 (1973).
117. J. C. Merle, C. Wecker, A. Daunois, J. L. Deiss, and S. Nikitine, *Surf. Sci., 37,* 347 (1973).
118. S. Kohn, Y. Petroff, P. Y. Uy, and Y. R. Shen, *J. Nonmetals, 1,* 147 (1973).
119. C. Y. Fong, Y. Petroff, S. Kohn, and Y. R. Shen, *Solid State Commun., 14,* 681 (1974).
120. H. A. Weakliem, R. Braunstein, and R. Stearns, *Solid State Commun., 15,* 5 (1974).
121. T. Itoh and S. Narita, *Proc. Int. Conf. Phys. Semicond., 12,* 977 (1974).
122. T. Itoh and S. Narita, *J. Phys. Soc. Jpn., 39,* 132 (1975).
123. J. Donecker and J. Kluge, *Phys. Status Solidi, 71,* 1 (1975).
124. G. Mondio, G. Saitta, and G. Vermiglio, *Can. J. Phys., 53,* 1664 (1975).
125. G. B. Abdullaev, T. G. Kerimova, S. S. Mamedov, T. R. Mechtiev, R. K. Nani, and E. Y. Salaer, *Phys. Status Solidi, 73,* K69 (1976).
126. H. Lange, J. Donecker, and H. Friedrich, *Phys. Status Solidi, 73,* 633 (1976).
127. J. Camassel, M. Schlueter, S. Kohn, J. P. Voitchovsky, Y. R. Shen, and M. L. Cohen, *Phys. Status Solidi, 75,* 303 (1976).
128. M. Yokogawa, K. Taniguchi, and C. Hamaguchi, *Solid State Commun., 19,* 261 (1976).
129. T. Ohnishi and S. Ebashi, *J. Biochem. (Tokyo), 54,* 506 (1965).
130. B. Chance and L. Mela, *J. Biol. Chem., 241,* 4588 (1966).
131. J. M. Goldstein, *Biochys. J., 10,* 445 (1970).
132. T. E. Cook, R. E. Santini, and H. L. Pardue, *Anal. Chem., 49,* 871 (1977).
133. A. R. Hawthorne and J. H. Thorngate, *Appl. Opt.* (in press).
134. S. Shibata, M. Furukawa, and K. Goto, *Anal. Chim. Acta, 46,* 271 (1969); *53,* 369 (1971); *62,* 305 (1972).
135. G. Bonfiglioli, P. Brovetto, G. Busca, S. Levialdi, G. Palmieri, and E. Wanke, *Appl. Opt., 6,* 447 (1967).
136. G. H. Haugen, B. A. Raby, and L. D. Rigdon, *Chem. Instrum., 6,* 205 (1975).
137. T. C. O'Haver, Modulation and derivative techniques in luminescence spectroscopy,

in: *Modern Fluorescence Spectroscopy* (E. L. Wehery, ed.), p. 65, Plenum, New York (1976).
138. P. L. Petrakis, *Aminco Lab. News, 31,* 11 (1975).
139. M. A. Fox and S. W. Staley, *Anal. Chem., 48,* 992 (1976).
140. D. A. Kolb and K. K. Shearin, Fingerprinting Petroleum Oils with Low Temperature Derivative Fluorometry, Paper 399, Twenty-Eighth Pittsburgh Conference, Cleveland, Ohio (March 3, 1977).
141. J. Funfschilling and D. F. Williams, *Appl. Spectrosc., 30,* 443 (1976).
142. R. D. Hunt and D. T. Williams, *Am. Lab., 9,* 10 (1977).
143. J. W. Strojek, D. Yates, and T. Kuwana, *Anal. Chem., 47,* 1050 (1975).

2

Apparatus and Methods for Laser Doppler Electrophoresis

Barton A. Smith and B. R. Ware

1. Introduction

1.1. Description of Technique

Laser Doppler electrophoresis is a descriptive name for a new analytical technique for the determination of the electrophoretic mobilities of particles in suspension or of macromolecules in solution. It is a combination of electrophoresis, the motion of charged particles in an electric field, with laser Doppler velocimetry, a technique for determining the velocities of particles by measuring the Doppler shift of laser light which is scattered from them. The laser Doppler approach differs from other types of electrophoresis in that the velocities of the particles are measured continuously, rather than being inferred from a measurement of displacement over a period of time. Several names have been applied to this new technique. The most common names in current use in the literature are laser Doppler spectroscopy and electrophoretic light scattering (ELS). We shall use primarily the latter name and the abbreviation ELS.

Electrophoretic light scattering measurements can usually be performed in a small fraction of the time required for equivalent measurements by other techniques. For solutions of macromolecules, both the electrophoretic mobilities and the diffusion coefficients of molecules in a sample can be determined simultaneously. In addition, since the measuring process does not require the separation of the different species in solution and no concentration gradients are established, it is possible to study in-

Barton A. Smith and B. R. Ware • Department of Chemistry, Harvard University, Cambridge, Massachusetts 02138

teracting systems in uniform concentrations. For suspensions of larger particles, ELS is often far superior to the standard microelectrophoresis method, since the measurement can be made much more quickly, objectively, and automatically. In many cases, ELS yields higher accuracy and better resolution than can be obtained by any classical technique.

The many advantages of ELS must be weighed against the limitations and difficulties introduced by the laser Doppler approach. One objective of this article is to provide a description of the technique in sufficient detail to allow a careful assessment by the reader of the applicability of ELS to whatever problem may be of specific interest. A second objective is to provide a systematic account of the current experimental methods and design considerations, which will hopefully be of use to the growing number of groups who are embarking on research programs in this area.

1.2. Applications

In this section we list the applications to date of electrophoretic light scattering. The reader should consult the review articles on the subject for a more thorough discussion of these applications.[1-3] The first theory and experiments were reported by Ware and Flygare,[4] who measured simultaneously the electrophoretic mobility and diffusion coefficient of bovine serum albumin. Several groups have reported various modifications and improvements in the apparatus, including adaptations of the technique for the study of different types of samples.[5-11] Further theoretical work has led to a better understanding of the quantities measured by ELS[12-15] and has predicted that ELS can be used as a probe of reaction kinetics.[16]

Suspensions of polystyrene latex spheres have been used extensively as a model system for the development of the technique.[5-9,17] Latex spheres have also been used for the study of interactions between charged particles,[18,19] for the study of polymer reactions,[20] and for the development of an immunological assay.[21] For the application of the technique to smaller macromolecules, the plasma proteins[10,22] and hemoglobin[11] have been used as model systems. Other sytems on which ELS experiments have been reported include suspensions of DNA,[23] bacteria,[5] viruses,[23,24] vesicles,[25] and secretory granules.[25]

The laser Doppler approach has been particularly useful for the study of living cells, for which the scattering intensity is high and the diffusion broadening is negligible. ELS has been used to characterize the electrophoretic distributions of blood cells[26,27] and to study the reactions of these cells with mitogens[28,29] and with antigens.[30] Differences in the electrophoretic mobility distributions of white blood cells from normal individuals and from patients with acute lymphocytic leukemia have been observed.[31] Correlations between charge and cell-surface marker characteristics[32] as well as the relationship between charge and cell aggregation[33] have been studied. ELS has also been used to characterize cell subpopulations.[34] Recently ELS and two other electrophoresis techniques have been applied to the study of red blood cells of differing age and have laid to rest the common belief that the surface charge density of red blood cells decreases with age.[35]

All of the above experiments have been performed on aqueous suspensions of uniform concentration. Application of this technique to the study of electrophoretic motion in supporting media such as gels has been attempted in our laboratory and elsewhere,[36] but only limited success has been achieved, primarily because of the intense signal which is produced by the internal motions of the supporting medium.

2. Principles

2.1. Electrophoresis

2.1.1. Electrophoresis Theory

The phenomenon of electrophoresis is the basis of a wide variety of analytical and preparative techniques used in research and industry. There are several reviews which describe the theory and apparatus in common use for solution electrophoresis.[37–39]

Electrophoresis is the migration of ions in an electric field. The velocity $\mathbf{v}$ of a given ion in a field $\mathbf{E}$ is given by

$$\mathbf{v} = U\mathbf{E} \tag{1}$$

where the constant of proportionality U is known as the electrophoretic mobility. In general, the electrophoretic mobility for a particular species will be a function of solution conditions and must be experimentally determined.

The electrophoretic mobility can be related theoretically to fundamental physical properties of the particles of interest. The mobility of a spherical macroion is given approximately by Henry's equation,[38]

$$U = \frac{Z\epsilon}{6\pi\eta R}\frac{X(\kappa R)}{(1 + \kappa R)} \tag{2}$$

where Z is the number of elementary charges on the particle, ϵ is the magnitude of a unit charge, η is the viscosity of the solution, R is the radius of the particle, and κ is the Debye-Hückel constant, which is proportional to the square root of the ionic strength of the solution. $X(\kappa R)$ is Henry's function,[38] which ranges from 1.0 for small values of κR to 1.5

for large values of κR. Thus, from the measured electrophoretic mobility of a macroion and its diffusion coefficient, from which the radius can be determined, the charge of the ion can be calculated. Equation (2) predicts that for values of κR much greater than one, the electrophoretic mobility of a particle is proportional to and uniquely determined by the net charge per unit surface area (the surface charge density) of the particle.

2.1.2. *Electroosmosis*

Electroosmosis is a phenomenon closely related to electrophoresis which often causes experimental difficulties in electrophoresis measurements. Electroosmosis is the flow of solution past a charged, stationary solid surface caused by an electric field parallel to the surface. In most electrophoresis sample chambers, electroosmosis causes a bulk flow of solution along the walls of the chamber in one direction, and a return flow along the center of the chamber in the opposite direction. This flow produces a velocity profile across the chamber which adds to the electrophoretic velocity of the particles of interest, complicating the analysis of the experimental data.

2.1.3. *Electrophoresis Techniques*

There are two common techniques for making electrophoresis measurements in solution. For macromolecules and other submicroscopic particles, moving-boundary electrophoresis[37] is used. In this technique, a sharp boundary is formed between a solution containing the particles of interest and a buffer solution. Electric current is passed through the chamber and the position of the moving boundary is monitored at later times by optical methods. Although this technique has made great contributions to biology and colloid research, it has been used less frequently in recent years because of a number of fundamental limitations. Each measurement generally requires several hours and a substantial quantity of material. Interpretation of the measurement is complicated, because gradients in conductivity, pH, macroion concentration, and salt concentration at the boundaries may cause the mobility of the boundaries to differ from the mobility of the macroion in dilute homogeneous solution. The interpretation becomes even more complicated for solutions of several species, particularly when interaction is present. The fundamental limitations of measurement time and quantity of material required are greatly alleviated by electrophoretic light scattering, and boundary anomalies are obviated entirely.

For larger particles, the most common technique is microelectrophoresis, in which the velocities of particles are determined by measuring the time required for each particle to traverse a predetermined distance. The particles are observed through a microscope with a calibrated reticle to measure the distance traveled, and the time is measured with a stop watch. The determination of the mobilities of a statistically significant number of particles in a sample by this method is obviously a laborious and time-consuming task.

2.2. Light-Scattering Velocimetry

The frequency spectrum of light scattered from a collection of moving particles contains information about the motion of these particles. The various applications of this principle have been reviewed in recent books.[40–42] We shall confine ourselves to a discussion of the application of this technique to velocimetry.[43] Stated simply, the frequency of the scattered light is Doppler-shifted from the frequency of the incident light by the motion of the particle. This frequency shift is usually very small with respect to the frequency of the light, so a special technique called optical mixing spectroscopy is employed to determine the frequency spectrum of the scattered light. In optical mixing spectroscopy, a coherent, monochromatic light source (laser) is used to illuminate the sample. Light scattered from the sample is mixed at the surface of a photodetector with light from the source which has not been shifted in frequency. The resulting signal from the photodetector has a frequency spectrum which corresponds to the difference between the frequencies of the incident light and of the scattered light. The use of a reference local oscillator is often called heterodyne detection, although some confusion exists in the literature because heterodyne detection in engineering literature implies that the reference local oscillator is shifted in frequency from the primary carrier frequency.

Figure 1 illustrates the geometry of a light-scattering experiment. All of the vectors in the figure lie in a plane which is perpendicular to the plane of polarization of the incident laser beam. The incident laser beam

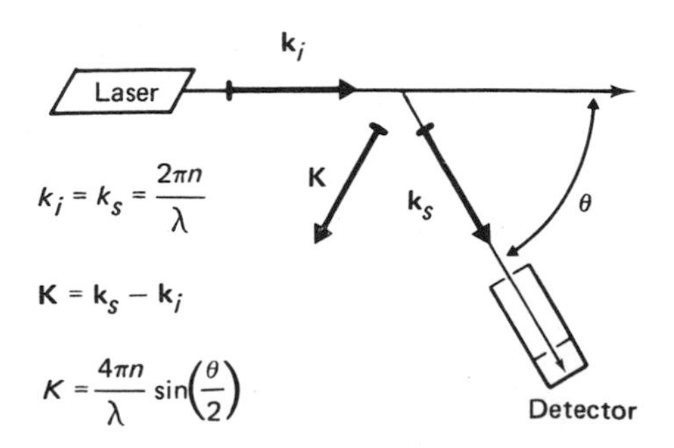

Figure 1. Geometry of a laser light-scattering experiment. The scattering vector **K**, the difference between the wave vectors of the scattered and incident light, represents the momentum imparted to the light by the scatterer. The scattering angle θ is that angle through which the light is scattered in the sample. Refraction at the windows of the sample chamber must be considered in determining the scattering angle from the relative orientation of the laser and detector.

is characterized by the wave vector $\mathbf{k}_i$ which is parallel to the direction of propagation of the beam and has magnitude $k_i = 2\pi n/\lambda_i$ where n is the index of refraction of the medium in which the scatterers are suspended and λ_i is the wavelength *in vacuo* of the incident light. Light is scattered in all directions by the sample, but only light scattered at a particular angle θ, which is called the scattering angle, is allowed to fall onto the detector. The light scattered at angle θ is characterized by wave vector $\mathbf{k}_s$ with magnitude $k_s = 2\pi n/\lambda_s$. The scattering vector $\mathbf{K}$ is defined by the equation

$$\mathbf{K} \equiv \mathbf{k}_s - \mathbf{k}_i \tag{3}$$

Since $\lambda_i \simeq \lambda_s$ because the Doppler shift is very small with respect to the incident frequency, the magnitude of the scattering vector is

$$K = \frac{4\pi n}{\lambda_i} \sin(\theta/2) \tag{4}$$

The power spectrum of the signal from the photodetector in an electrophoretic light scattering measurement of a collection of identical particles with diffusion coefficient D and electrophoretic velocity $\mathbf{v}$ is

$$S(\omega) = \frac{DK^2}{(\omega + \mathbf{K}\cdot\mathbf{v})^2 + (DK^2)^2} + \frac{DK^2}{(\omega - \mathbf{K}\cdot\mathbf{v})^2 + (DK^2)^2} \tag{5}$$

where $S(\omega)$ is the spectral intensity at angular frequency ω. Two terms appear in this expression because the optical mixing technique determines only the magnitude of the Doppler shift and not its sign. Equation (5) describes a Lorentzian frequency distribution centered at a frequency equal to $|\mathbf{K}\cdot\mathbf{v}|$ with half-width at half-height equal to DK^2. Thus from the frequency spectrum of the light scattered from a sample of macromolecules undergoing electrophoresis, one can determine simultaneously the diffusion coefficient and the electrophoretic mobility. The diffusion coefficient is inversely proportional to the friction constant, which for a sphere is given by $6\pi\eta R$, where η is the viscosity of the solvent and R is the radius of the sphere. From examination of equation (2), it is evident that a simultaneous determination of the electrophoretic mobility and the friction constant permits a calculation of the charge on the particles.

The ELS spectrum from a sample containing a mixture of noninteracting macroions is the sum of the spectra which would be obtained from each component measured separately. Each term in the sum is weighted by the intensity of light scattered by that component. If electrophoretically distinct species are interacting on a time scale which is of the same order as the correlation time of the measurement, then the spectrum will be modified, and in principle the kinetics of the interaction can be studied by proper manipulation of the scattering angle and electric field.[16]

The analytical resolution for macromolecules of identical electrophoretic mobility is defined to be the ratio of the frequency of the center of the peak to the width of the peak:

$$r \equiv \frac{|\mathbf{K}\cdot\mathbf{v}|}{DK^2} \tag{6}$$

For the case of particles which are sufficiently large that diffusion can be neglected, the Doppler frequency spectrum is a direct representation of the distribution of the velocities present in the sample. Thus, by use of the relationship $\omega = |\mathbf{K}\cdot\mathbf{v}|$ and equation (1), the frequency spectrum can be converted into the electrophoretic mobility distribution for the sample simply by renumbering the frequency axis in terms of mobility.

2.3. Signal Analysis

The signal from the photomultiplier tube can be processed to obtain either the autocorrelation function or the power spectrum, which is the Fourier transform of the autocorrelation function. The power spectrum can also be calculated as the square of the modulus of the Fourier transform of the time domain signal. The data from a laser Doppler velocimetry experiment are easiest to interpret in the frequency domain because of the linear relationship between frequency shift and velocity. We shall therefore discuss signal analysis in terms of the power spectrum.

According to the sampling theorem,[44] in order to determine the spectrum with a given resolution, we must record the signal for a period of time in seconds equal to the reciprocal of the desired resolution in hertz. This requirement sets a lower limit on the length of time that the electric field must be applied to the sample for a single measurement. The same lower limit applies to the length of time that each particle must remain under observation by the detector. Loss of resolution due to the particles traversing the incident beam in less than this lower time limit is known as transit-time broadening.

2.4. Implications of Theory for Experimental Design

2.4.1. *Direction and Magnitude of the Electric Field*

It can be seen from equation (6) that the resolution of the ELS measurement is maximized by maximizing the projection of the velocity **v** on the scattering vector **K**. The velocity is parallel to the applied electric field, so the apparatus should be arranged in a configuration in which the applied field is parallel or antiparallel to the vector **K**, which bisects the angle between $-\mathbf{k}_i$ and $\mathbf{k}_s$ as shown in Figure 1. Another way to increase the resolution is to increase the magnitude of $\mathbf{v}$, which can be done by applying the highest possible electric field to the sample, and by choosing solution conditions so as to maximize the electrophoretic mobility of the particles. However, the solution conditions are often dictated by the experiment to be performed, and the maximum possible electric field is limited by Joule heating as discussed in Section 2.5.

2.4.2. *Selection of Scattering Angle*

From equation (6) we can see that, other things being held constant, the resolution is proportional to $1/K$. Thus, to overcome diffusion broadening, it is desirable to work at the lowest possible scattering angle. There are, however, some difficulties associated with very low scattering angles. At low angles the relative angular uncertainty for a fixed detection aperture increases, and greater care is required to maintain a sufficiently precise definition of the scattering angle. A low magnitude of K also implies that the Doppler spectrum will be at very low frequencies, where mechanical noise tends to be a more serious problem. In addition low-frequency spectra require a longer duration of the electric field pulse to achieve sufficient spectral resolution, and the longer duration reduces the maximum electric field which can be applied. It is generally best to work in the spectral range above 1 or 2 Hz and to avoid electric field pulse durations greater than 4 s. Typical scattering angles for measurements on protein solutions range from 2° to 5°.

For suspensions of large particles such as blood cells, the diffusion broadening is negligible, and the linewidth is due primarily to electrophoretic heterogeneity. In this case, the optimal scattering angle may be quite high. The principal advantage of a higher scattering angle is an increase in the magnitude of the Doppler shift for a given field strength, while the typical disadvantages of a higher scattering angle are reduced scattered intensity and possibly some inconvenience in optical alignment. The resolution of measurements for these large particles is limited primarily by movement of the particles due to forces other than electrophoresis, such as convection.

2.5. Electric Field and Power Dissipation

The electric field in a solution with conductivity σ is determined by the current density $\mathbf{J}$:

$$\mathbf{E} = \mathbf{J}/\sigma \tag{7}$$

If the current density is constant throughout a specified region which has a uniform cross-sectional area A, the electric field is constant in magnitude throughout this region and can be calculated from the current I flowing through the region by the equation

$$E = I/A\sigma \tag{8}$$

The passage of current through a solution produces heat in the solution (Joule heating). The power dissipated per unit volume of solution is given by

$$P = J^2/\sigma \tag{9}$$

The maximum possible temperature rise which will be caused by a short current pulse can be estimated by assuming that none of the heat is removed from the sample during the pulse. Assuming that the solution has the specific heat of water, the temperature rise in °C for a current density J in A/cm² and a conductivity σ in mho/cm in a time t in seconds is

$$\Delta T = 0.24\, J^2 t/\sigma \tag{10}$$

The amount of heating must be limited since an excessive temperature increase will change the solution conditions and cause convection in the sample.

3. Apparatus

In this section we give a detailed list of the components which make up an electrophoretic light-scattering apparatus (Figure 2). In each case, the general considerations pertinent to design of the apparatus for various types of samples are given, followed by a description of the equipment currently in use in our laboratory for the measurement of electrophoretic mobilities of blood cells in physiological ionic strength media.

3.1. Laser

The two most important characteristics to consider in the choice of a laser are its wavelength and its power. The wavelength is usually chosen to minimize absorption of light by the sample, which can cause both convection and photochemical damage of the particles of interest. In addition, the wavelength dependences of the respective scattering cross sections of the particles and of the sensitivity of the photodetector should be considered. Both of these considerations generally favor shorter wavelengths. When diffusion broadening is a serious problem, the resolution (equation 6) is proportional to laser wavelength, resulting in a slight advantage at longer wavelengths. Typical power levels range from a few

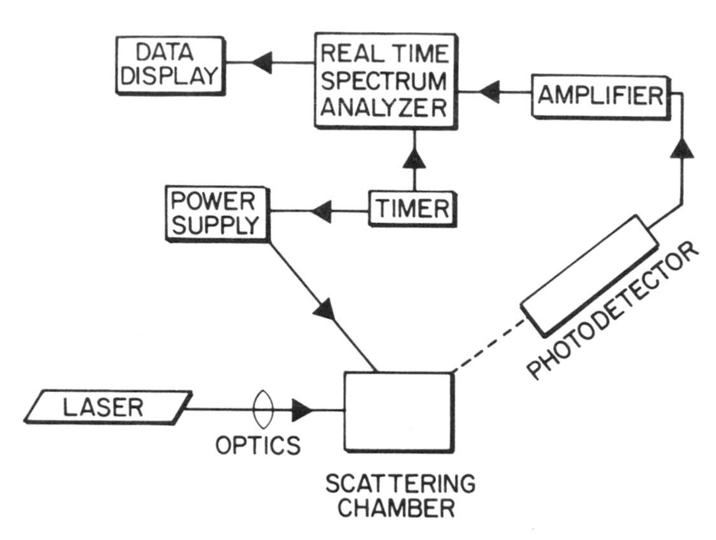

Figure 2. Diagram of the apparatus for electrophoretic light scattering. The sample is illuminated by a laser and the light scattered at a particular angle is detected. A timer controls the application of an electric field to the sample and synchronizes the data collection with the field pulses. The power spectrum of the signal from the photodetector is calculated and displayed by the spectrum analyzer, which also performs the signal averaging of the successive spectra collected during the respective electric field pulses.

milliwatts for strongly scattering samples to a hundred milliwatts for a dilute solution of macromolecules. The laser should be operated in the single transverse mode TEM_{00}. The polarization of the laser should be perpendicular to the plane determined by the incident and scattered wave vectors.

A further consideration in the selection of a laser is the magnitude of noise present in its output. Fluctuations in the intensity of the light at a frequency within the bandwidth being measured will appear as noise in the spectrum. Power-line-frequency ripple in the laser intensity is particularly troublesome. Helium–neon lasers are generally the best of the commonly available types with respect to noise, reliability, convenience of operation, and price. We use a Spectra Physics model 125A He–Ne laser operating at a wavelength of 632.8 nm at a power of 50 mW.

3.2. Optics

The optical system associated with the scattering chamber has two functions. The first is to focus the laser beam into the chamber, thereby defining the volume of the sample that is illuminated. Since the beam in the sample should be of approximately uniform intensity, spatial filtering may be desirable. The dimension of the illuminated volume along the direction of migration should be kept sufficiently large that the transit time of the particles during an electric field pulse is greater than the reciprocal of the frequency resolution of the Doppler spectrum.

The second function of the chamber-associated optics is to provide the reference beam (local oscillator) for optical mixing detection. In the simplest case, stray scattered light from the windows of the chamber will serve this purpose. This may be the method of choice for weakly scattering samples at low scattering angles. Two disadvantages of this method are that it does not allow easy adjustment of the relative intensity of the local oscillator and that it requires the scattering volume to include at least one chamber window. When the stray scattering from the windows will not serve as a suitable local oscillator, optical elements such as beam splitters and mirrors must be used to direct a part of the light from the incident laser beam onto the surface of the photodetector. Special effort must be made to minimize vibrations of these optical elements with respect to the scattering volume, since even very low amplitude vibrations, on the order of a fraction of the wavelength of light, will produce spurious signals in the output of the photodetector.

Figure 3 is a diagram of the optical arrangement used in our appa-

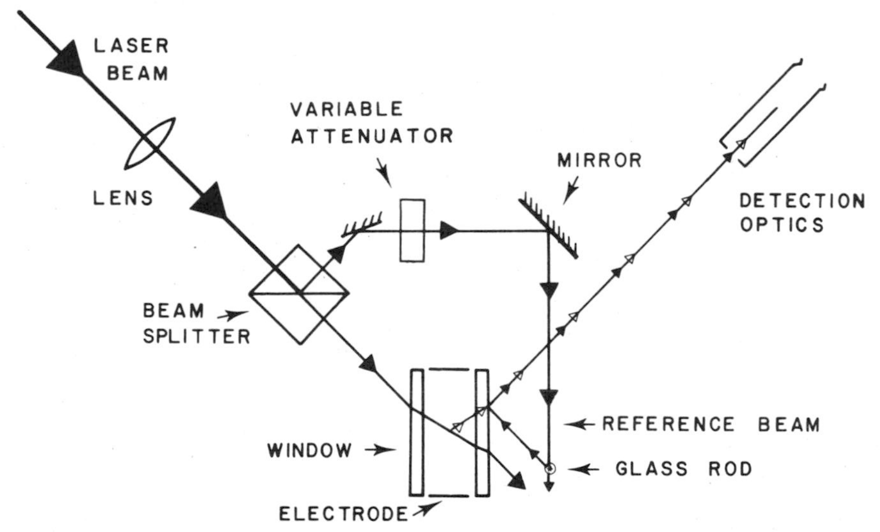

Figure 3. Optical system for electrophoretic light scattering at high scattering angle. The laser beam is divided into two parts, the reference beam and the sample-illuminating beam. The reference beam is combined at the photodetector with light scattered from the sample. The beam splitter, mirrors, attenuator, glass rod, and chamber are all mounted on a metal plate to prevent vibrations of these components with respect to each other.

ratus. The laser beam is focused into the chamber by a converging lens with a focal length of 20 cm. The beam diameter in the sample is approximately 0.15 mm. A cube beam splitter divides the beam into two parts, one of which goes through the sample and the other of which is directed, by means of front-surface mirrors, through a variable attenuator and onto a glass rod on the other side of the chamber. This rod, which has a diameter of 0.2 mm, is positioned so that light scattered from the rod will be reflected by the window of the scattering chamber into the detection optics. The chamber is oriented so that the electric field in the solution is parallel to the scattering vector. All of the optical components of the system, including the laser and photodetector, are mounted on a heavy table which is vibrationally isolated from the floor by foam rubber pads. In order to minimize relative vibration of these components, the beam splitter, mirrors, glass rod, and chamber are attached to a metal plate. The optical elements are mounted on small permanent magnets which attach them rigidly to the metal plate but allow the convenient adjustment of their positions when aligning the apparatus. This design also allows easy rearrangement of the optical system to different configurations. The optical arrangement in Figure 3 is only one of a variety of possible options which we have employed. Although we have found it to be an optimal design for many experiments, the selection of the optical configuration depends upon a number of specific factors such as chamber design, scattering angle, and available optical components.

3.3. Chamber

The electrophoretic light-scattering chamber must allow the application of an electric field to the sample, and it must have optical windows for the entrance and exit of light. Temperature control and stabilization are also important. Efficient heat transfer is required to minimize the rise in temperature of the solution due to Joule heating during the application of the electric current pulse. The chamber geometry should be designed to minimize convection in the scattering region, which can occur during the current pulse as a result of thermal gradients. It is desirable to reduce or eliminate electroosmosis. It is necessary to minimize the sample volume required for the measurement when only a small amount of material is available for the experiment. Of course, the materials of which the chamber is made must be chemically compatible with the samples being studied, and the chamber should be easy to clean. There are several quite different chamber designs in use,[5–11,22] and in general the chamber to use in a particular instance depends upon the type of sample being studied.

Uzgiris[7] has chosen the simplest possible chamber design, placing closely spaced parallel-plate electrodes in a standard cuvette. This design has the great advantage that it completely avoids the problem of electroosmosis, since there are no chamber walls in the region of the sample through which current passes. A disadvantage of this design is that square-wave modulation of the applied electric field must be used to reduce the effects of electrode polarization and current-induced concentration gradients. This modulation results in a complication of the observed spectrum, which makes data interpretation more difficult.[17]

Figure 4 is a diagram of the chamber which we use for the measurement of the electrophoretic mobility distribution of blood cell samples, as viewed looking through the optical windows (which are not shown). The two largest pieces of the chamber are made of silver-plated copper, chosen for excellent thermal and electrical conductivity. Water from a constant-temperature bath flows through these pieces to maintain the sample temperature. The electrodes are semicylindrical silver/silver chloride reversible electrodes. They are attached to the copper pieces by epoxy cement which conducts both heat and electricity (Emerson & Cuming, Inc. Eccobond Solder #56C). The incident laser beam is directed through the sample in

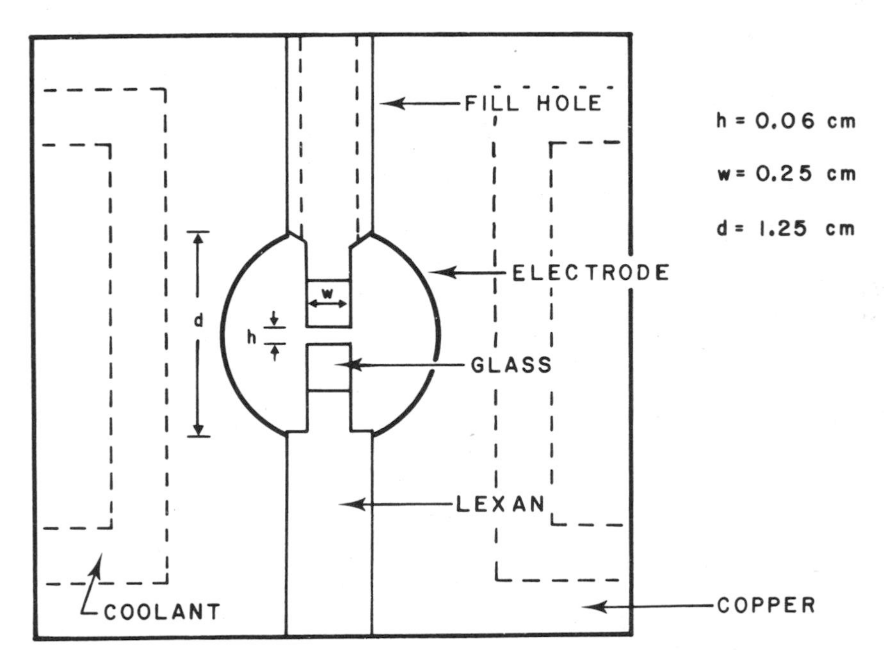

Figure 4. Electrophoretic light-scattering chamber, as seen looking through the optical windows. The body of the chamber is made of two pieces of silver-plated copper, to which the electrodes are cemented. The copper pieces contain channels through which water from a constant temperature bath flows. The two copper pieces are separated by a pair of insulators which restrict current flow through the sample to a small cross section. The distance between the inner surfaces of the windows is 0.37 cm.

the small region between the two glass spacers in the center of the chamber. These spacers and the plastic insulators restrict the current flow through the solution to a small cross-sectional area in which the scattering region is located. The current density is much greater in this region than at the electrodes; thus, a high electric field can be applied to the sample without exceeding maximum allowable current densities at the electrodes. The semicylindrical shape of the electrodes is chosen to establish a uniform current density at the electrodes in order to pass the maximum possible total current. Since significant Joule heating is confined to a small volume, the heat is quickly dissipated during the interval between electric field pulses. The narrow gap between the glass spacers serves also to inhibit convection in the scattering region. Although the solution outside the gap exhibits substantial convection, allowing rapid heat transfer from the gap to the thermostatted electrodes, no convection is observed in the gap due to the stabilizing effect of the closely spaced horizontal surfaces. This chamber is similar to that designed by Haas and Ware,[11] and the original account of that design includes a discussion of the electric field configuration and of the application of the chamber to experiments on protein solutions.

The windows and spacers are made of white crown glass. The windows are pressed onto the front and back of the chamber by plastic mounting flanges so that they lie flat against both the edges of the electrodes and the glass/Lexan spacers. The windows are sealed with a thin layer of stopcock grease. The distance between the inner surfaces of the windows is 0.37 cm. The interior glass surfaces of the chamber are coated with methylcellulose to reduce electroosmosis.

The sample is introduced into the chamber by means of a 23-gauge hypodermic needle through the fill holes in the top Lexan insulator. The chamber can be emptied by inserting the needle to the bottom. Thus the chamber can be rinsed and filled in place.

The cross-sectional area through which the current flows in the solution is constant over the region in which the light scattering is observed, so that the electric field is constant over this region. The electric field strength can be determined from a knowledge of the current flowing through the chamber, the solution conductivity, and the cross-sectional area of the gap (equation 8).

3.4. Power Supply and Switching

In almost all cases, a constant-current power supply is the best choice for ELS. If a constant voltage is applied to the electrodes, electrode polarization and other effects may cause variations of the electric field in the sample. However, if a constant current is passed through the sample, these electrode variations will not affect the electric field in the sample. Constant current has an additional advantage over constant voltage: typical current densities for electrophoresis result in Joule heating which raises the temperature of the solution significantly, often causing a substantial decrease in the viscosity of the solution. Since electrophoretic mobility is inversely proportional to viscosity (equation 2), a change in viscosity will result in a change in the mobilities of the particles being measured. However, the conductivity of the solution is also inversely proportional to viscosity. At a constant current density, the electric field will be reduced by a factor equal to the factor by which the electrophoretic mobility is increased (equation 8). Thus, the particles will have the same velocities as if no heating had occurred.

The power supply should provide a well-regulated, low-noise, constant current which is adjustable over the range necessary for various samples and chambers. In our laboratory, we have used currents in the range from 0.01 to 10 mA and employ an Electronic Measurements Inc. (Neptune, New Jersey) model C6112AL power supply.

An advantage of ELS over classical electrophoresis is that the polarity of the electric field in the sample can be alternated, thereby avoiding the formation of concentration gradients in the sample. In addition, the field can be pulsed with time for heat dissipation between the pulses, and the pulses can therefore be of higher field magnitude than can be employed with continuous current. A timing device is necessary to switch the field on and off and to alternate its polarity. This device must also synchronize the collection of data with the field pulses. The minimum duration of a single pulse is determined by the spectral resolution required for the measurement (Section 2.3). The ratio of field off-to-on time depends on the heat-dissipation time constant of the chamber and on the amount of heat produced during the pulse.

We use a custom-made digital electronic timing circuit which controls the application of the electric field to the sample and simultaneously triggers the spectrum analyzer. The timer actuates mercury-wetted contact relays which connect the constant-current supply to the chamber electrodes. The current is measured by a digital ammeter in series with the electrodes.

For measurements in physiological ionic strength media, we typically analyze the signal with a spectral resolution of 0.5 Hz. This requires an electric field pulse duration of 2.0 s. The time between pulses is typically 15 s.

3.5. Detection Optics

The optical arrangement used in connection with the photodetector determines the scattering angle and the scattering volume. The scattering volume is that volume of the sample from which scattered light reaches the photodetector. Light from other sources must be prevented from reaching the photodetector. In the simplest case, the detection optics can consist of a pair of iris diaphragms placed in the front of the detector so that only light scattered from the sample in a certain direction can enter the photodetector. More elaborate arrangements can be used to define the scattering volume more precisely, to limit the range and uncertainty in scattering angle, and to simplify the optical alignment procedure.

Figure 5 is a diagram of our photodetection apparatus, which is similar in function to a single-lens reflex camera. A lens of focal length 20 cm and diameter 2.5 cm forms a real image in the plane of the adjustable slit, which is directly in front of the photomultiplier tube. The distance from the lens to the real-image plane is approximately 75 cm, resulting in an image magnification of about a factor of three. By moving the reflex mirror into position, the image of the sample is diverted onto the viewing screen, which is a Nikon camera focusing screen with centering grid. The viewing screen and mirror are positioned so that the part of the image falling on the center of the grid will fall on the center of the slit when the mirror is moved aside. Thus we can, by looking at the viewing screen, determine exactly what will be "seen" by the detector during the measurement. The iris diaphragm in front of the lens allows the amount of light entering the detector to be adjusted. This adjustment determines both the depth of field of the image and the span of scattering angles collected by the lens. The position of the sliding tube in which the lens is mounted can be adjusted to bring the image into focus. The width of the slit determines

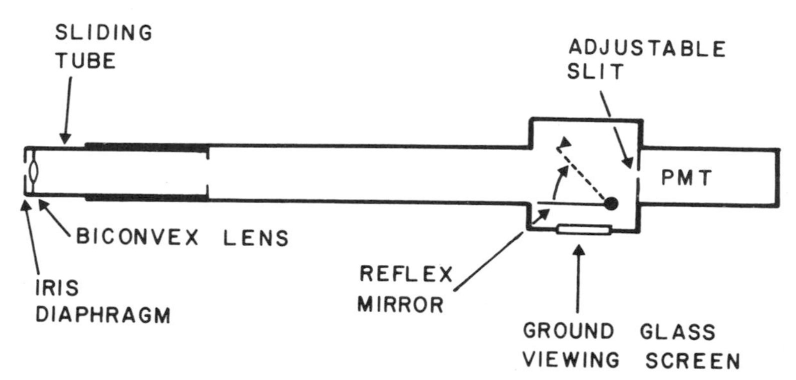

Figure 5. Reflex optical detection system. This system, which is similar in design to a single-lens reflex camera, allows the operator to determine exactly from which part of the sample scattered light will reach the photomultiplier. Light from the reference beam can be superimposed precisely upon the image of the sample to achieve proper optical mixing. The lens, which has a focal length of 20 cm, is approximately 75 cm from the adjustable slit onto which the real image of the sample is projected.

the size of the scattering volume. In addition to the adjustable slit shown in Figure 5, there is a fixed 0.7-mm-high horizontal slit which limits the vertical extent of the image allowed to fall onto the photomultiplier. The iris in front of the lens is usually set at a diameter of approximately 1 mm, and the slit is set at a width of about 0.5 mm.

The photomultiplier tube is an EMI model 9558B in a Products for Research housing. It is powered by Keithley Instruments model 244 high-voltage supply. This tube has a quantum efficiency of approximately 0.08 at 633 nm. We typically operate the tube at a gain of 10^4 with a photocurrent of 2 μA. The photocurrent is monitored with a digital ammeter.

3.6. Signal Analysis and Data Collection

The desired form of the data is the power spectrum of the photocurrent. It is important that the spectral components of the signal be determined with real-time efficiency, i.e., that all of the time-domain data be used in measuring each spectral intensity. Each calculated spectrum should be available to the experimenter as the measurement is in progress in order to facilitate any necessary adjustments or diagnosis of experimental problems. Because the frequencies of interest in ELS are generally in a very low frequency range (1–200 Hz), there are a number of efficient ways to calculate the spectrum.

Autocorrelators have become very popular among workers in the field of dynamic light scattering, and most autocorrelators are acceptable for ELS experiments, provided some means of fast Fourier transformation is available so that the data can be viewed in the frequency domain for immediate interpretation. However, clipped autocorrelators,[41] which have great advantages in other types of experiments, are not suitable for ELS measurements, because the local oscillator intensity will make it difficult to establish a suitable clipping level. In fact, any photon-counting device will require the incident laser intensity to be extremely low, since the photon counts per sample time increment in this experiment are typically far higher than those in most light-scattering experiments and far higher than the capabilities of any commercial photon-counting instrument. The required reduction of laser intensity would cause an unacceptable decrease in signal-to-noise ratio. The best signal-to-noise ratio will generally be achieved by treating the photocurrent as an analog signal, digitizing it, and performing digital operations to calculate the power spectrum.

A number of hard-wired fast Fourier transform audio spectrum analyzers have appeared on the market recently which are ideal for laser Doppler applications, including ELS. (Consult Princeton Applied Research, Princeton, New Jersey; Nicolet, Inc., Madison, Wisconsin; and Spectral Dynamics Corporation, San Diego, California.)

For those laboratories with a dedicated computer or access to rapid on-line processing, the method of choice will probably be to digitize the photocurrent and perform the Fourier transformation with the computer. Because the data are at very low frequency and there is a substantial off-time between pulses, any modern computer will be able to perform these operations with real-time efficiency.

In our apparatus, the photocurrent returns to ground through a 20-kΩ resistor. The voltage across this resistor is the signal, which is proportional to the intensity of light falling on the photocathode. This voltage is amplified by a Keithley Instruments model 103A nanovolt amplifier, which is an AC-coupled differential amplifier with variable high- and low-frequency cutoff filters, and gain variable from 10 to 1000.

The output of the amplifier is connected to a Honeywell-SAICOR SAI-51B real-time spectrum analyzer/digital integrator which computes the power spectrum of the signal and averages spectra computed during successive electric field pulses. The single and averaged spectra are displayed on CRT monitors. Averaged spectra can be recorded digitally on punched paper tape for later computer analysis or plotted on graph paper by an X–Y recorder.

4. Methods

4.1. Chamber Preparation

Prior to each series of experiments the chamber is disassembled and cleaned thoroughly. The glass and plastic parts are immersed in an ultrasonic cleaner with liquid detergent for about a half hour and then rinsed with distilled water. The metal parts are rinsed with distilled water. The interior glass surfaces are then coated with methylcellulose to minimize the surface charge and hence minimize electroosmosis.[45] The coating procedure is as follows:

A 2% solution of Dow Corning Z-6040 (glycidoxypropyltrimethoxysilane) is prepared in one part water, four parts methanol. This solution is acidified with a few drops of concentrated acetic acid and has a shelf life of 1 h. A 0.1% solution of methylcellulose (Polysciences #0846) is prepared in distilled water and centrifuged to remove particulate matter. This solution can be stored indefinitely at 4°C.

The glass is cleaned thoroughly, dried, and dipped into the Z-6040 solution for 1–2 min. It is allowed to dry at room temperature, placed in an oven at about 75°C for 1 h, then rinsed with distilled water. The glass is then dipped into the methylcellulose solution for 5 min, dried at room temperature, and placed in the oven for 1 h.

The methylcellulose coating will remain adsorbed to the walls for several days if the chamber is filled with distilled water and left undisturbed. However, the useful lifetime of a coating is dependent upon the particular experiment being done. A small amount of coating may be removed with each rinsing of the chamber. In addition, some samples may adhere to the wall coating and after a period of time the optical quality of the surfaces may be degraded and the wall charge may be increased. We have found the coatings to have an average working lifetime of about 10 h of operation.

The electrodes are made from 99.99% pure Ag sheet. The surface of the silver is cleaned with concentrated NH_4OH. The electrodes are then anodized in a 0.1 M HCl solution at a current of 5 mA/cm^2 for 5 min. This need not be repeated unless the electrodes are damaged either mechanically or by passage of excessive current.

4.2. Sample Preparation

There are two special considerations for the preparation of electrophoretic light-scattering samples. The first is that solution conditions which affect electrophoretic mobility, such as pH, ionic strength, and viscosity, must be carefully controlled and accurately known. The conductivity of the solution must also be measured accurately, so that the electric field can be determined from the current density.

The second consideration is that the sample must be free of extraneous particles which scatter a significant amount of light. The most common problems are dust and bubbles in the sample. Relatively dust-free solutions can usually be prepared by filtration through small-pore, e.g., 0.1-μm, filters. Filters which are coated with detergents or surfactants should be avoided, as these compounds can interfere with the measurements. The presence of bubbles can be minimized by degassing the sample and storing it at a temperature slightly above that at which the measurements will be made. Routine procedures will work for most samples, but for dilute solutions of macromolecules, extraordinary cleaning methods must sometimes be employed.[46]

4.3. The Measurement

Once the sample has been introduced into the chamber, the intensity of the reference beam is adjusted by means of a variable attenuator to be between 10 and 100 times the intensity of the light scattered from the sample. The relative intensities are determined by observing the photocurrent alternately with and without the reference beam blocked. If the reference beam intensity is too low, it will be insufficient for optical mixing. If it is too high, laser intensity noise will appear in the measured spectrum.

A low-power microscope is used to observe the scattering volume during the measurement to be sure that no foreign matter has drifted into the laser beam and to facilitate adjustment of the position of the beam in the chamber. The microscope also allows the sign of the electrophoretic mobility to be determined for microscopic particles by visual observation of the direction of their motion in the electric field.

For suspensions of large particles which sediment rapidly, it is sometimes necessary to stop data accumulation briefly and stir the sample. At the beginning of the measurement and after each stirring, the spectrum is observed in the absence of an applied electric field to determine that there is no undesirable motion, such as convection, in the sample. The measurement is complete when the averaged spectrum has an acceptably high signal-to-noise ratio and a statistically significant number of particles has been sampled, which usually requires about 5 min.

At the beginning of each experiment, the velocity is measured at various vertical positions in the chamber gap. The position is quantified using an external height gauge. If there is a significant deviation in the measurements at different gap positions, the usual interpretation is that electroosmosis is present. Small magnitudes of electroosmosis can be determined from the flow profile and corrected for in the interpretation of the data, or the scattering volume can be selected to be centered at the region in the chamber at which the electroosmotic velocity is zero, the so-called stationary layer.[39] The flow profile is checked periodically during the course of the experiment. If the electroosmotic velocity is greater than 10% of the electrophoretic velocities, it is best to disassemble the chamber and repeat the methylcellulose coating procedure.

4.4. Common Experimental Problems

The three major sources of instrumental noise in the light-scattering spectra are electromagnetic interference, laser noise, and vibration. The most common example of electromagnetic interference is the presence of peaks in the spectrum at the power-line frequency (60 Hz in the U.S.) and its harmonics. Noise from this source can be identified by its presence even when light is prevented from reaching the photodetector. Electromagnetic interference can be reduced to an acceptable level by proper shielding and grounding of the electronic components.

Laser noise will be present in the spectrum even when only the reference beam is allowed to fall onto the photodetector. Laser noise may consist of both broad-band components and harmonics (120 Hz, etc.) of the power-line frequency. If the laser is designed properly and in good operating condition, laser noise will be much smaller than the signal as long as the intensity of the reference beam is not excessive. Laser noise can thus be controlled by proper adjustment of the optics so that a minimum reference-beam intensity will suffice for optical mixing.

Vibrations of the optical components will introduce spurious sharp peaks into the spectrum. Peaks due to vibration can be recognized by the fact that they are not affected by the application of an electric field to the sample. Vibration noise can be eliminated only by preventing vibration of the components, which can be done by isolating the apparatus from sources of vibration and by mounting the components rigidly with respect to each other.

Obtaining proper optical mixing is often a major difficulty in electrophoretic light scattering experiments. A lack of proper optical mixing will result in the presence of a large signal at the low-frequency end of the spectrum, and a loss of intensity in the Doppler-shifted peaks.[40] It is sometimes difficult, however, to tell whether the absence of shifted peaks in the spectrum is due to poor optical mixing, or to the absence of electrophoretic motion in the sample. A good method to distinguish between these possibilities is to flow the sample slowly through the chamber while observing the spectrum. If no shifted peak due to the flow is observed, the optics are not properly aligned. Final adjustments of the positions of the optical components are best made while observing the quality of the spectra being recorded.

Bulk flow of the sample in the scattering region due to a difference in the level of the liquid on different sides of the chamber or to thermal convection will cause distortion of the spectra. In the absence of an applied electric field, the spectrum should consist of a single peak centered at zero frequency. The zero field spectrum should be observed at the beginning of each measurement to be sure that no bulk flow is present. Flow of the sample will also cause electrophoretically shifted peaks to alternate between two different positions as the direction of the electric field is alternated, since the flow will in one case add to, and in the other case subtract from, the electrophoretic velocity. Convection due to excessive Joule heating during the electric field pulse will broaden the peaks in the spectrum and introduce noise throughout the spectrum. This problem can be corrected by operation at a lower electric field or with shorter field pulses.

5. Examples of Electrophoretic Light-Scattering Spectra

5.1. Blood Plasma

As an example of an electrophoretic light-scattering spectrum from a protein solution, we have selected the spectrum of human blood plasma shown in Figure 6. This spectrum was collected under conditions of low

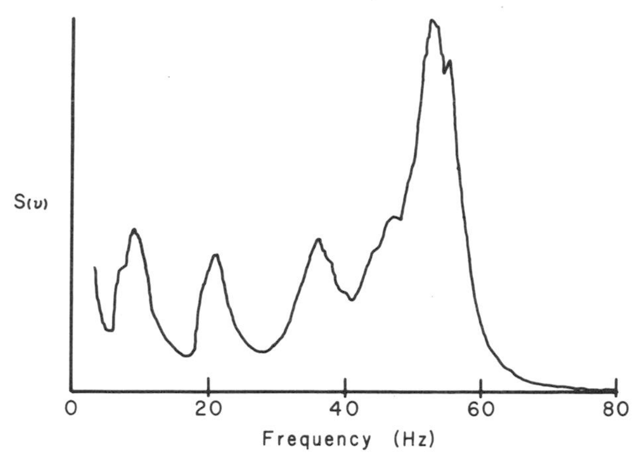

Figure 6. Electrophoretic light-scattering spectrum of human blood plasma. Fresh human plasma was dialyzed and then diluted several-fold to final solution conditions of pH 9.1 and ionic strength 0.004. This spectrum was taken with a high field strength (183 V/cm) to maximize the Doppler shift and at a low scattering angle (3.2°) to minimize the diffusion broadening of each peak. The large peak at the highest frequency can be identified as albumin from its relative magnitude and its electrophoretic mobility (3.9×10^{-4} cm^2/V-sec). Positive identification of the peaks at lower mobility cannot be made from the Doppler spectrum alone, but the form of the spectrum is similar to the known electrophoretic pattern of normal human plasma.

ionic strength, high electric field, and low scattering angle in order to maximize the resolution. The chamber used for this measurement was one of the original Ware–Flygare designs with a BeO cooling plate.[1] The most intense peak is also the highest mobility peak, and it can be identified as albumin both from its relative intensity and its mobility. The frequency shift of this peak was shown to depend linearly on electric field over the range from 25 V/cm to 250 V/cm. The peaks at lower mobility cannot be identified positively, but the form of the spectrum is similar to the known electrophoretic distribution of normal human plasma. Similar ELS spectra of blood plasma have been reported by Ware[1] and by Mohan *et al.*[10]

5.2. Blood Cells

As an example of electrophoretic light-scattering measurements on blood cells, we have chosen an experiment to determine the effect of the enzyme neuraminidase on the surface charge density of human lymphoblasts. Neuraminidase removes *N*-acetylneuraminic acid from glycoproteins and glycolipids on cell surfaces, thereby reducing the net negative charge of the cells.

A cryopreserved sample of mononuclear white blood cells from a patient with acute lymphocytic leukemia was thawed[31] and suspended in tissue culture medium 199 (Grand Island Biological Co. #235) at a concentration of 3×10^6 cells/ml. This suspension was divided into two samples. Neuraminidase (Sigma N-3001) was added to one sample at a concentration of 1.25 units/ml. The other sample served as a control. Both samples were incubated at 37°C for 70 min and then cooled to room temperature. Electrophoretic light-scattering spectra were then recorded for both samples. These spectra are shown in Figure 7. The spectrum from the neuraminidase-treated sample is labeled N and the spectrum from the control samples is labeled C.

The measurements were made at 20°C using the apparatus described in Section 3 of this chapter, at a scattering angle of 58° with an applied electric field of 25 V/cm. The horizontal axis in Figure 7 has been labeled in units of electrophoretic mobility. An electrophoretic mobility of 2.0×10^{-4} cm²/V-sec under these conditions corresponds to a frequency shift of approximately 100 Hz. The sign of the mobility is not indicated on this axis, since the optical mixing technique allows the measurement of only the magnitude of the frequency shift. These cells have a negative surface charge.

The two spectra in Figure 7 demonstrate the value of electrophoretic light scattering for measuring mobility distributions. The time required for

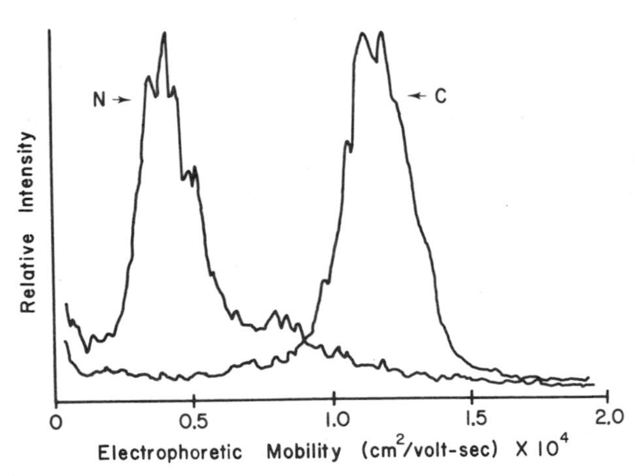

Figure 7. The effect of neuraminidase on the electrophoretic mobility of human lymphoblasts. The curve marked N is the mobility distribution for a sample of cells which have been treated with neuraminidase, an enzyme which reduces the net negative charge of the cells. C represents the control sample. Measurements were made in a physiological-ionic-strength medium at 20°C at a scattering angle of 58° with an applied electric field of 25 V/cm. Since the optical mixing technique does not determine the sign of the velocity of the scattering particles, the mobility axis in the figure specifies only the magnitude of the electrophoretic mobility and not its sign, which for these cells is negative.

data accumulation was about 8 min for each spectrum. Instrumental noise is not significant, and the width of the peaks is due to the heterogeneity of the samples, rather than to limited instrumental resolution. From these spectra, one can determine quantitatively the effects of the enzyme on the electrophoretic mobility of these cells.

ACKNOWLEDGMENTS

The development of ELS in this laboratory has been aided by a number of people whom we are pleased to acknowledge. Dan Haas had made numerous contributions to this effort and has served generously as reviewer of this manuscript. Other co-workers to whom we owe a great debt include Jack Josefowicz, Robert Mustacich, and David Siegel. Construction of various parts of the apparatus has been assisted by several skilled craftsmen. Charles Maheras, Emil Sefner, and Joe Ciampi have performed much of the machining and fabrication. George Pisiello has directed the machining, assisted in mechanical designs, and facilitated greatly the physical operation of our laboratories. Carlton D. Smith kindly designed and constructed the electronic switching circuit.

We are grateful to Prof. Gary Hieftje for editorial comments and assistance.

One of us (Ware) wishes to express his appreciation of the early contributions and continued support and enthusiasm of Prof. W. H. Flygare to the development of this technique.

We are most grateful for the generous financial support of the National Science Foundation. Professor Ware is an Alfred P. Sloan Research Fellow.

6. References

1. B. R. Ware, Electrophoretic light scattering, *Adv. Colloid Interface Sci.*, *4*, 1–44 (1974).
2. W. H. Flygare, B. R. Ware, and S. L. Hartford, in: *Molecular Electrooptics* (C. T. O'Konski, ed.), Part 1, pp. 321–366, Marcel Dekker, New York (1976).
3. B. R. Ware, in: *Chemical and Biochemical Applications of Lasers* (C. B. Moore, ed.), Vol. II, pp. 199–239, Academic Press, New York (1977).
4. B. R. Ware and W. H. Flygare, The simultaneous measurement of the electrophoretic mobility and diffusion coefficient in bovine serum albumin solutions by light scattering, *Chem. Phys. Lett.*, *12*, 81–85, (1971).
5. E. E. Uzgiris, Electrophoresis of particles and biological cells measured by the doppler shift of scattered laser light, *Opt. Commun.*, *6*, 55–57, (1972).
6. T. Yoshimura, A. Kikkawa, and N. Suzuki, Spectroscopy of light scattered by suspended charged particles, *Jpn. J. Appl. Phys.*, *11*, 1797–1804 (1972).

7. E. E. Uzgiris, Laser doppler spectrometer for study of electrokinetic phenomena, *Rev. Sci. Instrum.*, *45*(1), 74–80 (1974).
8. J. Josefowicz and F. R. Hallett, Homodyne electrophoretic light scattering of polystyrene spheres by laser cross-beam intensity correlation. *Appl. Opt.*, *14*, 740–742 (1975).
9. J. D. Harvey, D. F. Walls, and M. W. Woolford. Electrophoretic investigations by laser light scattering, *Opt. Commun.*, *18*, 367–370 (1976).
10. R. Mohan, R. Steiner, and R. Kaufmann, Laser doppler spectroscopy as applied to electrophoresis in protein solutions, *Anal. Biochem.*, *70*, 506–525 (1976).
11. D. D. Haas and B. R. Ware, Design and construction of a new electrophoretic light-scattering chamber and applications to solutions of hemoglobin, *Anal. Biochem.*, *74*, 175–188 (1976).
12. L. Friedhoff and B. J. Berne, Irreversible thermodynamic analysis of electrophoretic light scattering experiments, *Biopolymers*, *15*, 21–28 (1976).
13. G. D. J. Phillies, Effects of intermacromolecular interactions diffusion. III. Electrophoresis in three-component solutions, *J. Chem. Phys.*, *59*, 2613–2617 (1973).
14. M. J. Stephen, Doppler shifts in light scattering from macroions in solution, *J. Chem. Phys.*, *61*, 1598–1599 (1974).
15. M. B. Weissman, Investigations into the Theory and Applications of Fluctuation Spectroscopy, PhD thesis (physics), University of California, San Diego (1976).
16. B. J. Berne and R. Giniger, Electrophoretic light scattering as a probe of reaction kinetics, *Biopolymers*, *12*, 1161–1169 (1973).
17. A. J. Bennett and E. E. Uzgiris, Laser doppler spectroscopy in an oscillating electric field, *Phys. Rev. A*, *8*, 2662–2669 (1973).
18. T. Yoshimura, A. Kikkawa, and N. Suzuki, Measurements of electrophoretic movements with an optical beating spectrometer, *Jpn. J. Appl. Phys.*, *14*, 1853–1854 (1975).
19. T. Yoshimura, A. Kikkawa, and N. Suzuki, The spectral profile of light scattered by particles in electrophoretic movement, *Opt. Commun.*, *15*, 277–280 (1975).
20. E. E. Uzgiris and F. M. Costaschuk, Investigation of colloid stability in polyelectrolyte solutions by laser doppler spectroscopy, *Nature (London) Phys. Sci.*, *242*, 77–79 (1973).
21. E. E. Uzgiris, A laser doppler assay for the antigen-antibody reaction, *J. Immunol. Methods*, *10*, 85–96 (1976).
22. B. R. Ware and W. H. Flygare, Light scattering in mixtures of BSA, BSA dimers, and fibrinogen under the influence of electric fields, *J. Colloid Interface Sci.*, *39*, 670–675 (1972).
23. S. L. Hartford and W. H. Flygare, Electrophoretic light scattering on calf thymus deoxyribonucleic acid and tobacco mosaic virus, *Macromolecules*, *8*, 80–83 (1975).
24. L. Rimai, I. Salmeen, D. Hart, L. Liebes, M. A. Rich, and J. J. McCormick, Electrophoretic mobilities of RNA Tumor Viruses. Studies by doppler-shifted light scattering spectroscopy, *Biochemistry*, *14*, 4621–4627 (1975).
25. D. P. Siegel, B. R. Ware, D. J. Green, and E. W. Westhead, Electrophoretic light scattering spectra of chromaffin granules and a plasma membrane fraction from bovine adrenal medulla, *Biophys. J.* (May, 1978).
26. E. E. Uzgiris and J. H. Kaplan, Study of lymphocyte and erythrocyte electrophoretic mobility by laser doppler spectroscopy, *Anal. Biochem.*, *60*, 455–461 (1974).
27. E. E. Uzgiris and J. H. Kaplan, Laser doppler spectroscopic studies of the electrokinetic properties of human blood cells in dilute salt solutions, *J. Colloid Interface Sci.*, *55*, 148–155 (1976).
28. J. H. Kaplan and E. E. Uzgiris, The detection of phytomitogen-induced changes in human lymphocyte surfaces by laser doppler spectroscopy, *J. Immunol. Methods*, *7*, 337–346 (1975).

29. J. Josefowicz and F. R. Hallett, Cell surface effects of pokeweed observed by electrophoretic light scattering, *FEBS Lett., 60,* 62–65 (1975).
30. E. E. Uzgiris and J. H. Kaplan, Tuberculin-sensitized lymphocytes detected by altered electrophoretic mobility distributions after incubation with the antigen PPD, *J. Immunol., 117,* 2165–2170 (1976).
31. B. A. Smith, B. R. Ware, and R. S. Weiner, Electrophoretic distributions of human peripheral blood mononuclear white cells from normal subjects and from patients with acute lymphocytic leukemia, *Proc. Natl. Acad. Sci. U.S.A., 73,* 2388–2391 (1976).
32. J. H. Kaplan and E. E. Uzgiris, Identification of T and B cell subpopulations in human peripheral blood: Electrophoretic mobility distributions associated with surface marker characteristics, *J. Immunol., 117,* 115–123 (1976).
33. B. A. Smith, The Study of Cell Surface Charge by Electrophoretic Light Scattering, PhD thesis (chemistry), Harvard University (1977).
34. J. Y. Josefowicz, B. R. Ware, A. L. Griffith, and N. Catsimpoolas, Physical heterogeneity of mouse thymus lymphocytes, *Life Sci., 21,* 1483–1488 (1977).
35. S. J. Luner, D. Szklarek, R. J. Knox, G. V. F. Seaman, J. Y. Josefowicz, and B. R. Ware, Red cell charge is not a function of cell age, *Nature (London) 269,* 719–721 (1977).
36. M. W. Woolford, Application of Laser Light Beating Spectroscopy to Electrophoresis, MS thesis, p. 151, University of Waikato, New Zealand (July, 1975).
37. R. A. Alberty, An introduction to electrophoresis, *J. Chem. Educ., 25,* 426–433, 619–625 (1948).
38. C. Tanford, *Physical Chemistry of Macromolecules,* pp. 412–432. Wiley, New York (1961).
39. M. Bier, *Electrophoresis,* Academic Press, London (1959).
40. B. Chu, *Laser Light Scattering,* Academic Press, New York (1974).
41. H. Z. Cummins and E. R. Pike, eds., *Photon Correlation and Light Beating Spectroscopy,* Vol. I (1974) and Vol. II (1977), Plenum Press, New York.
42. B. J. Berne and R. Pecora, *Dynamic Light Scattering,* Wiley, New York (1976).
43. B. M. Watrasiewicz and M. J. Rudd, *Laser Doppler Measurements,* Butterworths, London (1976).
44. S. Goldman, *Information Theory,* Prentice-Hall, New York (1953).
45. S. Hjertén, Free zone electrophoresis, *Chromatogr. Rev., 9,* 122–219 (1967).
46. B. E. Tabor, in: *Light Scattering from Polymer Solutions* (M. B. Huglin, ed.), pp. 1–25, Academic Press, New York (1972).

Detectors for Trace Organic Analysis by Liquid Chromatography: Principles and Applications

Peter T. Kissinger, Lawrence J. Felice, David J. Miner, Carl R. Preddy, and Ronald E. Shoup

1. General Considerations

1.1. Introduction

Although liquid column chromatography (LC) had been used as a means of chemical separation for many years prior to 1969, it was not accepted as a method useful for rapid, routine analysis, due to the relatively long times required to achieve resolution. Significant theoretical and technological advances have since been made in this field, and excellent separations can now be achieved within a few minutes using high-efficiency LC column packings. Liquid chromatography offers many advantages over gas–liquid chromatography (GLC) in that no restrictions are placed on the size, volatility, or thermal stability of the sample molecules. In addition, LC offers tremendous flexibility in the choice of mobile and stationary phases such that most sample components can be conveniently resolved using some appropriate combination of mobile and stationary phases.

One area of widespread interest is the trace analysis of drugs, drug metabolites, pesticides, and other biologically important compounds in complicated matrices (e.g., body fluids, tissue samples, and foodstuffs). Often the thermal instability of these molecules makes their direct GLC analysis difficult, if not impossible. While derivatization is, in some cases,

Peter T. Kissinger, Lawrence J. Felice, David J. Miner, Carl R. Preddy, and Ronald E. Shoup • Department of Chemistry, Purdue University, West Lafayette, Indiana 47907

a satisfactory solution, this approach can be difficult to employ quantitatively on trace amounts of material. Another area of increasing concern is the detection of trace organic impurities in bulk industrial chemicals and commercial products (especially pharmaceuticals). There are now a number of examples in the literature where such determinations have been carried out by LC. Applications to trace (submicrogram) analysis have until recently been hampered by the lack of sufficiently sensitive detectors.

The objective of this chapter is to describe the variety of detectors in use for trace organic analysis by liquid column chromatography. Although many of the detectors discussed here have some applicability to inorganic ions and organometallics, this area will not be emphasized, since a very high percentage of trace LC work is focused on organic substances. This situation is accounted for primarily by the success of atomic spectroscopy techniques that are capable of multielement analysis without prior separation.

Following a brief introduction to the principles and instrumentation required for individual detectors, attention will be given to applications. The utility of chemical reactions to improve sensitivity, to increase specificity, or to extend the applicability of certain detectors will be described in some detail. We have attempted to give a realistic view of both the advantages and limitations of the various approaches currently in use as well as those which are under active development. A number of detectors which have been developed over the years are not reviewed because in our judgment they are not competitive with alternate methods.

Throughout this chapter we have avoided the use of the term HPLC, due to the misconceptions this nomenclature implies for inexperienced chemists. If this designation is used, the "P" refers to "performance" and not to "pressure." Very high pressure (a relatively meaningless term) is certainly not required in modern liquid chromatography nor is pressure a particularly important experimental variable. Frequently "high pressure" is construed as meaning very high cost. Excellent "HPLC" separations can be achieved with equipment operating at pressures below 500–1000 psi. While "high performance" (or "high efficiency") was a useful term a decade ago, today such characteristics of an LC system are commonly taken for granted.

1.2. Trace Analysis

1.2.1. Sample Preparation

The injection of raw biological or environmental samples directly into a chromatograph without prior work-up may at first thought seem attractive. In reality, this approach often leads to degradation of column performance, and places an unnecessary burden on the column to resolve many components which are likely to be of no interest. Even if the resolution is adequate for the component(s) of interest, direct injection of a raw sample will often extend the time required for a complete chromatographic run. Most applications of the apparatus described in this chapter will require the assay of very large numbers of samples. It is, therefore, very desirable to develop approaches which permit the most rapid analysis possible.

As a rule of thumb, trace analysis problems which are most often solved by LC are those involving aqueous solutions of very polar compounds. There are, of course, many exceptions which include fat-soluble substances of high molecular weight (e.g., aromatic hydrocarbons in air particulate samples). In general, low-molecular-weight volatile substances are best handled by GLC. Prior to the analysis of an organic compound in an aqueous tissue homogenate or biological fluid, several cleanup steps are usually necessary. The first of these commonly involves the removal of proteins from the sample. This can often be accomplished by the addition of an agent which causes the proteins to precipitate. A few of the common reagents used for this purpose are perchloric acid, ammonium sulfate, ethanol, and trichloroacetic acid.

After elimination of the protein, it is desirable to devise a method to isolate the class(es) of compounds of interest, while eliminating as many potential interferences as possible. Often this is accomplished by one or more solvent extractions. By careful pH adjustment and judicious selection of the solvent, it is possible to be quite selective in a single extraction step. For example, extraction of urine at pH 7 with ethyl acetate will remove most of the neutral aromatic compounds, while leaving the acidic and basic compounds in the aqueous phase. At pH 2, organic acids will be undissociated and will partition into ethyl acetate, while leaving bases and very strong acids (e.g., sulfuric acid esters) behind. Extractions of this type are most commonly carried out by mechanical shaking in small glass centrifuge tubes (15 ml or less). Brief centrifugation is often helpful to restore the two layers after the agitation step is complete. It is a very simple matter to extract a large number of samples in parallel, and isolate the separated phases using a pipette or syringe. The extracts are usually concentrated by evaporation under a steam of dry nitrogen, using a manifold of syringe needles to blow down several tubes simultaneously. Often the extract is taken to dryness and then accurately redissolved in a small volume of solvent for direct injection or TLC (see below). A preconcentration by a factor of 10 or more is often possible using solvent extraction followed by evaporation. This provides an obvious advantage in trace analysis.

In some cases, liquid–liquid extraction is not effective. Isolation may then be accomplished using a liquid–solid, liquid–gel, or ion-exchange extraction. In the analysis of urinary and tissue catecholamines using electrochemical detection, for example, advantage is taken of the fact that catechol compounds are selectively adsorbed onto alumina at pH 8.5 and then released at low pH. Sample work-up using one or more extraction techniques often seems hopelessly awkward and time-consuming to the novice. While the simplest procedures are most desirable, with appropriate equipment experienced personnel can rapidly carry out large numbers of extractions with good precision. A variety of mechanical pipettes, shakers, and evaporators are now commercially available, and total automation of the process is possible in some cases.

Thin-layer chromatography is another technique which is useful for LC sample preparation. Both specificity and sample preparation are often improved because large numbers of samples can be run in parallel and because the mechanism of separation in TLC is often very different (normal phase) from that used in the column (reverse phase). All components necessarily elute between R_f 0 and 1, whereas for LC there is no limitation on the retention time. This means that TLC–LC experiments on a large number of samples can often be carried out more quickly than LC alone, because uninteresting components eliminated in the TLC cleanup would be strongly retained on the LC column. Furthermore, a TLC cleanup will often prolong the life of very expensive LC packing materials. In the determination of a trace constituent in the presence of a large amount of a bulk chemical, TLC is a useful means of minimizing the amount of major constituent injected on the LC column. Avoiding an overload condition in this manner often makes it easier to resolve the trace component.

TLC zones chosen for examination by LC will normally cover 0.1 R_f units or less. The adsorbent is removed from the plate and treated with solvent in a small (1–3 ml) centrifuge tube. Before such a sample is injected into a modern liquid chromatograph, it should be passed through a filter with a submicron pore size to remove residual TLC stationary phase. Without this precaution, it is likely that the top frit on the LC column will become clogged.

1.2.2. Sample Injection

One of the most important factors in successful trace analysis is efficient utilization of the available sample without sacrifice of precision or accuracy. Gas chromatographic methods frequently require the injection of microliter amounts of derivatized sample contained in a volatile solvent. Difficulties with manipulating such small amounts of material often result in the adoption of procedures in which most of the available sample is wasted (i.e., not injected). The use of automatic sample injectors often involves additional sacrifice of precious sample. Fortunately these problems are greatly minimized in liquid chromatography because it is possible to inject rather large volumes of sample without sacrifice of resolution.[1]

Septum injection ports and stopped-flow injection techniques are a thing of the past in LC, and rotary injection valves (i.e., sample valves) are rapidly becoming the method of choice. When sample is quite limited, a microliter syringe can be used to partially load a sample loop. This technique avoids the waste normally encountered when a loop is filled by suction using a capillary tube immersed in the sample vial. Valves modified for syringe loading are also useful for development work, but they are not recommended for routine trace analysis. These devices lack the precision which can be expected from the conventional six-port valves and are more expensive. Many injection valves are equipped with needlessly long inlet lines and sample loops which are far too small. If the inlet line is shortened and the sample loop lengthened, a greater percentage of a precious sample can be injected while avoiding the sacrifice of precision and the possibility of contamination from a syringe. For commercial LC columns, there are few cases where any advantage will be gained from injecting less than about 20 μl of sample.

For practical work the injection volume can be *at least* as great as the peak standard deviation (σ_v) for the least-retained component of interest. In other words, if the response at the detector to an impulse injection (negligible volume) is known to be Gaussian with a baseline width expressed in volume of mobile phase, the injection volume may be as large as 25% of the peak width without significantly influencing the response. This means that if only a small amount of sample is available, it often may be diluted with mobile phase so as to fill a given injection loop. The important lesson here is that sample preparation techniques for trace components need not aim toward concentrating the final extract into the smallest possible volume. There is a common misconception that greater sample preconcentration leads to greater sensitivity. Beyond a certain point, dependent on the chromatographic efficiency, no benefit is derived from such an effort. It is frequently desirable to work with large volumes because of the improvement in precision and recovery which often results. To give one typical example, if a peak has a baseline width of 30 s and the flow rate is 1 ml/min, it is possible to inject 125 μl (or more) without seriously degrading performance! A complete mathematical treatment is described in Section 1.3.3.

There is one important caveat to the above recommendations. It is preferable that the sample be dissolved in the mobile phase prior to injection and that the unretained components do not overload the column. The influence of injection volume is most pronounced on nonretained substances ($k' = 0$) since these theoretically result in the narrowest peaks. For biological samples there are often numerous nonretained components at much higher concentrations than the trace components of interest. Injecting a very large volume (based on measurement of σ_v for the peak of interest) can overload the column with nonretained material, resulting in a "tail" which interferes with the determination of weakly retained substances. This is a problem which is much less important for detectors which can respond specifically to the analyte(s) under study.

The virtue of injecting the sample dissolved in the mobile phase extends to the fact that most detectors will respond to a change in the bulk properties of the effluent stream. For example, ultraviolet absorption detectors will often respond to a change in refractive index and electrochemical detectors will sense a change in ionic strength. If the components to be measured have small capacity factors (desirable for trace analysis), a detector response to the media injected can lead to serious problems. Because removal of solvent by evaporation or lyophilization is commonly the last step in sample preparation, dissolving the sample in the mobile phase is often convenient.

When trace analysis is the goal, impurities in the mobile phase can be a problem for two reasons. First there is the obvious difficulty of these components contributing to the baseline. More important in some cases is the problem of "vacancy peaks" which may occur when the sample is dissolved in a medium which does not contain the same impurities as the mobile phase. These "negative" peaks will appear at the retention time of the impurities. Those not familiar with "vacancy chromatography" may wish to refer to the literature for a detailed explanation.[2]

1.2.3. Chromatograph

Before beginning a discussion of individual detectors, it will be helpful to briefly consider the general arrangement of the four building blocks of LC: pump, injector, column, and detector. The arrangements schematically shown in Figure 1 will serve to illustrate several concepts which are useful in the design of chromatographs for analyzing both simple and complex samples.

Due to the large number and varied properties of components found in biological samples, there is a high probability that some components will be more-or-less permanently retained on the stationary phase, gradually altering its affinity for compounds of interest. The same problem can result from contamination by trace components in the reagents used to formulate the mobile phase. Since the mobile phase is pumped continuously (typically 1–2 ml/min) and the sample only injected periodically (typically 20–100 μl every 10–30 min), the latter problem can be more severe than one might expect. A "guard column" between the pump and the injection valve is a

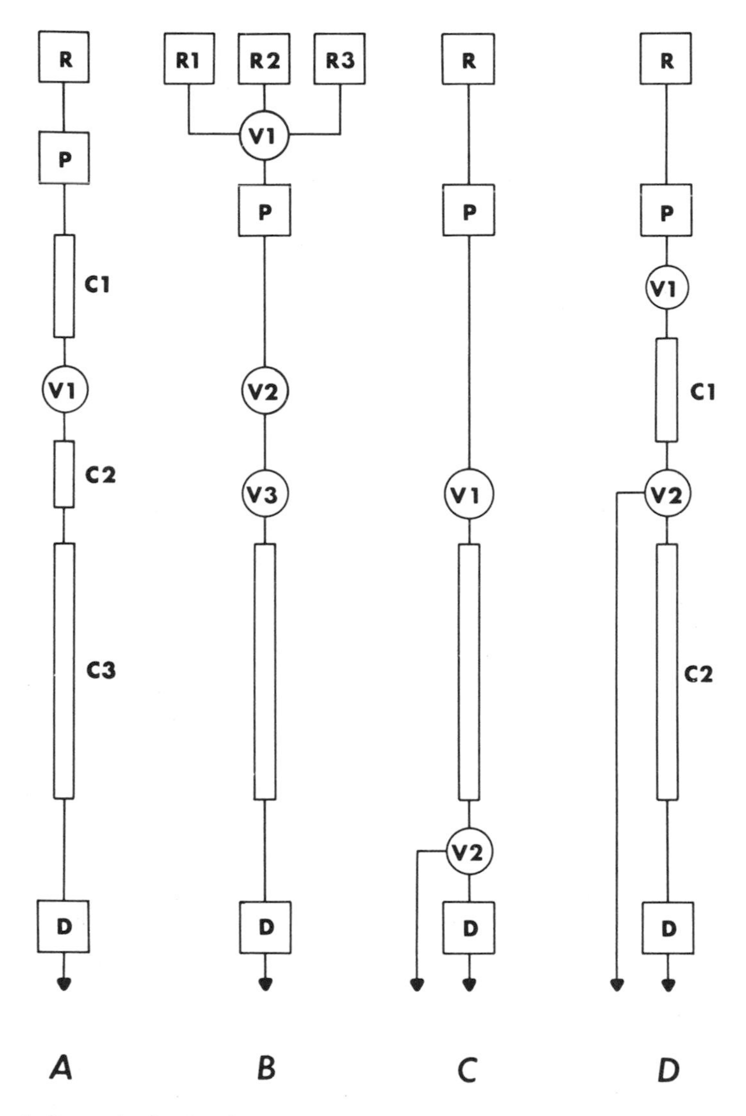

Figure 1. Strategies for development of LC systems for trace analysis. A: A guard column (C1) and pre-column (C2) following the sample injection valve (V1) enhance the life of the analytical column (C3). B: A selection of mobile phase reservoirs (R) prior to the pump (P) is used to select a periodic wash cycle whereas large volume injection valve (V2) is available to apply a gradient plus at some preset time following sample injection (V3). C: A low-dead-volume bypass valve (V2) prior to the detector (D) provides a shunt for concentrated components which degrade detector life. D: A split column permits diversion of high-capacity-factor components using a switching valve (V2) to bypass the major column segment (C2). (Reprinted from reference 3 by courtesy of the American Association for Clinical Chemistry.)

useful means of prepurifying the mobile phase. Guard columns are often packed with a relatively inexpensive, low-efficiency, high-capacity stationary phase that is chemically similar to the high-efficiency material used in the analytical column.

Contamination of the analytical stationary phase by sample constituents may be avoided by use of a short (2–5 cm) "precolumn" packed with high-efficiency material immediately after the injection valve. Such columns are essential when very complex samples such as urine or diluted serum are injected, but can often be avoided by the use of appropriate isolation procedures prior to injection. Precolumns are easily slurry-packed to give low plate heights, H, because they are short; however, some deterioration in over-all column efficiency is to be expected from the additional fittings required. Precolumns often can be used for several hundred samples before replacement is indicated (usually by an increase in back pressure for constant-flow systems).

The useful life of reverse-phase columns can sometimes be extended by periodically flushing them with one or more mobile phases containing a higher concentration of organic solvent. Several useful recipes have been described in manufacturers' literature (e.g., Waters Associates and Whatman have appropriate brochures). For routine work it is convenient to install a valve at the inlet to the pump (Figure 1B) to permit selection of a "cleaning solution." Solenoid-controlled pneumatically actuated valves permit the flush sequence to be carried out at night. The column can then be reequilibrated with the analytical mobile phase before arrival of the next batch of samples. A flush sequence should not be used with a guard column and/or precolumn in position.

A problem frequently encountered in clinical and environmental work is that uninteresting, strongly retained components limit the rate at which samples can be injected. This difficulty is frequently ameliorated by use of a step change in mobile-phase composition and/or flow rate. Step gradients can be accomplished either with a valve prior to the pump or with a six-port sampling valve with a large volume loop (V2 in Figure 1B) between the pump and sample injection valve (V3).

Some detectors perform best at the trace level (electrochemistry and fluorescence) and problems can occasionally result if a major component is permitted to pass through the detector due to adsorption of that component on the walls of the detector cell. In such cases, a bypass valve (V2 in Figure 1C) can be useful to shunt the major component(s) to waste. Figure 1D illustrates another approach ("column switching") which can be used to increase the sampling rate and/or shunt undesired constituents to waste. The chromatographic conditions are adjusted so that the component(s) of interest have an optimum k' (approx. 2–5) and components with a greater affinity for the stationary phase are diverted by V2 after the crucial peak(s) have eluted from the first column (C1). This scheme makes it unnecessary to wait for stongly retained components to pass through the entire column while avoiding a change in mobile-phase composition.

In some detergent-modified reverse-phase columns it is not possible to instantly reequilibrate the column following a step gradient due to the large k' for the modifying reagent. Adsorption and ion-exchange columns also reequilibrate slowly following a gradient elution. It is not always desirable to divert compounds with large k' to waste. In some complex problems the least-retained components are deposited on the top of C2; V2 is then switched to divert the more strongly retained components to the detector. After these peaks have been detected, V2 is switched back and the low k' components are detected after passage through C2. Due to the low diffusion coefficients of compounds in liquid media, this process can be accomplished with negligible loss in efficiency. The split-column technique requires careful adjustment of the mobile phase and relative column lengths, but once a procedure is established the savings in time can be substantial.

Most detectors respond to a change in the bulk properties of the mobile phase causing a baseline drift during the chromatographic run. It should now be clear that there are several excellent reasons to prefer column switching over gradient elution to solve problems in which capacity factors cover a wide range. At this writing the technique is just beginning to attract attention, and very few examples of its use have been published.

An alternative to the split-column idea is to back-flush a single column after the peaks of interest have been quantitated. When the appropriate valving is provided (not shown), it is possible, in effect, to invert the column so that mobile phase flushes strongly retained components off the "top" (now the bottom). There is some controversy about whether or not this procedure (not widely practiced to date) is good for microparticle columns. If a column is well packed (i.e., is not in a mechanically metastable state), most chromatographers believe that there is no detrimental effect from reversing the flow direction.

1.2.4. Peak Quantitation

Most detectors which have been developed for liquid chromatography respond directly to the concentration of a compound in the effluent stream and only indirectly to the amount of that compound which was injected on the column. Since LC experiments are reproducible, the relationship between the peak height (or area) and the amount injected is readily established by the use of calibration standards. Because a variety of factors influence the width and shape of an eluted zone, the calibration factor usually must be determined at least daily. Whenever a column or the mobile phase is changed, one often encounters significant differences in efficiency which will alter a calibration based on peak height but not on peak area. Some detectors (e.g., amperometric devices) are flow-sensitive and thus, if the mobile-phase velocity drifts with time, both peak height and area measurements will drift off calibration.

In practice, once a linear response has been established, quantitation is accomplished by the use of an external standard injected before and after a series of samples. If drift is likely, then it may be necessary to inject a standard quite frequently, perhaps after every tenth sample. When injection valves are used the precision of an LC measurement can be expected to be less than 2% coefficient of variation. This degree of precision is, of course, well within the requirements for most trace determinations. In some quality-control applications better precision is required.

Internal and external standards are often used in combination for LC methods. An internal standard (IS) is helpful in ascertaining whether or not the sample injection was made properly, since the IS peak height should be nearly the same from injection to injection while the sample peaks may vary widely. The ratio of the response for a sample component to that of an internal standard is often more precise than the response to the sample component alone. This advantage is, of course, more significant when the injection precision is poor (e.g., when small sample loops or syringe injection are used). If an internal standard is selected to improve the precision of the chromatography step alone, any readily available pure compound which has a capacity factor close to that of the peak(s) of interest will do the job nicely.

The use of an internal standard is far more important when the intention is to calibrate a sample preparation in a method where the absolute recovery is less precise than the final chromatographic quantitation. When a liquid–liquid or a liquid–solid extraction is carried out on a biological fluid such as urine, the extraction efficiency for the compounds of interest will vary somewhat from sample to sample due to matrix differences (e.g., in ionic strength). In such cases the internal standard must be carefully selected to meet two criteria: (1) variations in the extraction process should similarly effect the sample component(s) and the internal standard, and (2) the internal standard should elute from the column near the peak(s) of interest.

An example of a good internal standard is the use of 3,4-dihydroxybenzylamine to improve the precision of an assay for the catecholamine

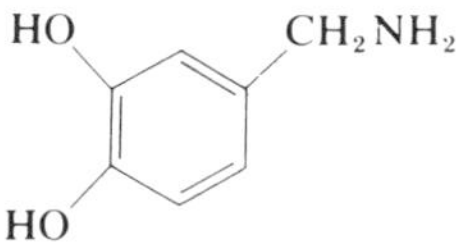

HO CH_2NH_2

HO

Internal standard

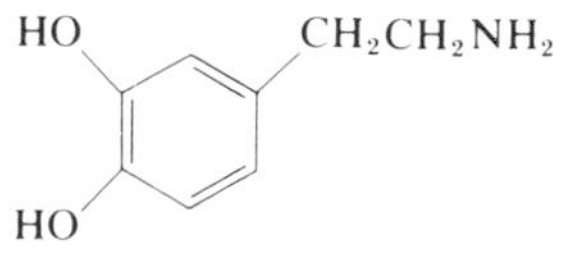

HO $CH_2CH_2NH_2$

HO

Dopamine

dopamine.[4] Both compounds are isolated from urine (spiked with a known amount of internal standard) by cation-exchange extraction and then by binding the catechol function to aluminum oxide. They are separated by reverse-phase ion-pair chromatography and detected electrochemically. For every step in the procedure the two molecules behave very similarly, although the hydrophobicity of the internal standard is sufficiently less than for dopamine to ensure adequate resolution of the two compounds. Even the electrochemical oxidation potential and kinetics are nearly identical, ensuring that the relative response factor for the detector will remain constant.

Some workers rely too heavily on an internal standard and assume that it will ensure adequate long-term calibration. This is sometimes not the case since small changes in the chromatographic conditions may change the relative peak heights for the internal standard and compounds of interest. With most detectors this problem can be alleviated by the measurement of peak areas.

While an internal standard is often reliable for quality-control applications where sample-to-sample variability is small, there is good reason to be cautious when very complex environmental or biological mixtures are under study. In the latter cases, limited peak capacity may make it very difficult to "find room" for an internal standard on the chromatogram. Furthermore, there is no guarantee that some future sample will not contain a compound with a capacity factor indistinguishable from that of the internal standard. In many cases it may therefore be desirable to forfeit the use of an internal standard.

1.2.5. Detector Sensitivity

The assignment of detector or method sensitivity is one of the most controversial subjects in chromatography. The reason for this is that there is *no completely general method of defining sensitivity which will permit convenient comparison between all detectors, compounds, and chromatographic conditions.* Much of the confusion could be eliminated if authors would carefully define what they mean by statements such as

> One ng could be detected
> It was possible to quantitate as little as 10 ng/ml of blood
> A detection limit of 0.2 ng was achieved and the method was linear from 1 ng to 10 μg
> Two ppm of the pesticide residue was detected
> The detector gave a signal-to-noise ratio of 5 at 10^{-7} g/ml

All of the above statements are ambiguous for various reasons. Attention to a few simple rules can clarify their significance:

1. *It is meaningless to state a sensitivity or detection limit figure without*

including some measure of precision. Does the quoted figure imply that "with some imagination I think we can see a peak" or that "good quantitation can be achieved at this level"? A signal-to-noise ratio of 2 is generally used to define the detection limit, but for many analytical objectives data at this level of precision are, of course, useless.

2. *When a sensitivity figure is quoted it should be clearly stated whether the figure refers to the concentration or amount in the original sample, in the injection valve, or in the detector cell.* In most cases it is best to express sensitivity *both* in terms of the original sample (before extractions, etc.) and in relation to the actual injection solution. The quantity or concentration *in a given volume of sample* should be stated (a concentration figure alone is not meaningful). Statement of the amount injected in grams or moles (be it for a sample or a standard solution) is more useful than the concentration injected, due to the relative independence of the LC response from the injection volume (see Section 1.2.2.).
3. The inevitable effects of band broadening on a sample slug as it travels through the chromatographic column necessarily cause dilution of that sample. The concentration in the detector may be an order of magnitude less than what was originally present prior to injection. For this reason, *impressive statements of sensitivity for a solution sitting in the detector cell are often unimpressive when couched in terms of an injected sample.* Another problem along these lines is the time constant of the detector. It can be very misleading to measure the sensitivity of the detector using a stagnant solution (where the time constant could be set at a very large value with no ill effects), if in fact the compound of interest eluted in a very sharp zone.
4. It is important to *give some information about the chromatographic efficiency and the compound for which the sensitivity figure is quoted.* All too often the compound in question is not even identified! Without some knowledge of k' and the column plate count, it is impossible to predict the sensitivity to be expected under different conditions. If detector sensitivity is an important issue, one should refer to the key molecular property (e.g., molar absorptivity at the wavelength used, quantum yield of fluorescence, specific refractive index, oxidation potential).

In general, the above suggestions have not been followed in published reports. It is therefore impossible to give an accurate comparative assessment of the sensitivity to be expected from the various detectors discussed in this chapter. When a number is quoted, it will be qualified as specifically as possible. Obviously, these detection limits are intended to be only a rough guide at the present state of the art. Uniformity in the citing of specifications of detector sensitivity would greatly aid future evaluations. We strongly urge that all reports refer to the amount injected on a column with about 2000 theoretical plates for a compound with a capacity factor between about 2 and 5.

There have been attempts in the past to compare the operating range of various detectors by means of a horizontal bar graph plotted as a function of concentration. Such plots are not meaningful due to the variety of molecular properties used to advantage in the different detectors. It is very difficult to generalize about detection limits and dynamic range without at least specifying the class of compound detected in addition to the chromatographic efficiency.

1.3. Instrumental Causes of Peak Distortion

1.3.1. Introduction

A great deal of attention has been given to the improvement of chromatographic efficiency over the past 20 years, with spectacular results. Modern column LC has earned the title "high performance" largely as a result of a steep reduction in attainable plate heights. The importance of over-all system efficiency in trace analysis cannot be overemphasized. As we examine sample components at lower levels, the number of compounds encountered sharply increases, and with it the probability of serious interference to quantitation. One countermeasure is the selection of a column and operating conditions leading to optimum selectivity (α), hence best resolution. There are both inherent and practical limitations on the ability to accomplish this aim, and the methods best employed are very much dependent on the nature of the analysis and analytes. A successful approach to a challenging trace determination using LC involves complex, semiempirical trade-offs of factors such as column length, stationary-phase composition, precolumn derivatization, mobile-phase modifiers, polarity, pH, ionic strength, and flow rate.

Highly efficient columns can provide the requisite resolution without need for great selectivity, lessening the analyst's burden *in every case*. Enhanced efficiency leads to increased system peak capacity, for the simple reason that each peak becomes narrower. Interference from neighboring peaks will be diminished, and shorter analysis times are possible. A second important benefit arises from decreased zone dilution, as evidenced by enhanced sensitivity. Efficient systems deliver separated components in more concentrated zones to the detector, yielding a relative increase in signal for a given amount of substance.

Because of dramatic improvements in column performance, extra-column contributions to zone-broadening can have a significant effect on resolution in some applications. The more efficient the column, the more concern must be given to reduction of all other detractors from total system efficiency. Extra-column peak distortions fall into two categories: those arising from mobile-phase hydrodynamics ("dead-volume" effects), and those of a purely instrumental nature. Both types can become critical, particularly in attempting to detect a minute amount of one component in the presence of a host of others.

The most common occurrence of distortion is introduced by a delay in response to change with a characteristic time constant. Both electronic time constants and extra-column dead volume manifest themselves in a tailing peak. To an experimentally adequate approximation, this distortion can be modeled by an "exponentially modified Gaussian" function. In this section we will examine the origin of instrumental distortion, the properties of the modified response functions, and the influence on system efficiency.

1.3.2. Effect of Finite Response Time

No electronic device is capable of responding instantaneously to a changing input or, in other words, to operate with signals of arbitrarily high frequency. Vanishing response times imply infinite bandwidth, suggesting a reciprocal relationship:

$$B = \frac{1}{\tau} \tag{1}$$

where B is the bandwidth in hertz and τ is the system response time, or time constant, in seconds. Limited bandwidth is an essential feature of amplifier design and may be nothing more than an intentional increase in the time constant (τ) to discriminate against high frequencies. A chromatogram is a low-frequency signal. Components above a certain frequency contribute no information and are observed as "noise." A low-pass circuit makes an effective noise filter as long as it does not reduce the bandwidth to the extent that frequencies carrying information are attenuated. When this occurs, distortion of the peak shape will become evident.

A completely general analysis of this type of distortion would involve application of the Fourier transform to the filter input waveform $E(t)$. The resulting function can be presented as the frequency spectrum $\bar{E}(\omega)$ equivalent to the input. The effect of the filter is completely described by its complex transfer function, which, when multiplied by $\bar{E}(\omega)$, gives $\bar{P}(\omega)$, the output frequency spectrum. The inverse Fourier transform of $\bar{P}(\omega)$ is $P(t)$, the output waveform. The power of the transform approach lies in its generality and the relative ease with which a result may be obtained from complex waveforms and transfer functions. The mathematical details are beyond the scope of this discussion.

The low-pass filter most often encountered in chromatography is the simple RC network or filtered amplifier stage, for which the former serves as an adequate model. Since a Gaussian excitation function $E(t)$ is a fair approximation of an eluting peak, a rigorously correct solution for the effect of $\tau = RC$ may be arrived at under these assumptions by solving a simple differential equation. This analysis, involving only elementary calculus, is detailed in Appendix 1, along with a graphical presentation of the results.

Excessive RC filtering (τ too large) produces a "tailing" peak which is lower, broader, and shifted to longer apparent retention times. These general characteristics, each of which is objectionable to the chromatographer, are observed regardless of the exact functional form assumed for a detector output. The extent of distortion is controlled by the ratio S of the undistorted peak standard deviation to the filter time constant ($S = \sigma/\tau$). Figure 2 depicts the influence of $1/S$ on relative peak amplitude, full-width at half-maximum, and displacement from the true retention time.

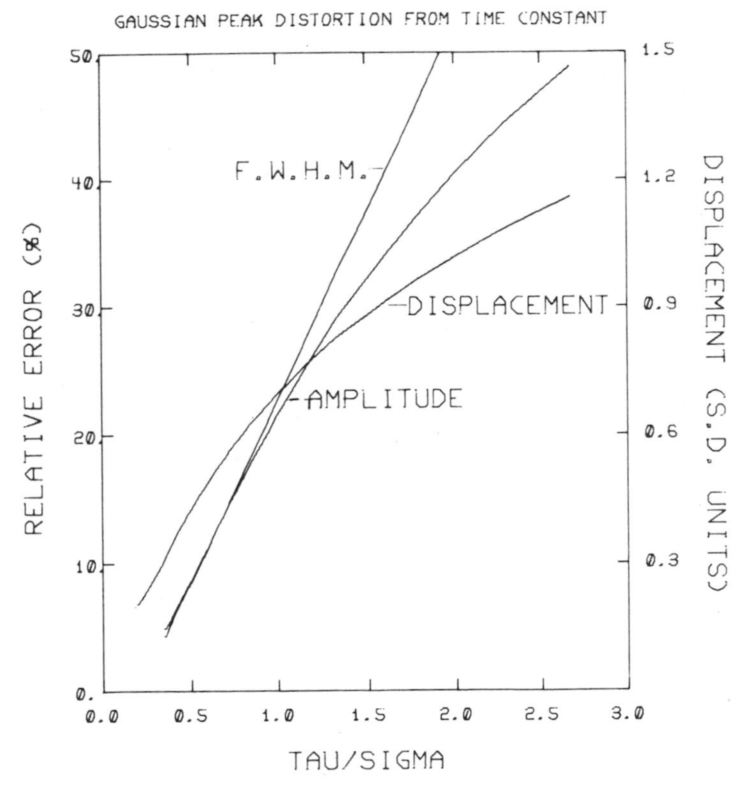

Figure 2. Relative error in (1) maximum amplitude, (2) retention time, and (3) full-width at half-maximum for increasing time-constant distortion. Note that $\tau/\sigma = 1/S$.

In the absence of complications, the peak standard deviation for a component of given capacity factor k' is related to the total number of theoretical plates N exhibited by the column by

$$\sigma \propto 1/\sqrt{N} \tag{2}$$

High-performance LC columns typically give large N values, thus "sharpened" peaks and lower S values. If the filter time constant is not also reduced, needless distortion may ensue. Unfortunately, it is rather difficult to recognize purely electronic distortion because quite similar asymmetric broadening is very often produced by column and flow-stream anomalies (see Section 1.3.4). Well-designed detector amplifiers should include a provision for time-constant adjustment so that the minimum distortion-free setting can be chosen. To avoid half-width broadening in excess of 5%, the time constant should be less than $\sigma/3$.

The deleterious effect of excessive filtering on resolution of neighboring peaks is quite pronounced. McWilliam and Bolton[5,6] have treated this problem in detail. Figure 3 illustrates the effect of "dialing in" a time constant on three Gaussian peaks of unity standard deviation. Two situations are illustrated, namely those of a minor component eluting prior to a major one and following a major one. These authors have shown that the latter case is more sensitive to the "swamping-out" effect. The overlap can become so severe as to obliterate the minimum between the two peaks.

1.3.3. Influence of Finite Detection and Injection Volume

Distortion can arise when the volume of mobile phase subject to detection at any instant is not negligible relative to the total peak volume. All commonly used detectors provide an output which is ideally proportional to the quantity of some component(s) of the mobile phase present in the detection zone. This zone must have a finite volume, so the detector output is actually representative of the average (integral) concentration across the zone. Rapid concentration changes passing through the zone will be "averaged out" to give a diminished response, whereas gradual changes will be unaffected. It is important to realize that this is entirely an instrumental artifact, separate from response time effects (Section 1.3.2.) and flow dynamics. Peak amplitude and dispersion, but not retention time, are affected. This symmetrical broadening can be shown to occur whether the zone of finite volume resides in the detector or the injector (see Appendix 2). Both effects are readily summarized as additive contributions to the peak second moment (variance):

$$\sigma_O^2 = \sigma_P^2 + \sigma_D^2 + \sigma_I^2 \tag{3}$$

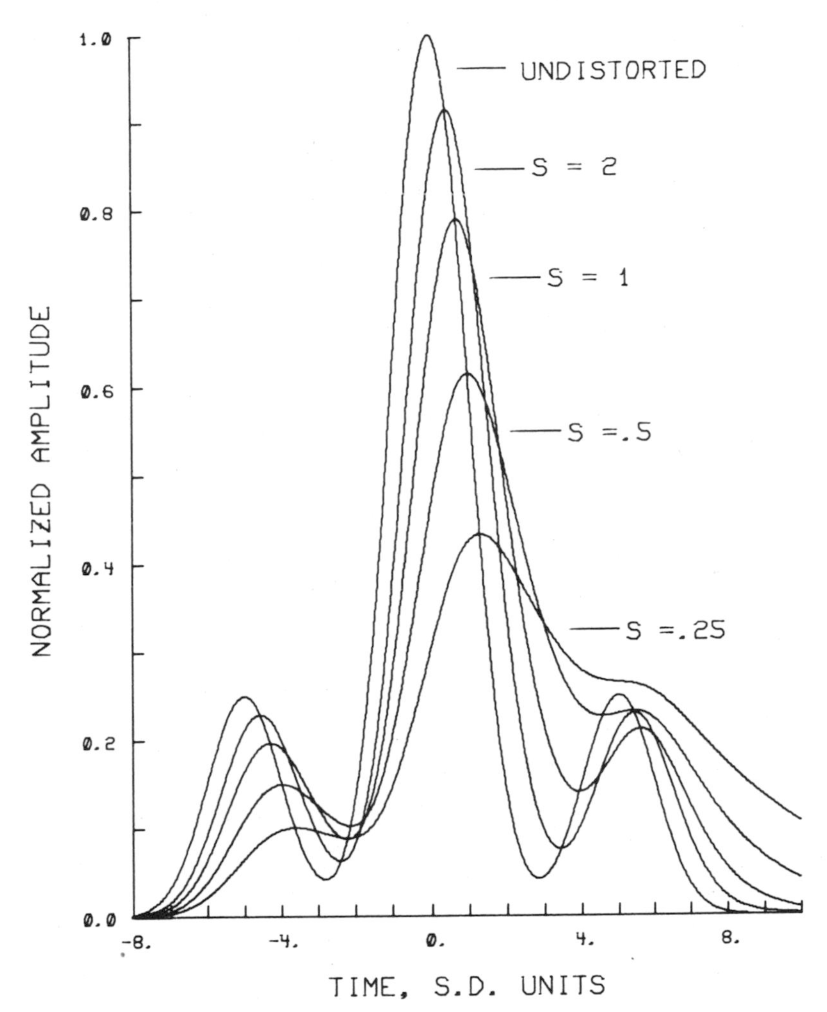

Figure 3. Artificial Gaussian chromatogram, showing effect of a time constant on shape and resolution. All peaks have unity standard deviation, and the time axis is arbitrarily shown as standard deviation units from the central peak.

where σ_O^2 = observed variance; σ_P^2 = inherent peak variance; σ_D^2 = contribution from the detector; and σ_I^2 = contribution from the injector.

The derivation of the modified peak-shape function and the dependence of peak amplitude, σ_D^2, and σ_I^2 on zone volume are given in Appendix 2. The results are summarized in Table 1, which expresses the maximum value of zone-to-peak volume (or time) ratio which can be tolerated for selected relative errors in peak amplitude and variance. The latter quantity is clearly the more susceptible to degradation. This will appear as

Table 1. Effect of Detection Zone Transit Time on Peak Amplitude and Variance[a]

Relative error in quantity (%)	Quantity (maximum τ/σ)	
	Amplitude	Variance
0.5	0.35	0.24
1	0.49	0.35
3	0.86	0.60
5	1.12	0.77
10	1.63	1.09

[a] The maximum τ/σ ratio allowed for fixed error limits on each quantity is shown. The two right-hand columns show the value to which τ/σ must be reduced in order to assure an error no greater than that in the left-hand column for the chosen quantity.

a decrease in realizable column efficiency and resolution, even though column processes are not involved. Furthermore, the situation is worse for early peaks in the chromatogram and for higher-efficiency columns, since the detector and injector volumes are fixed by their design. If the true peak-volume standard deviation equals the zone volume ($\tau/\sigma = 1$), the observed variance is increased by 10% and the standard deviation σ by about 5%. Since resolution is inversely proportional to observed standard deviation, resolution of neighboring peaks whose average standard deviation is less than either zone volume will be degraded by more than 5%. This statement supports the prior assertion (Section 1.2.2.) that the injected volume should be under 25% of the basewidth (4σ) if a 5% error margin is acceptable.

1.3.4. Fluid Time Constant Effects

Among the many influences on peak shape, only the electronic ones are well defined and lend themselves to straightforward, general analysis and correction. The most pervasive and obstinate detractions from system performance stem from the paths taken by the mobile phase between injection and detection. Flow anomalies engender two distinct types of band broadening; symmetric and nonsymmetric. Symmetric broadening phenomena in liquid chromatography have been extensively studied and appear to result largely from inhomogeneities in bed packing. Unlike for gas chromatography, longitudinal diffusion makes a very minor contribution.

Nonsymmetric distortion, or skewing, can arise from a variety of sources. Column overloading and other phenomena related to isotherm nonlinearity are common sources of skewing which are easily traceable and usually curable. Less amenable to correction and less well understood is band broadening due to extra-column flow disturbances. These are generally depicted as sharp transitions in tubing diameter and pockets of fluid not cleanly swept by the eluant stream, often lumped together under the term "extra-column dead volume." Gaps or channels in the column bed can also fall into this category.

A rigorous treatment of nonsymmetric broadening has been thwarted by the complexity of interacting convective and diffusive processes variant with pressure, temperature, viscosity, geometry, and flow rate. However, a good physical insight is offered by loosely applying the hydrodynamic analogies to electronics. For simple electronic circuits, one can envision an ideal fluid model in which pressure is analogous to voltage, flow rate to current, and diameter to resistance. One model for a capacitor is simply a "bulge" in a straight pipe which stores a quantity of fluid (charge) at a certain pressure (potential) for subsequent release. The presence of this "unswept pocket" introduces an element of delay into the flow stream. If the flow resistance due to tube walls ahead of the pocket is thought of as the fluid equivalent of a resistor, the analogy to the RC network (Figure 52) is complete. A fluid time constant is often intentionally introduced immediately following the pump in the form of a pulse damper. Its tremendous "dead volume" acts to partially absorb pressure impulses. In a like fashion, parasitic liquid time constants dampen impulses of solute concentration (peaks).

This analogy is obviously inexact, since it does not include liquid compressibility or velocity distributions, nor does it satisfactorily explain the situation at diameter transitions. However, its essential correctness is attested to by the remarkable degree to which the exponentially convoluted Gaussian model appears to fit experimental peaks.[7] Dead-volume broadening is therefore another manifestation of finite system response time, as discussed in Section 1.3.2.

The effect of several delaying elements in the system is cumulative; the broadening, attentuation, and displacement due to a series of time constants is approximately the sum of those from each acting separately, if no one is predominantly large.[5] Thus it is feasible to assess a chromatographic performance in terms of its over-all time constant. A novel method for extracting the time constant from an isolated peak profile has recently appeared.[8] Other studies[9,10] have demonstrated the validity of this model as a means of LC peak characterization.

Of particular relevance to trace analysis using LC is the direct relationship between system efficiency and detection limit. A detailed discussion of this point is given by Karger, Martin, and Guiochon,[11] who define the detection limit characteristic as

$$D^* = \sqrt{\frac{2\pi}{N}}\, V_r \cdot 5\, \frac{G}{S} \tag{4}$$

where N = theoretical plate number, V_r = retention volume of the sample component, and G/S is the noise-to-signal ratio characteristic of the instrument. N must be taken as the apparent plate number deduced from the peak width. Several alternatives exist for calculation of N through the definition of the plate height H of a column of length L:

$$H = \frac{L}{N} = \left(\frac{\sigma}{t_r}\right)^2 = 16\left(\frac{W_b}{t_r}\right)^2 \doteq 5.54\left(\frac{W_{1/2}}{t_r}\right)^2 \tag{5}$$

Here σ is the standard deviation in time units. For a Gaussian of standard deviation σ' convoluted with an exponential of time constant τ,

$$\sigma = \sqrt{\sigma'^2 + \tau^2} \tag{6}$$

(see Appendix 1). Both W_b and $W_{1/2}$, the baseline width and full-width at half-maximum, respectively, are also increased. W_b is usually found by triangulation at the inflection points, but examination of Figure 52 shows that location of the trailing inflection point becomes more troublesome with increasing τ. Regardless of the method used, a diminished N value is found, and so an elevated detection limit follows. The line of reasoning here is not entirely without complication, since increased RC filtering may help reduce G/S in some situations. No such virtue can be attributed to extra-column fluid time constants, whose chief effect is a dilution of the zone preceding its detection.

The cost and effort, often considerable, which must be expended to achieve very high efficiency (N) columns can easily be wasted by inattention to extra-column band broadening. Every portion of the system associated with the sample, injector, tubing, fittings, column bed, detector, amplifier, and recorder, may influence the output peak shape. Each can and should be examined independently to assess its contribution. Among the mathematical treatments of extra-column broadening, the enduring work of Sternberg[12] stands out. Although written with a focus on capillary gas chromatography, also a high-efficiency technique, its generality makes transference to LC virtually complete.

2. Optical Detectors

2.1. Infrared Detectors

The infrared spectrum of a molecule provides excellent information as to its identity. Thus a rapid-scanning IR detector for LC could aid in the identification of eluted compounds. A variable-wavelength IR detector would have adjustable selectivity: set to the C—H stretch region it would be a universal detector for organics; tuned to other regions it might primarily detect compounds having a given functional group.

The infrared region is not widely used for LC detection for several reasons. A major problem is IR absorption by the mobile phase. One is limited to relatively nonpolar aprotic solvents, and from these a solvent must be chosen which transmits significantly in the region of interest. With these limitations on solvent selection, chromatographic optimization is difficult (e.g., most reverse-phase separations are ruled out). To overcome this, Griffiths and Kuehl have studied several techniques for eliminating the mobile phase in an on-line fashion.[13] They had greatest success with an FT–IR system in which the effluent is sprayed into a gold-coated light pipe. The solvent is evaporated off and the reflection–absorption spectrum of the residual sample is measured. More work is required before this approach becomes practical.

A second limitation of IR detectors is that absorption coefficients are lower in the IR than in the UV, thus sensitivity is poor. The use of Fourier transform techniques could alleviate this problem, as it has in GC–IR, if solvent absorption problems can be overcome.

IR detection has been successfully used in gel permeation chromatography of polymers. In this application, high sensitivity is not usually required. Furthermore the choice of mobile phase is less critical in size exclusion separations than in other forms of chromatography. In comparison with the refractive-index detectors also employed in polymer work, IR detectors can be more sensitive, their response varies less with molecular weight, and they are much less temperature sensitive.[14] A quality instrument designed for GPC is available from Wilks Scientific. The Miran I operates over the full IR region from 2.5 to 14.5 μm with good photometric accuracy and low noise and drift. Several low-volume flow cells are available, providing different path lengths for solvents of different opacities. Minimum detectable quantities are on the order of 1 μg.

2.2. UV–Vis Absorbance

2.2.1. Recent Advances in Instrumentation

General. Ultraviolet absorption detectors are the most widely used detectors for LC. Absorption of UV radiation by solutes is often characterized by broad bands with large extinction coefficients. Therefore UV absorption detectors provide relatively low specificity and high sensitivity. Because the concepts of electronic spectroscopy are widely appreciated in connection with other applications, we will devote our attention here to several aspects of major importance to the LC application.

Several types of UV detectors are commercially available. The simplest and most common detector is the fixed-wavelength photometer. This instrument utilizes the 254 nm line from a low-pressure mercury lamp. The line is sharp and strong, and the detector is quite sensitive and stable. Another wavelength used alternately or in tandem with 254 is 280 nm, which is obtained from a phosphor excited by 254 nm light. Multiwavelength filter photometers employ a medium-pressure mercury lamp, which has lines of good intensity at 254, 280, 313, 334, and 365 nm. The line of interest is isolated with a narrow bandpass interference filter. Variable wavelength spectrophotometers employ continuous sources such as deuterium or xenon-arc lamps. A single wavelength of interest is selected with a grating monochromator. These instruments are much more flexible, although less stable, less sensitive, and more expensive.

Variable-wavelength instruments with two new capabilities have recently become available from several manufacturers. These detectors use a deuterium arc source and are stabilized by double-beam design. Their first advantage is that they can be operated down to 190 nm, opening up the far-UV region. This was made possible by minimizing stray light problems which normally interfere due to low source intensity in this region. Many compounds absorb strongly in the far UV, including aromatics and unsaturated or heteroatom-containing compounds. Many common solvents which are used as mobile phases will also absorb in the far UV, but these can be successfully avoided. Solvents which are used must be of very high purity, and their cost is high.

The focus of much recent work is on the acquisition of more spectral information from a single chromatographic run. Several commercial instruments now have the capability to obtain complete spectra of species in the cell. This is accomplished by stopping the flow through the chromatograph. A complete spectrum of the column effluent trapped in the cell is then obtained by switching on a wavelength drive mechanism. This is a less-than-ideal method for obtaining spectra since the chromatographic process will be more or less disturbed, but the spectra obtained can be quite useful for identification of unknown peaks. Simply collecting fractions and using a conventional spectrophotometer would appear to be a more realistic approach.

A simple means of acquiring limited qualitative information involves obtaining ratios of absorbances at two wavelengths. These ratios are characteristic of a given species, and several ratios matching those of a standard constitute good evidence for the identity of a peak. Ratios can be obtained from two detectors operating in tandem or from a variable-wavelength instrument by stopping the flow through the cell and manually adjusting the wavelength dial. Several authors have discussed the value of absorbance ratios for peak identification and for quantitative determination of unresolved peaks.[15–17]

Flow Cells. The standard flow cell used for absorption detection is schematically depicted in Figure 4. Recent work in the literature describes two modifications of the standard absorption cell. The use of a tapered cell, whose radius increases toward the detector end, was suggested as a means for eliminating refractive index effects.[18] Some light is always refracted at the entrance window–solution interface and strikes the walls of a conventional cell. Changes in the refractive index of the solution in the cell will change the amount of light striking the walls. If the cell is constructed such that the walls are strongly absorbing, the amount of light reaching the detector will also change. Refractive index changes can occur during gradient elution or when a high concentration of sample passes through the cell. The tapered cell minimizes the amount of refracted light which strikes the walls and thus stabilizes the baseline response of an instrument.

A second cell modification involves the reconnection of the two cells of a dual-beam instrument such that the solution from the column flows

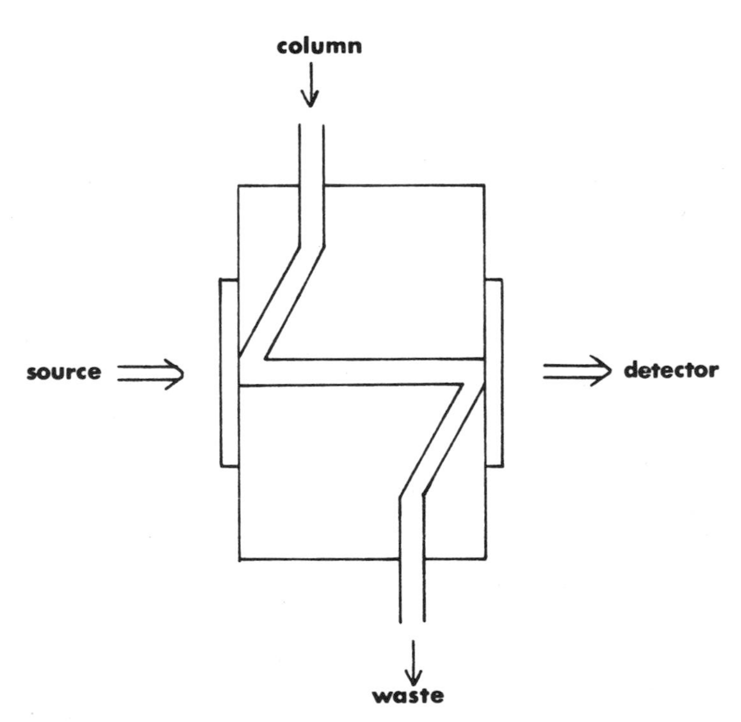

Figure 4. "Z" cell used for UV detection.

first through the sample cell and then in the same direction through the reference cell.[19] The result is that the derivative signal dA/dt is obtained from the instrument. This is easily done and does not affect the stability of the dual-beam instrument, but the derivative signal is generally of little interest.

Simultaneous Multiwavelength Instruments. Currently there is some interest in the development of simultaneous multiwavelength UV detectors. On the slow time scale of a liquid chromatographic separation, these detectors provide essentially instantaneous scans of the absorption spectra of the contents of the detection cell. Thus they can provide a wealth of information about the chromatographic process. (It should be clear that such an instrument is only practical when interfaced to a computer for processing the enormous amount of data generated.)

Since the original work in this area done by Bylina and co-workers,[20] three instrumental approaches to the problem have been studied. An oscillating-mirror rapid-scanning spectrometer has been used as a detector.[21] The 3-dimensional chromatograms generated by this instrument (see Figure 5) serve to illustrate the potential of multiwavelength detection

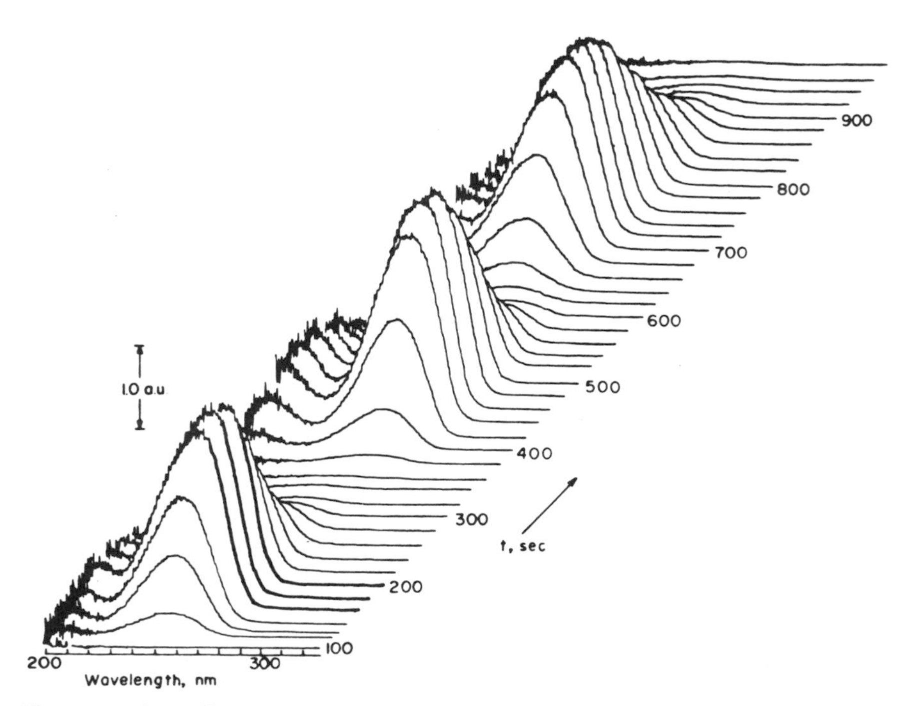

Figure 5. Three-dimensional chromatogram obtained with rapid scanning spectrometer for the separation of uracil, cytosine, and adenine on Aminex A-4 cation exchange resin. (Reprinted from reference 21 by courtesy of the American Chemical Society.)

for LC. The cost and complexity of this dispersive spectrometer precludes its routine use in the near future. Available repetition rates, for example (218 Hz) are much greater than necessary.

Two other approaches to multiwavelength detection, in principle, provide a multichannel advantage in signal-to-noise ratio over the oscillating-mirror detector by simultaneously monitoring multiple wavelengths. In these instruments the total output of a UV source (deuterium or xenon arc) passes through the flow cell and is dispersed afterwards. A multichannel detector is placed in the exit focal plane of the monochromator (see Figure 6). Two types of array detectors have been tested for this application. Pardue and McDowell used a vidicon tube in the simple single-beam configuration.[22] In the vidicon the light beam impinges on a photoconductive target. The incident radiation creates electron-hole pairs which discharge the capacitive elements. The signal is extracted by restoring the target potential with an electron beam. Several groups have used solid-state linear photodiode arrays in both single- and double-beam configurations.[23–25] These arrays consist of a series of photosensitive *p–n* junction diodes together with address and control circuitry and an amplifier.

The relative merits of the two detectors are such that for liquid chromatography no clear choice can be made between them at this time. The vidicon is capable of being operated in a dual-beam mode without having two sets of optics and two detectors as is required for the linear diode arrays. However, their integration time is limited. The linear arrays are less expensive and should become even less costly in the future. Both detectors are subject to systematic pattern noise, but this can easily be computer-corrected. A limitation of both array detectors is that compared with single-wavelength instruments they perform poorly below 210 nm, a problem which is not readily solved.

A number of advantages to obtaining complete multiwavelength data have been demonstrated or postulated. Complete spectral data allows selection of detection wavelengths for each peak such that maximum sensitivity and best resolution are obtained while interferences are tuned out. Spectral information is available to complement retention times in deter-

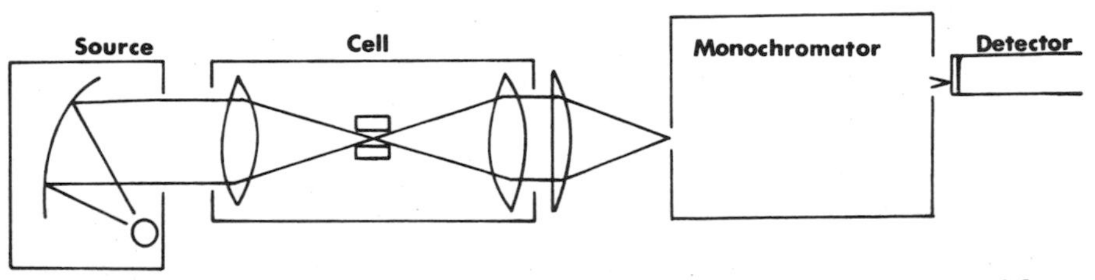

Figure 6. Schematic representation of optical layout of array detectors. (Reprinted from reference 22 by courtesy of the American Chemical Society.)

mining peak identities. By the same token, impurities can be detected by comparing spectra at different points across a peak. If the spectra of overlapped peaks are known or are at least partially chromatographically resolved, they can be deconvoluted. Ultimate sensitivity for well-resolved peaks should be enhanced by integration over the entire absorption spectrum. This advantage over single-wavelength detectors has not yet been achieved due to the superior performance of photomultiplier tubes compared with the individual elements making up an array detector.

An interesting treatment of multiwavelength data was carried out by Milano and Grushka.[26] They demonstrated two uses for derivative spectral monitoring ($dA/d\lambda$ vs. time). Two overlapping peaks can be deconvoluted by monitoring $dA/d\lambda$ at a wavelength where the derivative for one of the components is zero. Derivative chromatograms are reportedly more sensitive for single-beam instruments in that the derivative reduces the noise due to lamp fluctuations.

Utilization of all the data generated by an array detector requires a computerized data-handling system. Sophisticated algorithms are currently being developed for efficient data reduction. Ultimately, comparison with libraries of spectral data for compound identification should be possible; however, the molecular specificity of UV absorption is rather limited when compared to other techniques such as mass spectrometry.

The cost of a complete simultaneous multiwavelength instrument (if one were available) would be at least $20,000, as compared to $5,000 for the dual-beam variable-wavelength spectrometers discussed earlier. Are the capabilities of the simultaneous multiwavelength detector worth the cost? For routine applications, the answer is no. They should, however, be of considerable value in the development of new LC assays and may solve intractable resolution problems in complex cases by virtue of their deconvolution potential. If the technology improves somewhat, as seems likely, array detectors will certainly be a very useful tool in liquid chromatography.

2.2.2. *Direct UV Detection of Column Effluent*

Applicability. It should be clear from the above discussion that there has been a great deal of activity in the optimization of UV absorption measurements in small-volume flow cells. A few years ago the performance of most commercial UV detectors was not satisfactory when the injected amount of a simple aromatic compound was less than about 0.2 μg. In those days fluorescence and amperometric detectors were the only choices when quantitative data were required at the low-nanogram level. It is important to recognize that the UV detector is currently a very viable approach to trace analysis, and the advent of sensitive variable-wavelength detectors has provided a degree of selectivity which was previously unavailable at the trace level. At this writing, the single-wavelength UV detector at 254 nm is the most generally satisfactory LC detector for a wide variety of problems.

The authors estimate that better than 95% of all submicrogram LC methods are currently carried out using UV detection. Before examining pre- and post-column derivatization techniques, it is appropriate to consider the power of UV detection for directly detecting sample constituents. A large number of complex organic molecules contain conjugated systems which strongly absorb UV radiation. Examples in the literature of direct UV methods are numerous and thus will not be covered here, instead the reader is referred to Chapter 5 of this volume, where examples in the area of clinical analysis are discussed in detail.

In assessing the applicability of UV detectors it is useful to examine data such as those listed in Table 2. The effect of numerous functional groups on the extinction coefficient and absorption maximum for monosubstituted benzenes is summarized. Starting with data such as these, it might be expected that maximum sensitivity would be attained using a

Table 2. Absorption Maxima of the Substituted Benzene Rings Ph—R[a]

R	λ_{max} nm (ϵ) (solvent H_2O or MeOH)			
—H	203.5	(7,400)	254	(204)
$—NH_3^+$	203	(7,500)	254	(160)
—Me	206.5	(7,000)	261	(225)
—I	207	(7,000)	257	(700)
—Cl	209.5	(7,400)	263.5	(190)
—Br	210	(7,900)	261	(192)
—OH	210.5	(6,200)	270	(1,450)
—OMe	217	(6,400)	269	(1,480)
$—SO_2NH_2$	217.5	(9,700)	264.5	(740)
—CN	224	(13,000)	271	(1,000)
$—CO_2^-$	224	(8,700)	268	(560)
$—CO_2H$	230	(11,600)	273	(970)
$—NH_2$	230	(8,600)	280	(1,430)
$—O^-$	235	(9,400)	287	(2,600)
—NHAc	238	(10,500)		
—COMe	245.5	(9,800)		
—CHO	249.5	(11,400)		
—Ph	251.5	(18,300)		
$—NO_2$	268.5	(7,800)		
$—CH\overset{t}{=}CHCO_2H$	273	(21,000)		
$—CH\overset{t}{=}CHPh$	295.5	(29,000)		

[a] From: D. Williams and I. Fleming, *Spectroscopic Methods in Organic Chemistry*, McGraw-Hill, London (1973), by permission. Most values from: H. H. Jaffe and M. Orchin, *Theory and Applications of Ultraviolet Spectroscopy*, Wiley, New York (1962), by permission.

variable-wavelength instrument adjusted to λ_{max}. This is not necessarily the case, however. For compounds with maxima around 254 nm, the fixed-wavelength instrument is the one of choice. When using a variable-wavelength instrument it must be remembered that S/N is a function of source intensity and detector response as well as ϵ_{max}. In addition, the need to minimize interferences may be a factor in selection of an optimum wavelength.

Use of the Far UV. The above-mentioned considerations are especially important when operating in the far UV. In this region source intensity is low and falling off rapidly. Most organics absorb very strongly in this region (see Table 3), which is both an advantage and a serious drawback. It is an advantage since compounds such as alkyl halides, nitro compounds, and sugars, which are difficult to detect by other means, will absorb strongly in this region. However, this is also true of numerous interferences in complex samples and of most mobile-phase constituents; these two factors make chromatographic optimization more difficult. Solvents must be of very high purity and therefore are expensive. Nevertheless, Figure 7 illus-

Table 3. Characteristics of Simple Chromophoric Groups[a]

Chromophore	Example	λ_{max}nm	ϵ_{max}	Solvent
$>C=C<$	1-Octene	177	12,600	Heptane
$-C\equiv C-$	2-Octyne	178	10,000	Heptane
		196	*ca*2,100	Heptane
		223	160	Heptane
$>C=O$	Acetone	189	900	Hexane
		279	15	Hexane
$-CO_2H$	Acetic acid	208	32	Ethanol
$-COCl$	Acetyl chloride	220	100	Hexane
$-CONH_2$	Acetamide	178	9,500	Hexane
		220	63	Water
$-CO_2R$	Ethyl acetate	211	57	Ethanol
$-NO_2$	Nitromethane	201	5,000	Methanol
		274	17	Methanol
$-ONO_2$	Butyl nitrate	270	17	Ethanol
$-ONO$	Butyl nitrite	220	14,500	Hexane
		356	87	Hexane
$-NO$	Nitrosobutane	300	100	Ether
		665	20	Ether
$>C-N<$	Neopentylidene *n*-butylamine	235	100	Ethanol

[a] From: Dyer, *Applications of Absorption Spectroscopy of Organic Compounds*, Prentice-Hall, Englewood Cliffs, New Jersey (1965), by permission. Most values from *Organic Spectral Data*, Vols. I, II, and IV, Interscience, New York.

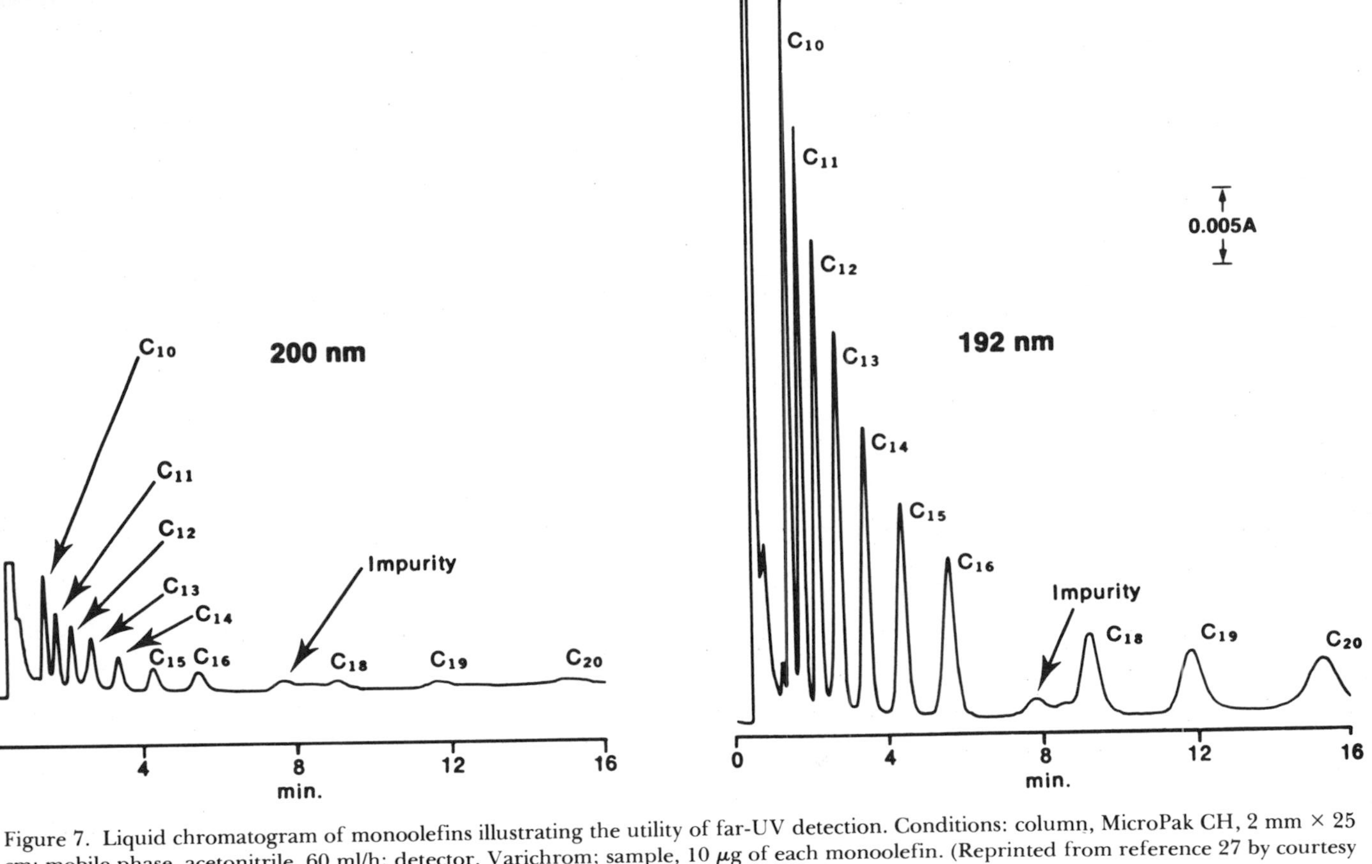

Figure 7. Liquid chromatogram of monoolefins illustrating the utility of far-UV detection. Conditions: column, MicroPak CH, 2 mm × 25 cm; mobile phase, acetonitrile, 60 ml/h; detector, Varichrom; sample, 10 μg of each monoolefin. (Reprinted from reference 27 by courtesy of Varian Instrument Division.)

trates the impressive sensitivity advantage which can sometimes be realized.

PTH Amino Acids. Classical spectrophotometric methods of analysis can often be modified to LC/UV procedures, thereby gaining specificity and sensitivity. A related example is the analysis of PTH amino acids. These derivatives are formed in the course of the Edman degradation and can also be directly detected in the UV. The Edman degradation is a powerful technique used for the determination of the amino acid sequence in polypeptides. In the Edman degradation, phenylisothiocyanate is coupled to the terminal amino group of a peptide, resulting in a phenylthiocarbamoyl peptide (Figure 8). Treatment with mild acid results in the cleavage of the *N*-terminal amino acid as a cyclic thiazolinone derivative, which rearranges to form a phenylthiohydantoin. The parent peptide remains intact and is again subjected to the Edman reaction. The process is repeated until the entire peptide has been degraded. This process has been automated and can be carried out very rapidly and efficiently. The PTH derivative resulting from each reaction sequence is then analyzed and the amino acid sequence determined.

The PTH derivatives are often analyzed by TLC, GC, or by an amino acid analyzer following the conversion of the PTH derivative to its parent

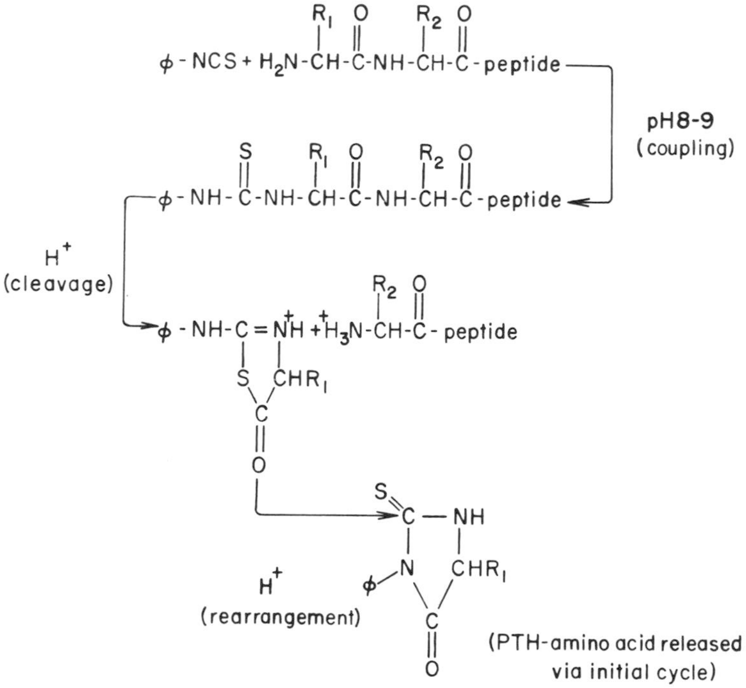

ϕ-NCS + H_2N-CH(R_1)-C(=O)-NH-CH(R_2)-C(=O)-peptide

pH8-9
(coupling)

ϕ-NH-C(=S)-NH-CH(R_1)-C(=O)-NH-CH(R_2)-C(=O)-peptide

H^+
(cleavage)

ϕ-NH-C=$\overset{+}{N}$H (S, CHR$_1$, C=O ring) + ^+H_3N-CH(R_2)-C(=O)-peptide

H^+
(rearrangement)

S=C—NH, ϕ—N, CHR$_1$, C=O (ring)

(PTH-amino acid released via initial cycle)

Figure 8. Edman degradation reaction.

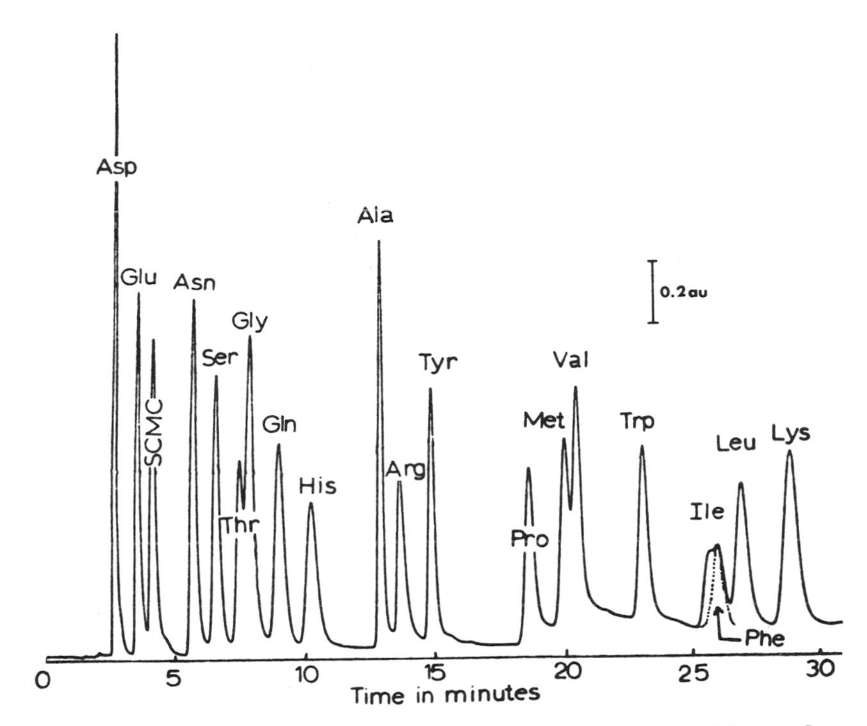

Figure 9. Reverse-phase (C_{18}) separation of PTH amino acids. (Reprinted from reference 30 by courtesy of the Federation of American Societies for Experimental Biology.)

amino acid. Recently modern LC has been applied to the analysis of the PTHs with great success.[28–30] The PTHs have been resolved by a number of normal and reverse-phase systems using linear or exponential gradient elution. An alternate to gradient LC is offered by Zimmerman *et al.*[29] These authors use two columns simultaneously under isocratic conditions. Polar PTHs are resolved on a C_{18} column and the less polar PTHs are separated on an alkyl ether bonded phase. All of the PTHs are eluted from either column in less than 40 min. Unfortunately, this approach requires two detectors.

An alternative to either continuous gradients or dual columns has been developed in order to avoid the need for expensive pumping systems or dual detectors.[30] All 20 PTH derivatives can be separated on a single reverse-phase column by means of a step-gradient elution profile. In this manner it is possible to process the results from a single Edman step in approximately 30 min (Figure 9). Over-all instrument precision ultimately limits the Edman method when it is no longer possible to distinguish the residue released in the current cycle from the background of residues released in previous cycles. The sample injection, column stability, and detector performance must be optimized if reliable results are to be obtained over a long period of time.

2.2.3. Pre-Column Derivatization and Post-Column Reactions

Pre-Column Derivatization. In liquid chromatography derivatization can be performed prior to separation or in a post-column mode.

In pre-column derivatization there is considerably more latitude in the choice of reaction conditions. In this approach the reaction medium need not be compatible with the chromatographic system. If necessary, the derivatives can be isolated from the reaction solvent prior to chromatographic separation. Much longer reaction times and higher reaction temperatures are also tolerable. Pre-column derivatization may also facilitate the separation of a class of compounds by changing their properties (e.g., reduce polarity). However, the advantages of pre-column derivatization are gained at the expense of convenience, since sample manipulation is usually increased.

A number of pre-column and post-column reagents are now available which can be used to enhance the UV absorption or fluorescence of various classes of compounds. The number of reagents, however, is small, and the use of derivatization in liquid chromatography remains underdeveloped. Those reagents used for UV enhancement are mainly of the pre-column type and can be used with several different functional groups. The few fluorescence derivatizing agents, discussed in a later section, are usually employed in a post-column reactor. A recent book by Frei describes some aspects of derivatization in LC and presents recipes for the use of a number of promising reagents.[31]

Post-Column Reactions. In the post-column or on-line approach the eluted compounds are mixed with a derivatizing reagent as they pass from the column. Reaction takes place in a length of capillary tubing or in a column packed with glass beads. Alternatively, the column effluent may be segmented by air bubbles, with the reaction occurring in each segment. The various "on-line" approaches are discussed in Chapter 5 of this volume.

The attractive feature of the on-line "reaction detector" approach is the reduced sample manipulation and analysis time. However, this approach imposes many restraints upon the derivatization reaction, making it difficult to find a suitable reagent. The reagent must be transparent to the detector and should react to near completion in only minutes. Long reaction times are undesirable, since they require long reaction coils which can lead to considerable band broadening. Incomplete reactions require precise temperature control since the reaction kinetics are critical to the conversion efficiency. High temperatures may improve the efficiency, but limitations imposed by solvent boiling point, detector noise, and air-bubble formation may be too severe. In addition the reaction solvent must be compatible with the mobile phase. A derivatizing agent which has a very specific solvent requirement will be usable with only a limited number of chromatographic systems.

Ninhydrin. Ninhydrin remains the most widely used reagent for the determination of amino acids. The ninhydrin is dissolved in a DMSO/acetate buffer mixture and reacts in a post-column mode producing a colored species. This product, "Ruheman's purple," has an absorption maximum at 570 nm and results from the reaction of any primary α-amino acid. Reaction times of about 15 min at pH 5 and 95°C are required.

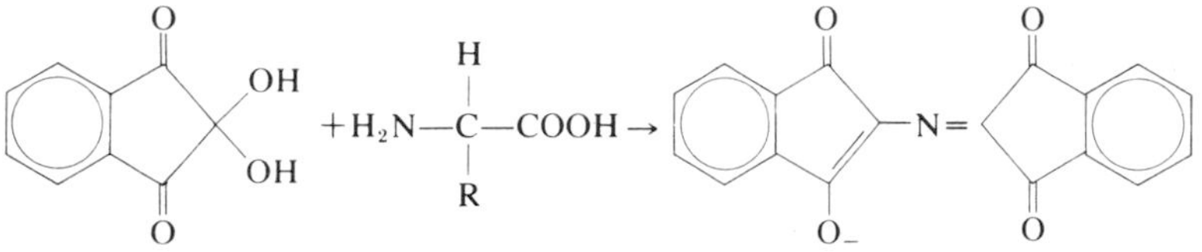

$+ H_2N—C(H)(R)—COOH \rightarrow$

Ninhydrin   Amino acid   "Ruheman's purple"

Anderson *et al.*[32] have developed a pressurized reaction cell which allowed operating temperatures of over 135°C and reaction times of less than 2 min. The reaction mechanism is complex and appears to be somewhat controversial. Hydrindantin, a reduced ninhydrin product, facilitates color formation. It is added to the ninhydrin solution or generated *in situ*

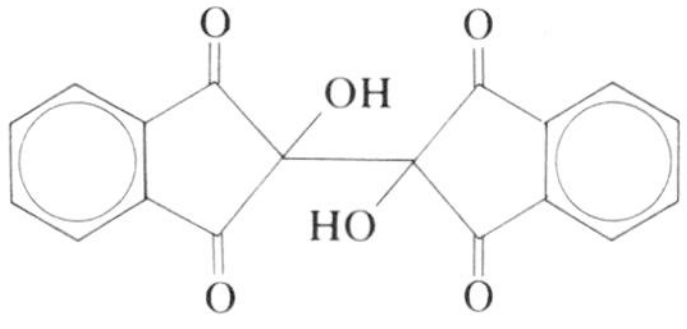

Hydrindantin

by the addition of a reducing agent. In one of the most recent studies of the reaction presented by Lamothe and McCormick,[33] the kinetic dependence of color formation on hydrindantin concentration is demonstrated. These authors review some earlier work and present a detailed mechanism of the ninhydrin reaction. The interested reader is referred to this paper for details.

The secondary amino acids, proline and hydroxyproline, also react with ninhydrin, but the reaction products vary with the conditions. Under those conditions normally used for the primary amino acids, a yellow product is formed. The absorption of this product at 440 nm is used for quantitation of the prolines. Ninhydrin also reacts with peptides, ammonia, and primary amines.

DNPH Derivatives. In addition to those reagents designed specifically for use in liquid chromatography, there are a variety of classical functional-group reagents used for qualitative organic analysis which may be valuable for LC derivatization. An excellent example is the derivatization of carbonyls using 2,4-dinitrophenylhydrazine (DNPH). Henry *et al.*[34] prepared a

$$\text{DNPH } (2,4\text{-}(O_2N)_2C_6H_3NHNH_2) + R_1R_2C{=}O \rightarrow 2,4\text{-}(O_2N)_2C_6H_3NHN{=}CR_1R_2 + H_2O$$

variety of steroid dinitrophenylhydrazones and resolved them on a number of chromatographic systems. Normal-phase, reverse-phase, and ion-exchange chromatography were used to separate the DNPH derivatives of estrogens, androgens, progestogens, and adrenocortical hormones, as well as the insect hormone ecdysone and several steroid conjugates. The derivatization was carried out in acidic methanol heated to 50°C for several minutes. The resulting derivatives had extinction coefficients on the order of 10^4 at 254 nm. Siggia and co-workers utilized DNPH derivatives for the analysis of 17-ketosteroids in human urine and plasma.[35] The separation and detection of simple aliphatic aldehydes and ketones has also been accomplished using DNPH.[36,37]

Carboxylic Acid Derivatives. Fatty acids are a class of biologically important molecules well suited to analysis by LC. Unfortunately the detection limits for free fatty acids are severely limited by the lack of a strongly absorbing chromophore. To increase the utility of LC for fatty acid analysis, several reagents have been designed to enhance the UV absorption of carboxylic acids. The reagent benzyl-3-*p*-tolytriazine can be used to form the benzyl esters of fatty acids.[38] The acids are derivatized in diethyl ether at 36°C for 3 h. The 1-*p*-nitrobenzyl-3-*p*-tolyltriazine converts carboxylic acids to their nitrobenzyl esters. With this reagent, esterification is accom-

$$H_3C\text{—}C_6H_4\text{—}N{=}N\text{—}NH\text{—}CH_2\text{—}C_6H_4\text{—}NO_2 + RCOOH \rightarrow$$

$$R\text{—}\overset{\overset{\displaystyle O}{\|}}{C}\text{—}OCH_2\text{—}C_6H_4\text{—}NO_2 + CH_3\text{—}C_6H_4\text{—}NH_2 + N_2$$

plished in ethanol at 56°C for 1 h. The resulting nitrobenzyl esters have extinction coefficients of approx. 5×10^3 at 254 nm. An alternate reagent for the formation of nitrobenzyl esters is *O*-*p*-nitrobenzyl-*N*,*N*-diisopropylisourea. Esterification of fatty acids in methylene chloride requires 2 h at 80°C.[39]

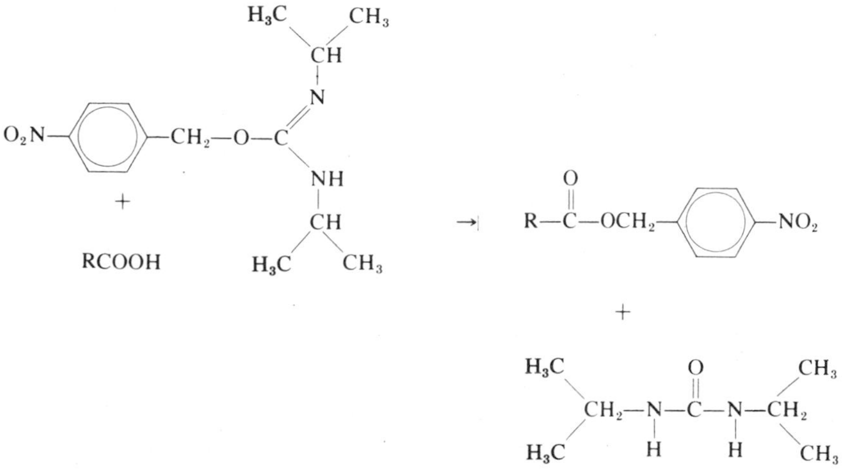

$$O_2N-C_6H_4-CH_2-O-C(=N-CH(CH_3)_2)(NH-CH(CH_3)_2) + RCOOH \rightarrow R-C(=O)-OCH_2-C_6H_4-NO_2 + (CH_3)_2CH_2-NH-C(=O)-NH-CH_2(CH_3)_2$$

Grushka and co-workers[40] have developed a unique approach to the formation of a variety of esters of carboxylic acids. The esterification is carried out in an aprotic solvent using a crown ether catalyst. The crown ether complexes with the salt of the acid to be derivatized, increasing its solubility and reactivity in the aprotic solvent. The resulting acid–crown

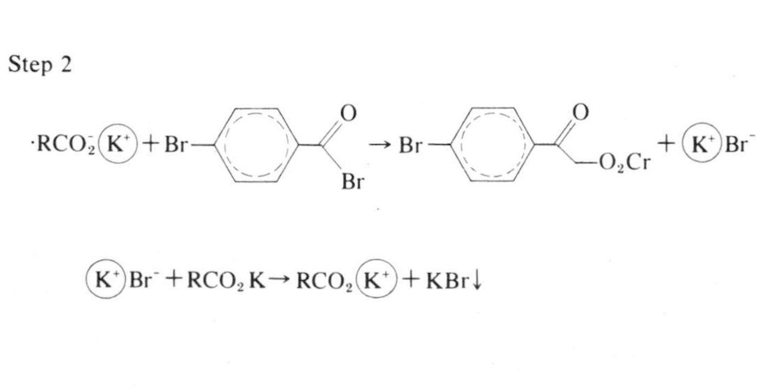

Step 1

$$K^+ {}^-O_2CR + \text{(crown ether)} \rightarrow \text{(crown ether}\cdot K^+) {}^-O_2CR \equiv (K^+) {}^-O_2CR$$

Step 2

$$\cdot RCO_2^-(K^+) + Br-C_6H_4-C(=O)-CH_2Br \rightarrow Br-C_6H_4-C(=O)-CH_2-O_2CR + (K^+)Br^-$$

$$(K^+)Br^- + RCO_2K \rightarrow RCO_2(K^+) + KBr\downarrow$$

ether complex will then react with a variety of esterifying agents. This procedure was used to esterify fatty acids with α-*p*-dibromoacetophenone in acetonitrile or benzene. The reaction time was 15 min at 80°C. The resulting phenacyl esters had extinction coefficients of greater than 17,000 at 254 nm. In addition to acetonitrile and benzene, the reaction will proceed in other aprotic solvents such as cyclohexane, methylene chloride, or carbon tetrachloride. The crown ether approach can also be used to form benzyl, nitrobenzyl, *p*-chlorophenacyl, *p*-phenylphenacyl, and 2-napthacyl esters. Grushka *et al.*[41] have used the crown ether approach to prepare phenacyl and benzyl derivatives of some biologically significant dicarboxylic acids. These esters were readily resolved on a C_9 reverse-phase system and detected at the 5- to 15-ng level.

The utility of the *p*-bromophenacyl esters for fatty acid analysis is demonstrated in a recent paper by Borch.[42] Phenacyl derivatives of 24 C_{12}–C_{24} acids were resolved using a microparticulate C_{18} column and an acetonitrile–water step gradient (see Figure 10). The author reports the successful application of this system to the analysis of fatty acids in chick fibroblasts and platelets.

Fatty acids have also been derivatized with *p*-methoxyaniline to form the corresponding anilides.[43] Two derivatization procedures have been developed and applied to C_6–C_{24} fatty acids. One procedure involves the heating of the acid in carbon tetrachloride in the presence of triphenylphosphine at 80°C for 5 min to form the acyl chloride. *p*-Methoxyaniline in ethyl acetate is then added, and the mixture is heated for 1 h. The alternate method employs triphenylphosphine bound to polystyrene, which decreases the reaction time from 1 h to 10 min. The use of the bound reagent also prevents sample contamination with triphenylphosphine or triphenylphosphine oxide. The anilides are resolved on a microparticulate C_{18} column using a methanol–water or an acetonitrile–water gradient (see Figure 11). The methoxyanilides have extinction coefficients of approx. 2.4×10^4 at 254 nm.

$$\mathrm{R{-}\overset{\overset{\displaystyle O}{\|}}{C}{-}OH + (C_6H_5)_3P + CCl_4 \rightarrow R{-}\overset{\overset{\displaystyle O}{\|}}{C}{-}Cl + (C_6H_5)_3PO + CHCl_3}$$

$$\mathrm{R{-}\overset{\overset{\displaystyle O}{\|}}{C}{-}Cl + CH_3O{-}(C_6H_4){-}NH_2 \rightarrow R{-}\overset{\overset{\displaystyle O}{\|}}{C}{-}NH{-}(C_6H_4){-}OCH_3}$$

The prostaglandins are a class of oxygenated fatty acids that are found in most bodily secretions. Their biochemical role is not well understood and is currently the object of intensive research. The prostaglandins are

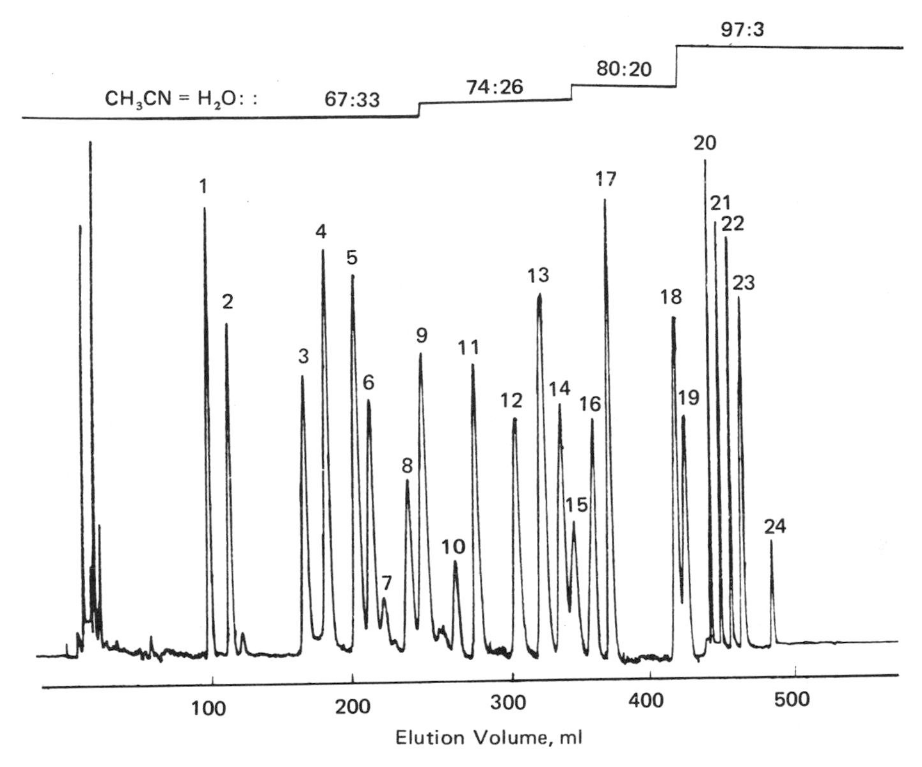

Figure 10. LC of fatty acid phenacyl esters. Peak 1, lauric (12:0); 2, myristoleic (14:1); 3, α- and γ-linolenic (18:3); 4, myristic (14:0); 5, palmitoleic (16:1); 6, arachidonic (12:4); 7, *trans*-palmitoleic (*trans* 16:1); 8, linoleic (18:2); 9, pentadecanoic (15:0); 10, linolelaidic (*trans* 18:2); 11, licosatrienoic (20:3); 12, palmitic (16:0); 13, oleic (18:1, Δ^9) and vaccenic (18:1, Δ^{11}); 14, petroselinic (18:1, Δ^6); 15, elaidic (*trans* 18:1); 16, eicosadienoic (20:2, $\Delta^{11,14}$); 17, heptadecanoic (17:0); 18, stearic (18:0); 19, eicosaenoic (20:1, Δ^{11}); 20, non-adecanoic (19:0); 21, arachidic (20:0) and reucic (22:1); 22, heneicosanoic (21:0); 23, behemic (22:0) and nervonic (24:1); 24, lignoceric (24:0). Column: 90 cm × 0.64 cm μ-Bondapak C_{18}; eluent: acetonitrile–water; flow rate: 2.0 ml/min. (Reprinted from reference 42 by courtesy of the American Chemical Society.)

extremely potent and elicit a number of very diverse physiological responses. Their action is quite dependent on structure. Only a minor change in structure can cause a dramatic change in response. The analysis of prostaglandins in biological fluids and pharmaceutical preparations requires efficient separation and very sensitive detection methods.

Several LC assays based on derivatization and UV detection have been developed, but they have been primarily used for pharmaceutical analysis. The *p*-nitrophenacyl esters of over 20 prostaglandins have been prepared.[44] The derivatization is carried out in acetonitrile in the presence of

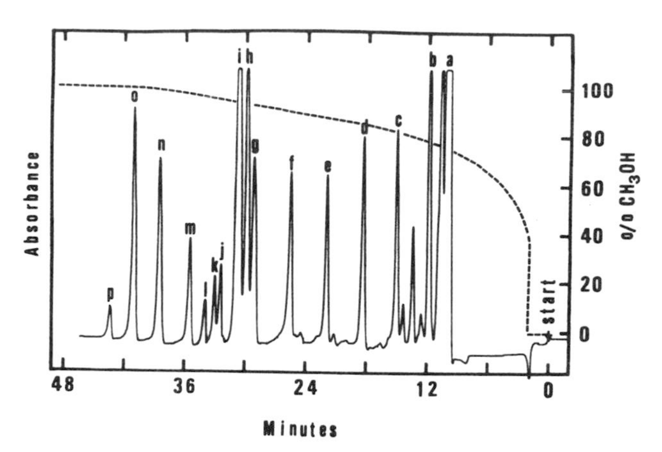

Figure 11. Water–methanol gradient elution LC of fatty acid *p*-methoxyanilides. Peaks: (a) *p*-methoxyaniline, (b) unknown, (c) C_6, (d) C_8, (e) C_{10}, (f) C_{12}, (g) C_{14} + $C_{18:3}$, (h) $C_{22:6}$, (i) C_{15} + $C_{16:1}$—$C_{20:4}$, (j) C_{16}, (k) $C_{18:1}$, (l) C_{17}, (m) C_{18}, (n) C_{20} + $C_{22:1}$, (o) C_{22} + $C_{24:1}$, (p) C_{24}. (Reprinted from reference 43 by courtesy of the American Chemical Society.)

N,N-diisopropylethylamine and *p*-nitrophenacyl bromide. Only 15 min at room temperature is required. The extinction coefficient of the *p*-nitrophenacyl ester of prostaglandin $F_{2\alpha}$ at 254 nm was 1.3×10^4. The phenacyl esters were separated with a microparticulate silica column (see Figure 12). A silver-ion-loaded microparticulate cation-exchange resin has also been shown to be effective for the phenacyl derivatives of very similar prostaglandins.[45]

A very imaginative approach to the derivatization of prostaglandins has recently been developed by Morozowich and Douglas.[46] The carboxylic acid moiety of the prostaglandin is reacted with *p*-(9-anthroyloxy)phenacyl bromide to yield the corresponding *p*-(9-anthroyloxy)phenacyl ester or so-called "panacyl ester" (see Figure 13). The resulting esters have absorption maxima at 253 nm with extinction coefficients in excess of 174,000. This allows the detection of less than 100 pg of prostaglandin E_2 by LC/UV. This is an order of magnitude better sensitivity than that achieved with phenacyl derivatives. The reagent should also be applicable to other carboxylic acids, as well as amines. The reaction requires 40–100 min at room temperature in acetonitrile–THF (4 : 1). The resolution of a number of panacyl derivatives of prostaglandins on a silica-gel system was similar to that of the phenacyl derivatives (see Figure 14).

2.3. Fluorescence Detection in Liquid Chromatography (LC/F)

2.3.1. Recent Advances in Instrumentation for LC/F

Introduction. The use of fluorescence for LC detection provides high sensitivity and specificity. Compounds which display native fluorescence or

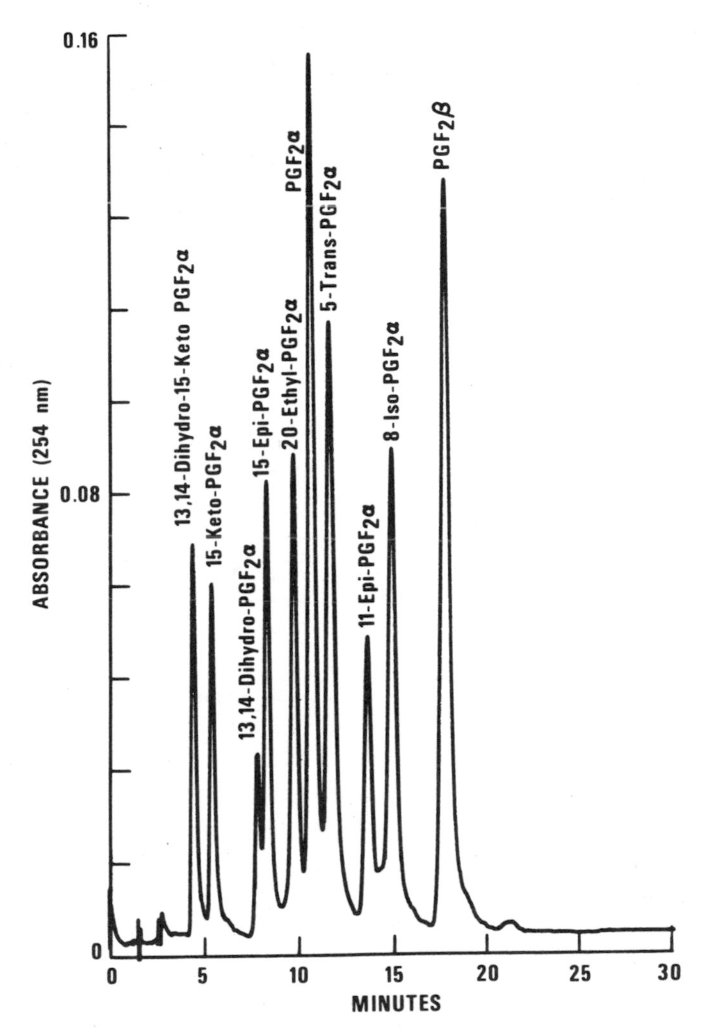

Figure 12. LC separation of a mixture of F-series prostaglandin *p*-nitrophenacyl esters on two series couples 2.1 mm (ID) × 25 cm microparticulate silica gel columns. Conditons: mobile phase, methylene chloride–hexane–methanol (55:45:5); 3000 psi; 0.3 ml/min. (Reprinted from reference 44 by courtesy of Dr. P. W. Ramwell.)

which can be derivatized to form fluorescent species can sometimes be detected at the picogram level, two or more orders of magnitude lower than is possible at present with UV absorption detection.

Fluorescence detection of LC effluents was originally carried out by incorporating simple flow cells into existing fluorometers. Since 1972 when Thacker introduced the miniature flow fluorometer,[47] a number of instruments have been designed specifically for LC. Although some room for improvement remains, good equipment is now commerically available

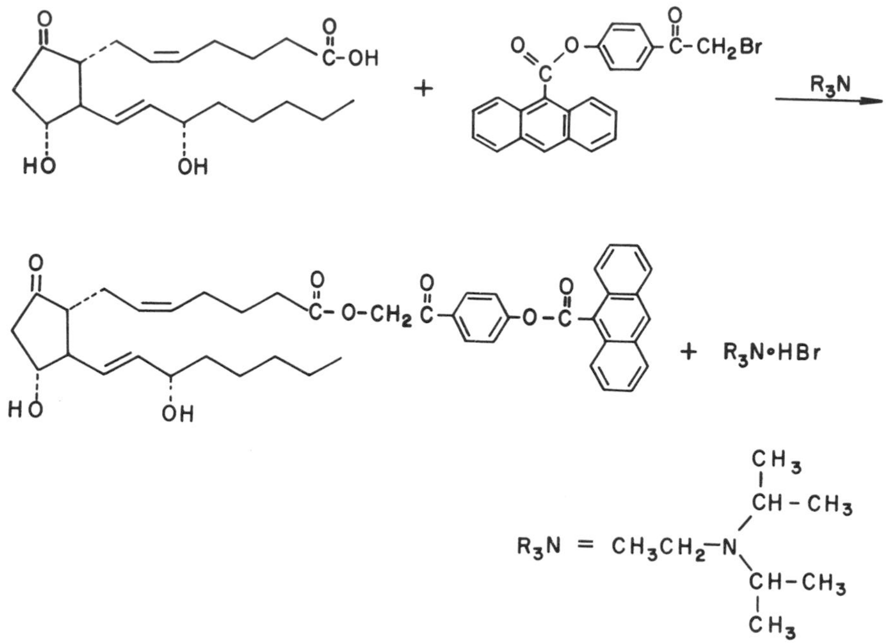

$R_3N = CH_3CH_2-N(CH(CH_3)CH_3)_2$

Figure 13. Synthesis of PGE_2 panacyl ester. (Reprinted from reference 46 by courtesy of the American Chemical Society.)

and the emphasis of recent work has shifted to the development of applications and means for producing a signal from molecules which are nonfluorescent.

Flow Cells. In designing a detector for LC, the volume of the flow cell is kept small to minimize post-column band broadening. This necessarily means that for a given sample concentration, the quantity of fluorescing material in the cell at any given moment is small in comparison with classical cuvettes. In addition, the small volume requirement makes it difficult to focus the excitation beam and to collect the emitted photons. The inherent advantage of fluorescence is not easily achieved in small-volume flow cells, especially when chromatographic performance demands a low time constant for the photometer.

The most straightforward approach is to utilize a piece of quartz tubing. Exciting radiation is focused on the tube at right angles to the direction of flow and fluorescence is monitored at right angles to both. Light scattered by the tubing can be a serious problem in cells of this type, therefore a filter is normally placed between the cell and the photomultiplier tube. The flow cell in the Du Pont model 836 is a twist on the simple approach.[48] This cleverly designed cell permits both absorbance and fluorescence detection (see Figure 15). The 20-mm × 1-mm ID quartz tube is capped at the ends with quartz windows. The excitation beam enters through the planar windows, traveling in the direction of the flow, and the light transmitted is monitored. Simultaneously, fluorescence radiation is collected and monitored at right angles to the flow. To minimize detection of scattered light, the photomultiplier "sees" only photons produced in the middle 8 μl of the 16-μl cell. This means, however, that much of the fluorescence produced in the cell is not collected. In addition, absorbance of some of the exciting beam before it reaches the middle of the cell can accentuate problems arising from strongly absorbing components. Absorbance data available from the detector allow the experimenter to determine if the absorbance in the cell is too high for linear fluorescence measurements to be made.

A more recent design is the flow cell employed in the Schoeffel model FS-970 (see Figure 16). The excitation beam enters from the left, exciting the sample in the 5-μl cuvette. Fluorescent radiation emitted in the direction of the hemispherical mirror is reflected back parallel to the excitation beam. The light not striking the narrow chamber bar is filtered and detected at the right. This design is attractive from several points of view. Relatively efficient optical collection is obtained from the silvered quartz mirror mounted opposite the emission filter and photomultiplier. The detector primarily collects front surface fluorescence, minimizing inner

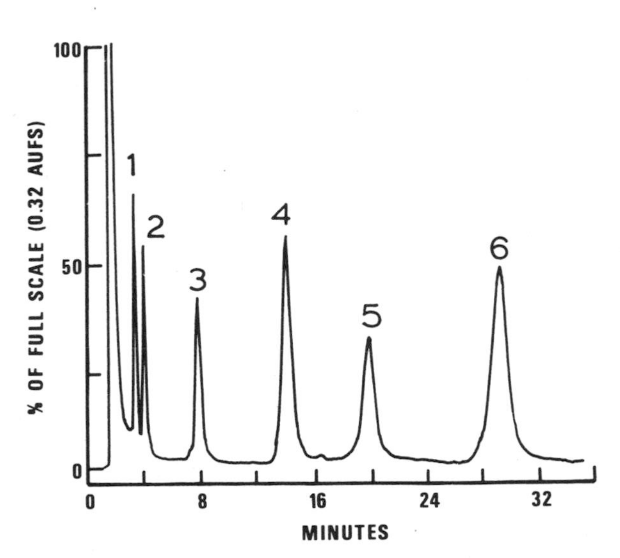

Figure 14. LC separation of the panacyl esters of 13,14-dihydro-15-keto-$PGE_{2\alpha}$ (3); $P6F_{2\alpha}$ (4); 8-iso-$PGF_{2\alpha}$ (5); and $PGF_{2\beta}$ (6) (70–350 ng each). LC conditions: ZORBAX-SIL, 2.1 mm (ID) × 25 cm; methylene chloride–methanol (97 : 3); 2200 psi; 0.5 ml/min. (Reprinted from reference 46 by courtesy of the American Chemical Society.)

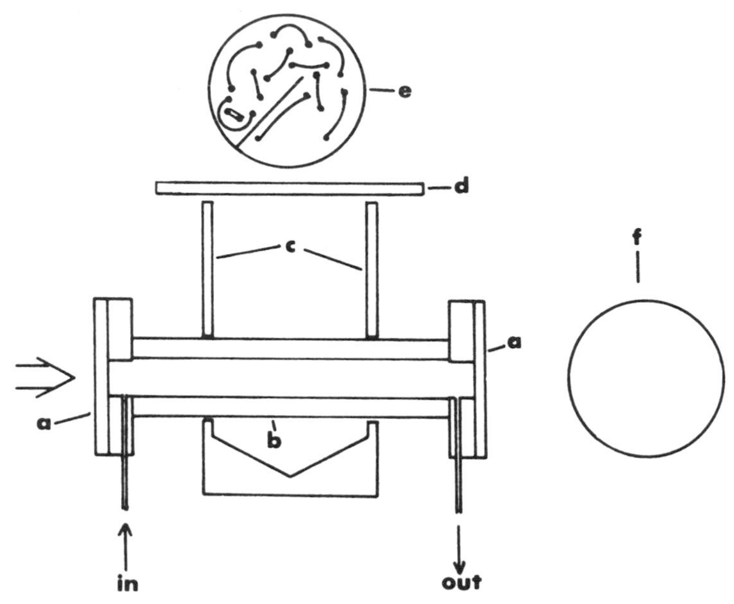

Figure 15. Absorbance fluorescence flow cell: (a) quartz window, (b) quartz tube, (c) reflective fluorescence chamber, (d) emission filter, (e) photomultiplier, (f) absorbance phototube. (Redrawn from reference 48 by courtesy of Elsevier Scientific Publishing Corp.)

filter effect problems. One problem is that scattered light is collected as well, but this is virtually eliminated by the emission filter.

A novel approach to fluorescence detection for LC was recently reported by Martin *et al.*[49] The authors developed a system which eliminates the need for a flow cell as such. The column effluent is formed into drops by a needle embedded in the top of a brass sphere. The falling drops pass through a horizontal light beam. Scattered light is detected by a photomultiplier. The output of this photomultiplier creates timing pulses for processing the output of a second photomultiplier which receives the fluorescence at right angles to the excitation radiation. This simple approach presumably means that the detector is easily and inexpensively constructed. There are problems with this design, however. Changes in the viscosity or surface tension of the mobile phase may change the size of the drops, altering response to a given concentration. Also, scattered light may be worse from a moving spherical drop than from a planar quartz window as in a flow cell. As a result, the sensitivity of the "falling-drop detector" is not comparable to commercially available detectors.

Along the same line, a windowless cell was recently described where the drop does not "fall" (see Figure 17). In this design, the column effluent flows continuously from a 1.6-mm stainless-steel capillary tube down onto a 1.6-mm stainless rod 2 mm below it, forming a 4-μl cell in the gap. This unit is further discussed in the following section.

Sources and Optics. The factors involved in the evaluation of light sources include spectral output, cost, and stability. Up to a point, increased fluorescence sensitivity will be obtained if the intensity of the exciting radiation can be increased. Thacker's miniature flow fluorometer used lines isolated from a mercury arc lamp for excitation. The bands are strong and sharp, but the choice of wavelength is limited. Deuterium lamps provide a continuum of light, with useful power down to 190 nm. Increased intensity in a continuous source is provided by the use of xenon arc lamps. However, xenon arcs are bulky, less stable, and more costly. These problems are somewhat overcome by the use of a repetitive xenon flash lamp.[50] Instruments employing all of the above are commercially available.

The tremendous potential of LC fluorescence for quantitation of small amounts of materials has been recently demonstrated with a laser fluorescence detector described by Diebold and Zare.[51] A He–Cd ion laser which produces 325-nm radiation was used for excitation. The laser had an internal feedback loop, and its output was amplitude modulated at 50 kHz through the use of an oscillator and an acousto-optic light modulator (see Figure 17). Fluorescence was viewed by a photomultiplier whose signal was detected in a phase-sensitive mode using a lock-in amplifier. The high intensity of the laser together with the noise reduction from the associated electronics allowed chromatographic quantitation orders of magnitude lower than had been previously attained.

Another detector employing a laser has been described.[52] It is based

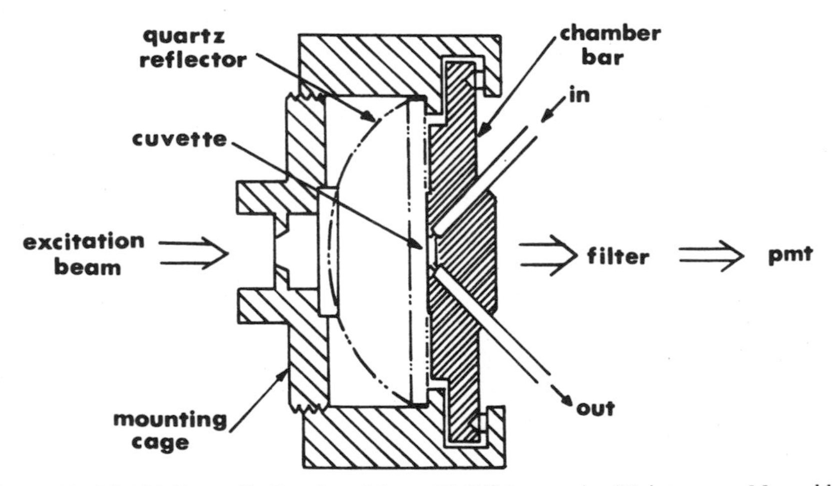

Figure 16. FS-970 flow cell. (Reprinted from FS-970 Instruction/Maintenance Manual by courtesy of Schoeffel Instrument Corp.)

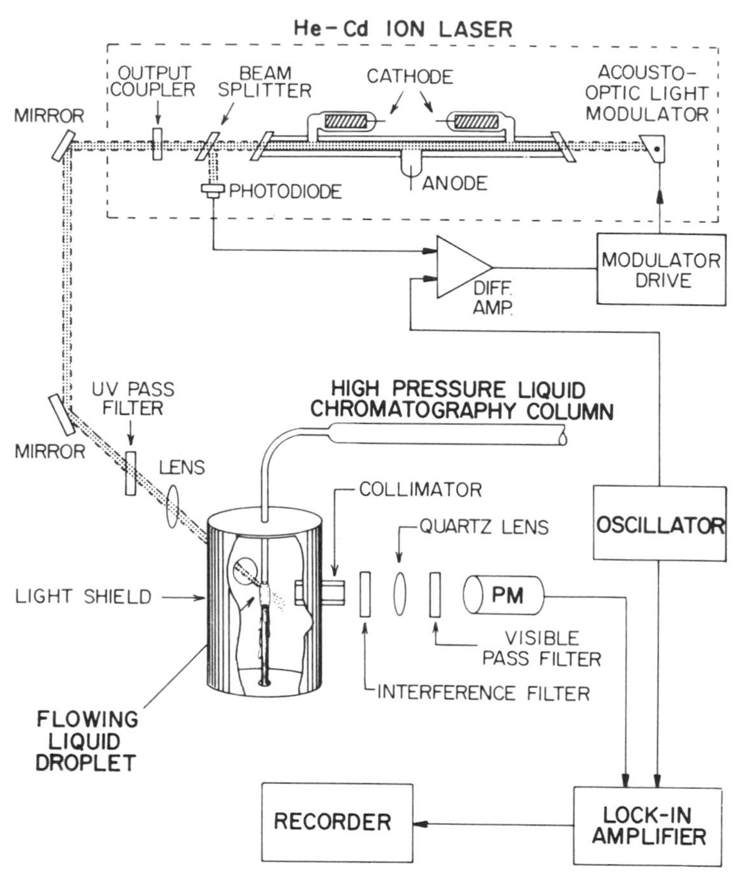

Figure 17. LC–laser fluorescence detector. (Reprinted from reference 51, copyright 1977 by the American Association for the Advancement of Science.)

on the process whereby a molecule absorbs two photons simultaneously to achieve an excited state, from which state it fluoresces as in conventional single-photon fluorescence. While the idea is intellectually interesting, the value of two-photon fluorescence for analytical chemistry lies elsewhere.[53] This detector could only detect highly fluorescent standards at the ng level, without demonstrating any advantage over conventional LC/F detectors.

The choice of optics for selection of excitation and emission wavelengths involves a trade-off between selectivity (resolution) and sensitivity (light throughput). Commercial instruments are available which employ filters for both excitation and emission, a monochromator for excitation or emission only, or monochromators for both excitation and emission. The use of wide-band filters is the simplest and least expensive approach.

Monochromators improve selectivity but also increase cost and decrease light throughput. In most routine applications the selectivity gained should be unnecessary since fluorescence coupled to LC is inherently selective. A good compromise is the continuously graded interference filter employed in the emission portion of the Perkin Elmer LC-1000. It is wavelength-tunable yet has good transmittance.[50]

In addition to devices employing single excitation and emission wavelengths, some instruments have been developed to scan the excitation or emission spectra of compounds eluted from an LC column. The additional spectral information allows qualitative confirmation of the identity of eluted peaks, and together with computer processing power could allow the quantitation of superimposed peaks. Two straightforward approaches to scanning were described in 1973.[54,55] One approach was to stop the flow while scanning. The second approach was to record spectra (12–20 s required) while compounds passed through the cell. More recently, a multichannel detector similar to those discussed on pages 78–80 has been described.[56] This instrument continuously recorded emission spectra of chromatographed petroleum fractions. For most purposes, however, collecting fractions and using conventional instrumentation is adequate.

A promising new approach to acquiring complete spectra, which could be applied to LC, was recently described by Christian *et al.*[57] Their "video fluorometer" allows rapid acquisition of three-dimensional fluorescence intensity plots. The output of a xenon arc lamp is spectrally dispersed by a monochromator and impinges on the sample cuvette (see Figure 18). Perpendicular to the excitation light plane, emission is monitored. Fluorescent light from the entire cuvette is dispersed using a second monochromator. The result is a two-dimensional image giving the fluorescence intensity as a function of both excitation and emission wavelengths (see

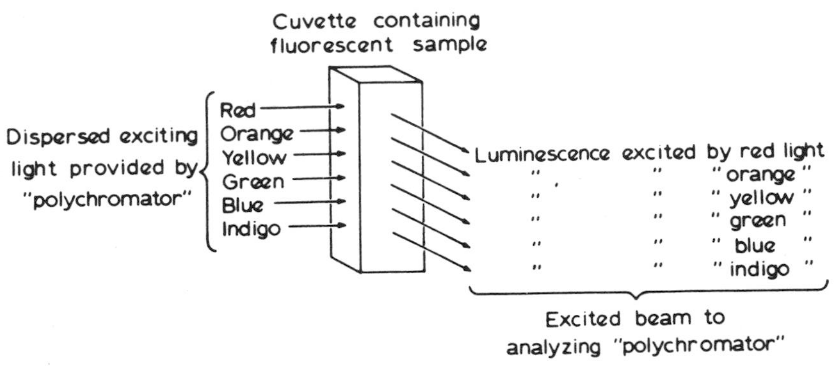

Figure 18. Geometry of illumination of the sample in the video fluorometer. (Reprinted from reference 57 by courtesy of the American Association for Clinical Chemistry.)

Figure 19). This image is monitored with an intensified-silicon intensified-target vidicon, a low-light-level TV camera. Spectra can be recorded in under 100 μs, and integration and time-averaging capabilities are built into the instrument. The video fluorometer generates a vast amount of data. At present a large computer is required for processing, but the cost and size should come down in the future as more efficient data-reduction algorithms are developed. The ability to quantitate five components simultaneously, given standard spectra, has already been demonstrated.

Applied to LC, the rapid scanning capability of the video fluorometer would allow signal averaging and complete fluorescence characterization of all peaks. Comparisons with standards would give positive identification

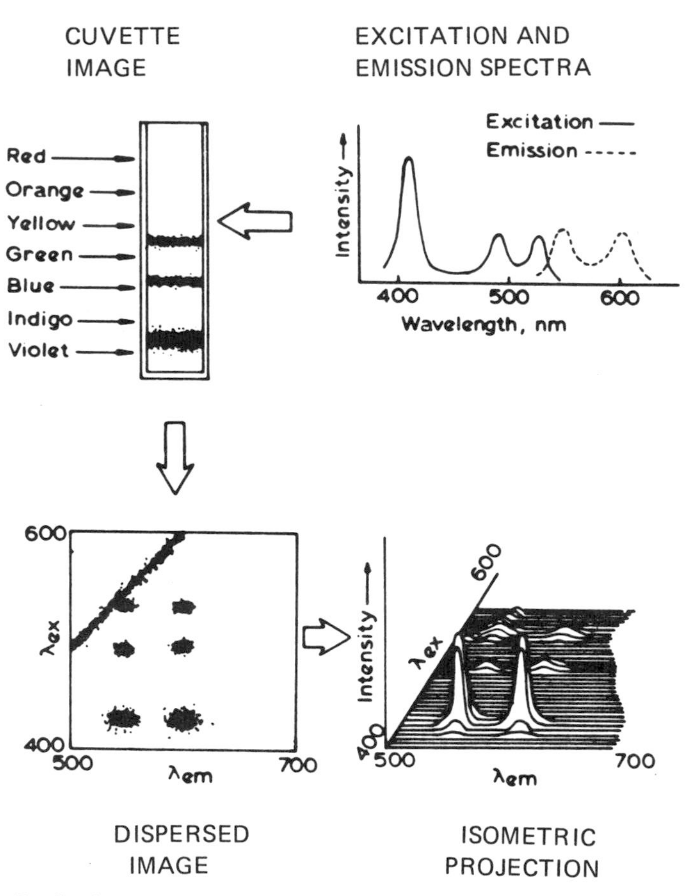

Figure 19. Production of the emission excitation matrix for a hypothetical compound. (Reprinted from reference 57 by courtesy of the American Association for Clinical Chemistry.)

of species eluted. Also, resolution of two or more overlapping peaks, even in cases where one component is only weakly fluorescent, should be possible given the wealth of spectral information available. The additional data obtained by taking complete spectra during an LC separation will be superfluous for most routine determinations but could be of some value during the development of assay procedures to optimize wavelength selection. At the present time the instrumentation required for such experiments is not cost effective considering the relatively meager advantages.

2.3.2. *Direct Fluorescence Detection*

Characteristics of Fluorescent Molecules. Relatively few organic molecules exhibit intense native fluorescence. The emission characteristics of molecules are difficult to predict, although some generalizations can be made. Extended conjugation favors a high fluorescence quantum yield. Although most fluorescent compounds are aromatic, aromaticity is by no means a guarantee that a molecule will fluoresce. Fluorescence is particularly sensitive to small structural changes in a molecule. Changes in the substituents on an aromatic ring can significantly alter the fluorescence quantum yield. Substitution of —OH, —OCH_3, —NH_2, —$NHCH_3$, —$N(CH_3)_2$, —F, and —CN generally results in fluorescent compounds, while —$NHCOCH_3$, —Cl, —Br, —I, —CO, —NO_2, —SO_3H, and —COOH substituents result in weak or nonfluorescent compounds. The emission characteristics of some representative nonsubstituted benzenes are listed in Table 4. When more than one substituent is added to the ring, the result is not always easily predicted.

Other structural features such as the planarity and rigidity of the molecule are also important. Any deviation from planarity that restricts the mobility of electrons in a π system will, in general, result in a decreased quantum yield. Rigidity will decrease the loss of energy through vibrational modes (i.e., radiationless transitions).

In addition to the structural features of a molecule, fluorescence is very dependent on the sample matrix. Solvent properties, pH, and sample contaminants all may influence the emission characteristics of a compound. Bromine and iodine atoms, for example, reduce the quantum efficiency when substituted on a fluorescent molecule and in addition adversely effect the analyte's fluorescence when present in the solvent molecule as well. Solvents such as ethyl iodide would quench fluorescent molecules. Dissolved oxygen promotes intersystem crossing of an excited singlet to its triplet state and is therefore a potent quenching agent. The polycyclic aromatic hydrocarbons are particularly sensitive to oxygen quenching.

Matrix effects are not always detrimental to measurement of fluorescence and in some instances can be used to an advantage. The emission

Table 4. Fluorescence of Monosubstituted Benzenes,[a] C_5H_5R

Compound	Substituent R	ϵ_{max}	Excitation (λ_{max} nm)	Fluorescence (μ_{max} nm)	Relative intensity
Ortho–paradirecting					
Benzene	H	204	260	291	1
Aniline	NH_2	1,430	280	345	46
Monomethylaniline	$NHCH_3$		280	360	
Dimethylaniline	$N(CH_3)_2$		283	363	114
Acetanilide	$NHCOCH_3$			None	0
Fluorobenzene	F		262	285	13
Chlorobenzene	Cl	190	265	294	0.02
Bromobenzene	Br	192		None	0
Iodoébenzene	I	190		None	0
Toluene	CH_3	225	265	292	3.8
Ethylbenzene	C_2H_5	240	265	292	1.2
Phenol	OH	1,450	272	320	112
Anisole	OCH_3	1,480	269	302	92
Phenoxide ion	O^-	2,600	289	345	~1
Metadirecting					
Benzoic acid	COOH	970		None	0
Benzoate ion	COO^-	560		None	0
Nitrobenzene	NO_2	7,800		None	0
Benzene sulfonic acid	SO_3H			None	0
Benzenesulfonamide	SO_2NH_2	740		None	0
Benzaldehyde	CHO	11,400		None	0
Benzenearsonic acid	AsO_3H_2			None	0
Benzonitrile	CN	1,000	273	294	45

[a] From: E. J. Bowen, *Luminescence in Chemistry*, p. 85, Van Nostrand, New York (1968), by permission.

properties of ionizable compounds are often pH-dependent. In some instances the pH can be used to selectively enhance the fluorescence of an analyte relative to other potentially interfering compounds. As discussed on pages 111–112, even the quenching phenomenon itself can be used to detect the nonfluorescent quenchers. The successful utilization of fluorescent methods demands that careful control be maintained over the many factors which may influence such methods.

The use of fluorescence in LC has not yet gained widespread popularity. Most of the published applications of direct fluorimetric detection have been directed to the determination of the polycyclic aromatic hydrocarbons and the aflatoxins. The measurement of these two classes of compounds is discussed in detail in the following sections.

Aromatic Hydrocarbons. Polycyclic aromatic hydrocarbons are constituents of the particulate matter in air. They are the result of a variety of combustion processes such as the burning of gasoline in automobiles and the use of coal for fuel. Many PAHs are known carcinogens and have been implicated in the cause of lung cancer. Currently much effort is under way to more fully evaluate the role of airborne PAHs in cancer.

The PAHs are usually collected from air by filtration and then subjected to a variety of chromatographic procedures. The large number of individual PAHs and their small quantities make identification and quantitation difficult. After a number of cleanup steps, the PAHs are frequently detected by their native fluorescence. The fluorescence spectra of over 50 PAHs were measured by McKay and Latham.[58] Due to the diversity of this class of compounds, the excitation and emission spectra occur over a wide range of wavelengths. Due to their rigid structure, most of the compounds exhibit several maxima in both their excitation and emission spectra.

Recently fluorescence has been coupled to LC and successfully applied to the analysis of PAHs in airborne particulates. Air-filter samples were extracted with benzene and subjected to LC/F. Over 35 components were resolved on a microparticulate C_{18} column (see Figure 20). Fractions were collected and subjected to further analysis by fluorescence and mass spectrometry. Eleven of the 35 components were identified and measured in a number of samples. Numerous examples of the LC analysis of PAHs are

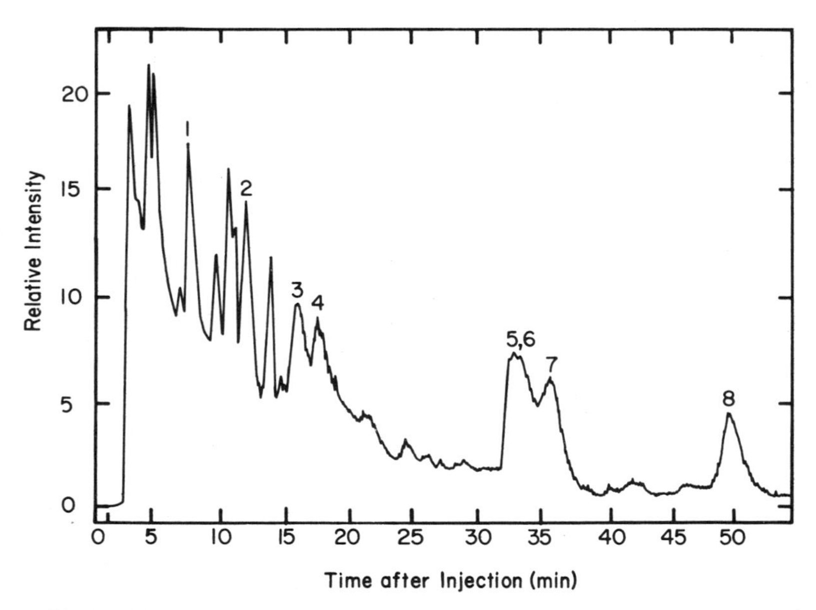

Figure 20. Partial LC/fluorescence trace of a benzene extract of an atmospheric particulate matter sample collected on the roof of Wing 1 of the Department of Chemistry, University of Maryland, College Park, Maryland. Conditions: 0.25 m × 8 mm (ID) ZORBAX ODS column, 7:3 (v/v methanol–water), 65°C, 1600 psi, flow rate 1.7 ml/min. (Reprinted from reference 59 by courtesy of the American Chemical Society.)

cited by Fox and Staley.[59] It is apparent that PAH analysis is an ideal application for LC/F.

Aflatoxins. The aflatoxins are a group of structurally related toxins produced by several species of the fungus *Aspergillus* (Figure 21). Under certain conditions of storage, these fungi will grow in a variety of grains, seeds, and nuts and can eventually find their way into food products. The extreme hepatocarcinogenic activity of these compounds makes them a serious threat to human and animal populations. Sensitive assays are required to screen agricultural and food products. Since these toxins are found in a variety of matrices, the methods of isolation are quite variable. However, most approaches utilize the natural fluorescence of the aflatoxins for their detection.

The fluorescent properties of the aflatoxins have been well characterized.[60,61] The long-wavelength absorption maxima of the aflatoxins (B_1, B_2, G_1, and G_2) in methanol occur between 335 and 360 nm, with a hypsochromic shift of the maxima as the solvent polarity is decreased. Both the emission maxima and fluorescent intensities are significantly altered by solvent polarity (see Table 5). The fluorescent intensities of aflatoxins B_1 and G_1 in polar solvents are relatively low. In methanol the relative intensities of G_2, B_2, G_1, and B_1 are 14.5 : 8.8 : 1.7 : 1.0, respectively. Treatment of the B_1 and G_1 with a mild acid converts these compounds to their

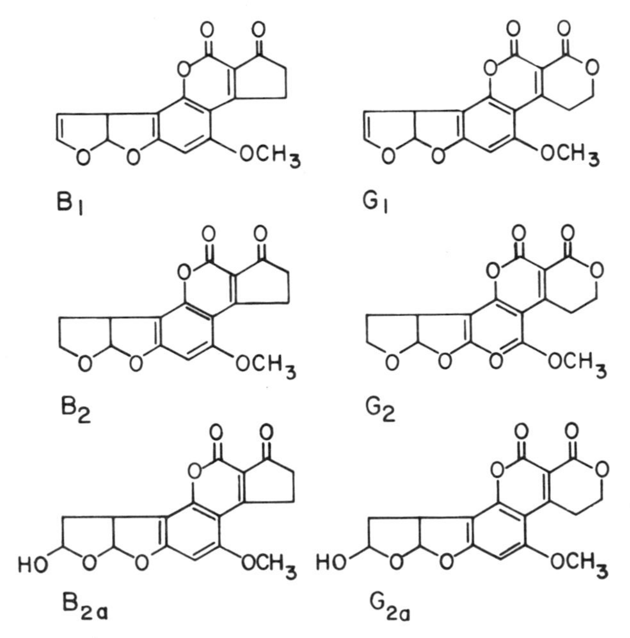

Figure 21. Chemical structures of six aflatoxins.

Table 5. Fluorescence of the Aflatoxins in Different Solvents

Aflatoxin	Excitation wavelength (nm)	Emission wavelength (nm)			KQ^a		
		Methanol	Ethanol	Chloroform	Methanol	Ethanol 95%	Chloroform
B_1	365	430	430	413	0.6	1.0	0.2
B_2	365	430	430	413	5.3	2.7	0.25
G_1	365	450	450	430	1.0	1.4	6,2
G_2	365	450	450	430	8.7	4.7	6.8

[a] $KQ = \dfrac{\text{Fluorescent intensity of aflatoxin solution}}{\text{Fluorescent intensity of standard quinine sulfate solution } (\mu\text{g/ml}) \times \text{concentration of aflatoxin solutions } (0.2\ \mu\text{g/ml})}$

hemiacetal derivatives B_{2a} and G_{2a} (see Figure 21), which have much higher fluorescent intensities. This reaction is incorporated into many analytical schemes to decrease the detection limit for B_1 and G_1.[51,62]

Most methods of analysis rely on TLC for the final separation of the aflatoxins. Recently a number of LC separations of aflatoxins using microparticulate silica have appeared in the literature.[63–65] Unfortunately, the majority of these LC methods have not been applied to real samples. Takahashi[66] has compared the use of silica, bonded —CN, and C_{18} columns for the analysis of aflatoxins in wine. Six aflatoxins were resolved in less than 15 min and could be detected in wine at the 0.02-μg/liter level. Both a UV and fluorescence detector were used in this work, and the advantage of a selective detector such as the fluorescence detector is beau-

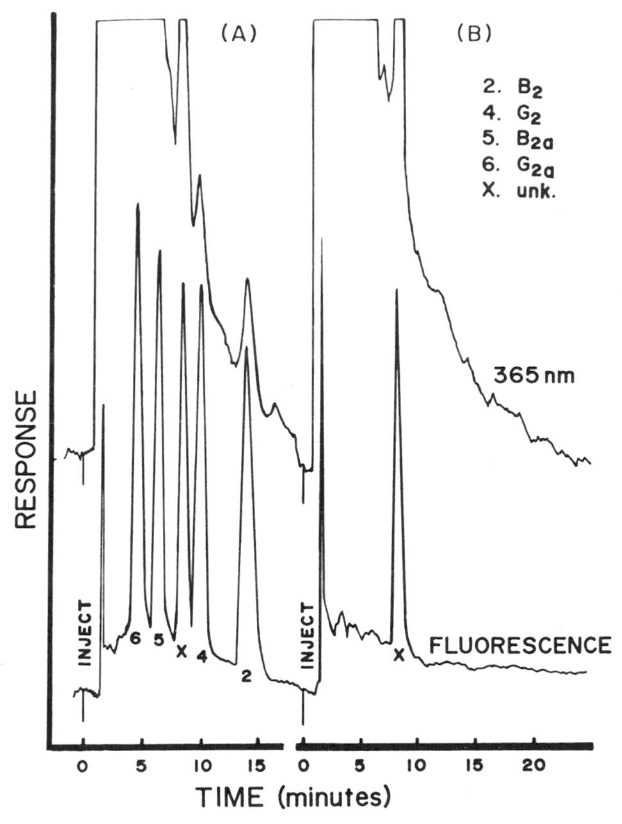

Figure 22. Madeira wine extract. (A) spiked with aflatoxins at 1 μg/l level; (B) unspiked sample extract. Conditions: 25 cm × 3.2 mm (ID) 10-μm Spherisorb-ODS, water–methanol (3:2); pressure, 2300 psi; 2.1 ml/min; 2 = aflatoxin B_2; 4 = G_2, 5 = B_{2a}, 6 = G_{2a}, X = unknown. (Reprinted from reference 66 by courtesy of Elsevier Scientific Publishing Co., Inc.)

tifully demonstrated (see Figure 22). The potential of LC/F for trace analysis is clearly shown by Diebold and Zare.[51] Development of the laser fluorimetry system discussed on page 97 allowed the determination of four aflatoxins in less than 15 min, with the detection limit of 750 fg. This instrument was applied to the analysis of aflatoxins in yellow corn extract, where aflatoxin B_1 could be determined at the 2-ppb level after conversion to B_{2a} (see Figure 23).

2.3.3. *Derivatization and Indirect Fluorescent Detection*

Fluorescamine. Fluorescamine is a reagent used for the conversion of primary amines to fluorescent products.[67] The reaction proceeds to near completion in less than a second at room temperature. The resulting fluorophors have excitation maxima at 390–410 nm and emit at 475–500 nm. Fluorescamine itself is nonfluorescent and is hydrolyzed in water to nonfluorescent products. The reaction with amino acids and aliphatic

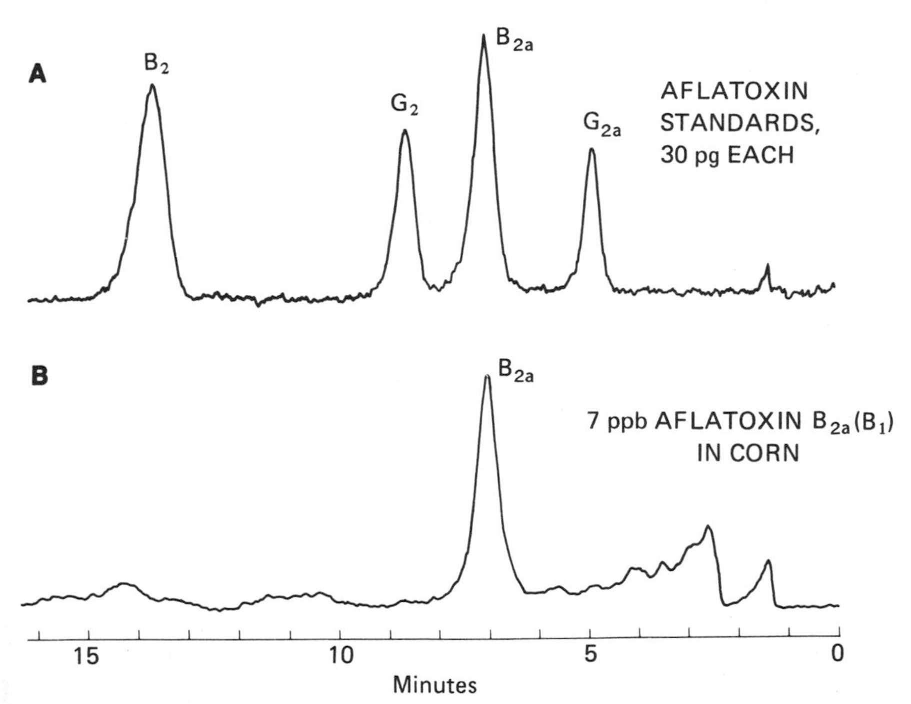

Figure 23. Aflatoxin chromatograms: (A) detection of the four aflatoxin standards following addition of HCl and (B) detection of aflatoxin B, in corn extract, by the same procedure, using a 10-μl sample injection. The chromatogram represents 45 pg of B_1 in the injected sample. (Reprinted from reference 51. Copyright 1977 by the American Association for the Advancement of Science.)

amines is generally carried out in alkaline buffer, while many aromatic amines favor acidic conditions. Since the reagent is unstable in water, it is dissolved in a water-soluble nonhydroxylic solvent such as acetone or dioxane. This solution is then mixed with a buffer containing the analyte of interest. Reaction with primary amines occurs rapidly, followed by hydrolysis of fluorescamine to nonfluorescent products.

$\xrightarrow[t_{1/2} = 100\text{–}500\ \text{ms}]{R\text{—}NH_2}$

Fluorescamine — Fluorescent product

$t_{1/2}$ = 5–10 s H_2O

Nonfluorescent products

The rapidity of the reaction makes it ideally suited for on-line detection in LC. The fluorescamine reaction has been studied in detail and both the kinetics with primary amines and the properties of the resulting fluorophors have been well characterized.[68]

Although fluorescamine is suitable for the detection of a wide variety of compounds, the analysis of amino acids has received the most attention. The combination of high-performance ion-exchange chromatography with fluorescamine detection is the basis for several of the commercial amino acid analyzers. The analysis of picomole amounts of the naturally occurring α-amino acids is accomplished in about 2 h. The secondary amino acids proline and hydroxyproline can be detected after oxidation to primary amines by *N*-chlorosuccinimide. The advantages of fluorescamine over the use of ninhydrin are its more rapid reaction time and its relative insensitivity toward ammonia. Also, fluorescence techniques are generally more sensitive than absorption techniques.

Several recent publications illustrate the general applicability of the fluorescamine reagent. Frei *et al.*[69] developed an assay for the nonapeptides oxytocin, lysine–vasopressin, and ornipressin in pharmaceutical injectables, using the fluorescamine approach. A thorough study of the reaction parameters of fluorescamine with the peptides is presented. Using a reverse-phase system, the peptides elute in 10 min or less and are detectable at 5- to 10-ng levels. Compatibility of the fluorescamine reaction with a number of other pharmaceuticals, both aromatic and aliphatic amines, has also been recently tested.[70] Standard mixtures of diamines and polyamines have been determined by LC–fluorescamine.[71] Several biogenic amines have also been analyzed by derivatization with fluorescamine *prior* to LC.[72]

o-Phthalaldehyde. Another relatively new fluorescence reagent is *o*-phthalaldehyde which was first proposed by Roth in 1974.[73] Like fluorescamine, it has primarily been used for the detection of amino acids and is the reagent used in some commercial amino acid analyzers. Although *o*-phthalaldehyde is reported to form fluorescent products with primary amines, the nature and scope of the reaction have not been fully investigated.

At room temperature *o*-phthalaldehyde reacts with amino acids and primary amines in just seconds to form a fluorescent isoindole product.[74]

$$C_6H_4(CHO)_2 + HOCH_2CH_2SH + RNH_2 \rightarrow \text{isoindole } (SCH_2CH_2OH, N\text{—}R)$$

The reaction is usually carried out at alkaline pH and requires a thiol, such as mercaptoethanol. The excitation maximum of the substituted isoindole is at 340 nm with an emission maximum at about 455 nm. This product slowly decays and measurements should be accomplished within 30 min of its formation.

The detection of proline and hydroxyproline is possible only after oxidation of these compounds with sodium hypochlorite. Cysteine gives a very low fluorescent yield with *o*-phthalaldehyde. In order to achieve sensitivity equivalent to that obtained for the other amino acids, cysteine must be oxidized to cysteic acid, prior to the coupling reaction. The sensitivity of this approach for amino acid analysis relative to that of ninhydrin and fluorescamine remains to be accurately determined.

Meek[75] has utilized the *o*-phthalaldehyde reaction for the analysis of the putative neurotransmitters taurine and γ-aminobutyric acid in small regions of the rat brain. The brain tissue homogenate could be analyzed with no work-up other than centrifugation of the proteins. Only 3–7 min is required per sample, and the amines were detectable at the 50-pmol level.

Dansyl Chloride. Dansyl chloride deserves mention because of its widespread use as a fluorescent derivatizing agent for primary and secondary amines, phenols, and alcohols. Long reaction times and the fact that dansyl chloride itself fluoresces has limited its use to pre-column derivatization.

Reactions are usually carried out in an alkaline buffer–acetone solution and require at least 1 h at room temperature. An added drawback of this reagent for on-line use is the low fluorescent yield obtained in solvents of high dielectric constant such as water. This limits the sensitivity in ion-exchange or reverse-phase chromatography, since these are both typically used for the separation of amines.

Cl
SO_2
$\xrightarrow{RNH_2}$
R
SO_2
N
CH_3 CH_3
N
CH_3 CH_3

One rather impressive approach to the analysis of amino acids involves derivatization with dansyl chloride prior to separation by LC.[76] With the use of microparticulate silica and reverse-phase columns, the dansyl derivatives of 20 amino acids were resolved in about 30 min (see Figure 24). Using an exciting wavelength of 340 nm and monitoring the fluorescence at 510, femtomole quantities were detectable.

Frei *et al.*[77] have used dansyl chloride for the analysis of alkaloids in complex pharmaceutical mixtures. The use of fluorescence eliminates many UV-absorbing interferences. The dansyl derivatives of a large number of carbamate pesticides are separated under a variety of LC conditions and allow the analysis of these pesticides in soil and water samples.[78] Other interesting applications of the dansyl reagent are the analysis of therapeutic levels of barbiturates in blood[79] and the determination of diamines and polyamines in tissue samples.[80]

Cerate Oxidative Monitor. The so-called "cerate oxidative monitor" is a reaction detector in which compounds such as carbohydrates, phenols, and aromatic acids are oxidized by cerium IV in an acidic medium.[81] The cerium III produced by the one-electron transfer reaction is measured by fluorescence (excite 260 nm, emit 350 nm). This approach has been used in combination with the traditional UV absorption detector. The combined detection system has been successfully applied to complex urine, serum, and environmental samples.[82,83] The reduction potential for cerium IV varies considerably with the medium, and thus the conditions under which the oxidation is carried out can be varied to control specificity.

The eluate from the column of a combined UV–fluorescence system flows first through a UV photometer and then to a mixer where it is diluted with cerium IV in strong acid solution (see Figure 25). The mixed

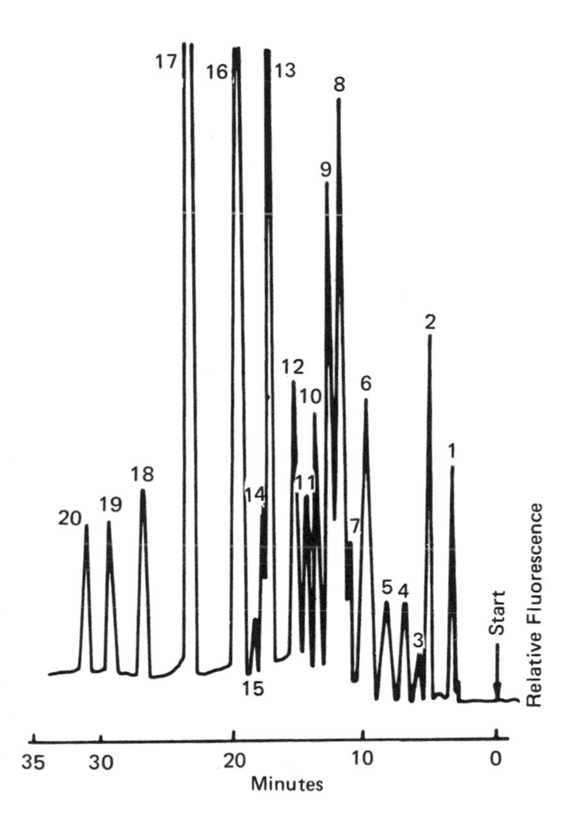

Figure 24. Reversed-phase chromatography of dansyl amino acids on Li Chrosorb RP8 (Merck A.G.), 10 μm, 50 cm × 0.3 cm (ID), 45°C, 140 atm, 1.5 ml/min; eluent: aqueous 0.01 M Na_2HPO_4 buffer–methanol (50:20) to which 1.5 ml methanol/min is added. Peak (1): dimethylaminonaphthalene-5-sulfonic acid; dansyl derivatives of (2) histidine, (3) serine, (4) glycine, (5) threonine, (6) alanine, (7) arginine, (8) proline, (9) valine, (10) methionine, (11) phenylalanine, (12) isoleucine, (13) leucine, (14) cystine, (15) aspartic acid, (16) tryptophan, (17) lysine (didansyl derivative), (18) tyrosine (didansyl derivative), (19) dansyl amide, and (20) dansyl chloride. (Reprinted from reference 76, courtesy of the American Chemical Society.)

eluate passes through a reaction coil in a boiling-water bath and on to a fluorometer. Although dilution and diffusional band spreading in the reaction coil result in some loss of sensitivity, the system is capable of detecting as little as 10 ng of injected phenol.[83] As yet there have been no published reports on the application of this reaction detector to separations on silica-based microparticle columns.

Fluorescence Enhancement and Quenching. Fluorescence is undoubtedly one of the more sensitive modes of LC detection. However, the dependence of fluorescence intensity on the solvent can become a problem. The mobile-phase requirements of a particular separation may result in poor fluorescence intensities of the compounds to be separated and thus lower sensitivity. This dependence of fluorescence intensity on environment is used to advantage by Novotny and co-workers.[84] The dependence of the fluorescence intensity of a dye on its environment is used as the basis for an LC detector. The fluorescence intensity of the dye 1-anilinonaphthalene-8-sulfonic acid (ANS) in an aqueous buffer is significantly enhanced by the presence of phospholipids. This is presumably due to formation of micelles by the lipids and the partitioning of ANS into the hydrophobic micelles.

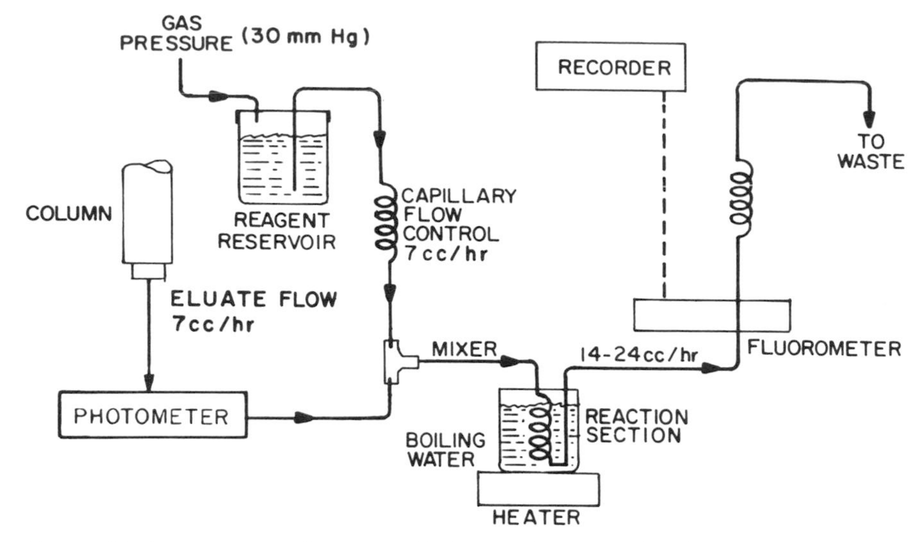

Figure 25. Arrangement of cerate oxidative monitor. (Reprinted from reference 81, courtesy of the American Association for Clinical Chemistry.)

For LC detection, a solution of ANS is mixed with the effluent of an LC column and passed through a fluorescence detector. The fluorescence of the ANS is continuously monitored. As a phospholipid elutes from the column, it is recorded as an enhancement of the steady-state fluorescence. Using this principle, several lipids were separated on the microgram level. The detection limit for phosphatidylcholine is given as 500 ng. Hardly a detector for trace analysis! As the authors point out, the system has not been optimized. With further refinements this clever approach may become a viable method for difficult-to-detect molecules such as phospholipids. At present, it is at least competitive with such detectors as the refractive-index and moving-wire detectors.

3. Electrochemical Detectors

3.1. Introduction

Many organic molecules of practical importance are electrochemically reactive at modest potentials, whereas many others are not. It is therefore sometimes possible to use electrode reactions as the basis for direct detection of one or two compounds in very complex samples without prior separation. While electrochemical techniques are inherently quite sensitive, the resolution of mixtures of electroactive compounds is very difficult. At most only about 3 V is available for analytical work between those potentials at which the medium itself begins to oxidize and reduce. Because a single electrode is rarely stable itself over this range, a practical working range may be less than 1.5 V for a given electrode material. Within this limited "potential window," it is only possible to reliably determine components of a mixture if the separation of half-wave potentials is at least 200–400 mV, depending on the kinetics of the reaction and the relative amount of each component present. It is therefore very unusual when more than 2 or 3 substances can be determined in a single electrochemical experiment. Because electrochemical reactions are surface reactions confined to a very small volume of adjacent solution, the resolution problem can often be overcome by placing an electrode in the effluent stream from a liquid chromatography column. The low dead volume required, the great sensitivity of hydrodynamic electrochemistry, and the tunability afforded by the electrode potential are desirable features which make the match of liquid chromatography to electrochemistry(LC/EC) ideal for many trace analysis problems.

In the following sections the principles, limitations, and major applications of the LC/EC technique will be described. Our discussion will be limited to finite-current electrochemical measurements in which a net conversion of material occurs. Although potentiometric measurements (e.g., selective electrodes) may ultimately have some limited utility in LC, at present potential measurements at zero current do not appear promising for this application.

While conductivity measurements are certainly within the purview of modern electrochemistry, they are not well suited to trace organic analysis, particularly in the presence of a relatively large concentration of background electrolyte. The innovative technique "ion chromatography" provides an exception to this point of view. In this case, a two-stage ion-exchange system is used. Figure 26 illustrates the principles of this technique for cation separations, using dilute hydrochloric acid as eluant. Sample constituents (denoted as M^+X^-) are first separated on a low-capacity pellicular column with fixed anionic sites (R^-) by undergoing the following reaction:

$$M^+X^- + R^-H^+ \rightarrow R^-M^+ + H^+X^-$$

Effluent from the separating column is then passed into a high-capacity "stripper column" with fixed cationic sites (R^+) in the hydroxide form (R^+OM^-). This latter stage suppresses the background electrolyte by two simple neutralization steps:

$$M^+X^- + R^+OH^- \rightarrow M^+OH^- + R^+X^-$$

$$H^+Cl^- + R^+OH^- \rightarrow R^+Cl^- + H_2O$$

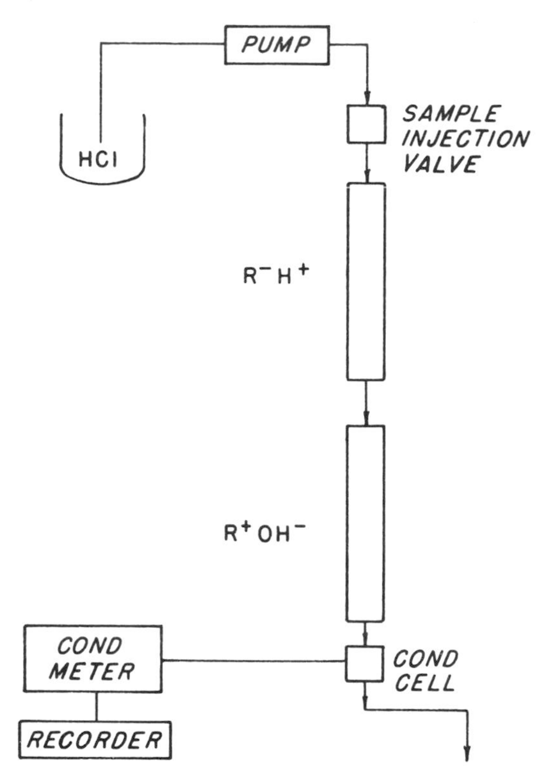

Figure 26. Schematic diagram of ion-chromatography apparatus (illustrated for cation analysis). (Reprinted from reference 85, courtesy of the American Chemical Society.)

In essence, the only electrolyte remaining to pass through the conductivity detector is that due to the analyte, transformed to the hydroxide form. The mobile phase is effectively neutralized, provided that the capacity of the "stripper" is not exceeded. In this clever manner a conductivity detector is made to be quite useful. The technique performs best for inorganic ions, although some organic ions have also been determined with good sensitivity.[85,86] At present ion chromatography is being most widely employed for the determination of ions in aqueous samples (waste water, boiler feedstock, municipal water supplies).

3.2. Applicability of Finite-Current Electrochemical Detectors

The applicability of electrochemical detection to a given trace-analysis problem ultimately depends on the voltammetric characteristics of the molecule(s) of interest in a suitable mobile phase and at a suitable electrode surface. It is therefore important to evaluate the voltammetric behavior under conditions meaningful to the chromatographic experiment. This is normally carried out using the cyclic voltammetry (CV) technique which is convenient for quickly assessing the thermodynamic and kinetic properties of the electrode reaction *per se* and the stability of the initial product. Hydrodynamic voltammetry (next section) is more closely related to the actual situation in an LC/EC experiment; however, CV is far more convenient to carry out in practice.

All detectors place limitations on mobile-phase composition to some degree; however, in electrochemical detection one must be conscious of the fact that a complex surface reaction is involved which depends on both the physical and chemical properties of the medium. Considerable effort may be required to optimize an LC/EC determination. Both the column and detector must be considered together. In fact the entire chromatograph may be considered part of the electrochemical cell. Fortunately, it is possible to make a few generalizations. For all practical purposes *direct* electrochemical detection is not likely to be useful in normal phase adsorption (e.g., silica gel) separations since nonpolar organic solvents are not well suited to many electrochemical reactions. The preferred LC stationary phases for LC/EC detection clearly include all ion-exchange and reverse-phase materials since these are compatible with polar solvents containing dissolved ions. The ionic strength, pH, electrochemical reactivity of the solvent and electrolyte, and the presence of electroactive impurities (e.g., dissolved oxygen, halides, trace metals, etc.) are all important variables that must be adjusted simultaneously for optimum separation and detection.

The choice of electrode material is more critical in LC/EC than for the usual electroanalytical experiment, primarily because of the ruggedness and long-term stability required. While electrochemists are accustomed to mechanically awkward devices such as dropping-mercury or rotating-disk electrodes, such gadgets will try the patience of the most stalwart chromatographer. Electrodes subject to complicated surface-renewal problems (e.g., platinum, glassy carbon, and mercury films) work well in some cases and are disastrous in others. In spite of these difficulties, it is possible to routinely achieve quantitation at the picomole level (or far below in some cases) using the LC/EC approach.

3.3. Principles of Amperometric and Coulometric Detection

3.3.1. Some Basic Considerations

In order to fully understand the performance of electrochemical detectors, it is necessary to appreciate hydrodynamic voltammetry. For chromatographic detection it is most advantageous to operate an electrode at a fixed applied potential. If the potential were to be scanned (either stepwise or as a slow ramp) and the current plotted as a function of potential, then one would observe a voltammogram as illustrated in Figure 27A for the mobile phase alone. The addition of a dilute oxidizable solute to the medium would result in the conventional "voltammetric wave" shown in Figure 27B. The "voltammogram" is characterized by the half-wave potential and limiting current for the oxidation wave. It is important to recognize that the hydrodynamic voltammogram represents a steady-state experiment. Under ideal conditions the scan rate and direction will have no influence on the recorded curve.

Quantitative determinations based on voltammetry depend on the degree to which the limiting current can be distinguished from the background current. At trace levels this ultimately becomes very difficult on the basis of signal-to-noise considerations. More significant, however, is the

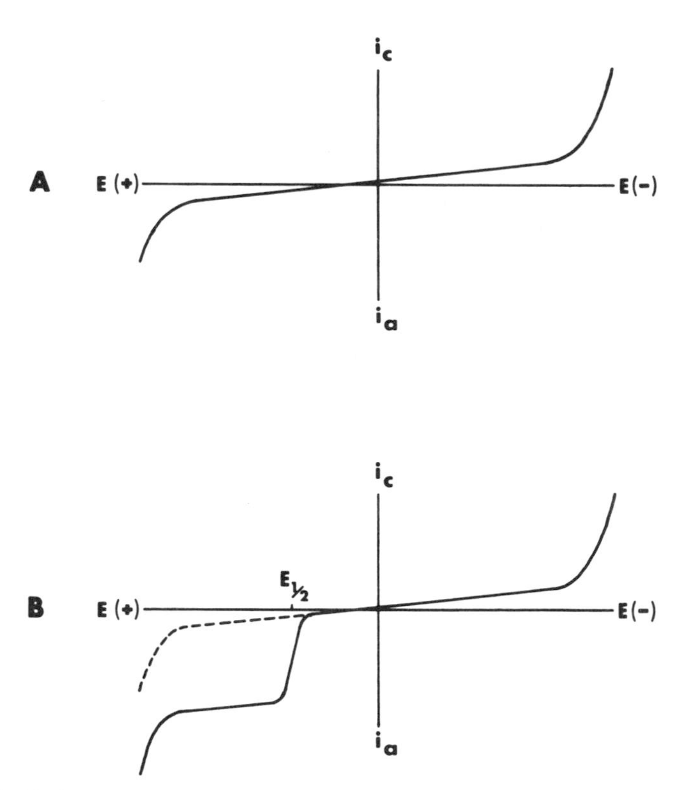

Figure 27. Hydrodynamic voltammetry. A: Current–potential curve for a mobile phase containing an inert electrolyte. B: Current–potential curve for the same solution following addition of an electrooxidizable substance.

problem of reproducing the background trace. This difficulty results from the fact that very slow processes alter the character of the electrode surface during a potential sweep. The background current will frequently depend on such variables as the initial potential and the elapsed time prior to initiation of the sweep. The anomalous "memory" effects, which apparently are related to previous experiments carried out using the same surface, are very difficult to deal with in routine practice. The heterogeneous electron-transfer kinetics for the sample molecule is generally fast and therefore the limiting current is related more to mass transport than to surface chemistry. In summary, hydrodynamic voltammetry is a good quantitative tool when the background current is small relative to the limiting current.

The situation can be improved dramatically by operating the electrode at a fixed potential (usually on the limiting current plateau) and recording the current as a function of time. The use of a constant potential improves the stability of the background current sufficiently to permit an enhancement of sensitivity by as much as two or three orders of magnitude. Most LC/EC experiments are carried out in this amperometric mode.

Once the operating potential is chosen, LC/EC chromatograms are obtained by plotting current as a function of time. The chromatographic baseline consists of the background current discussed above. The data are recorded with this "bucked out" to zero. In view of the fact that excellent current-to-voltage converters can be constructed from rugged integrated circuit operational amplifiers with field-effect inputs, it is not surprising that electrochemical detection is quite inexpensive, even for work at the picoequivalent level and below (where nanoamp currents are often encountered).

All of Faradaic electrochemistry ultimately depends on Faraday's law:

$$Q = nFN \tag{7}$$

For each chromatographic zone the number of coulombs (Q) passed will be proportional to the number of moles (N) converted to product in the cell, the number of electrons (n) involved in the reaction, and the Faraday constant (F). Coulometric detectors are defined as those where the conversion efficiency (extent of reaction) is 100%. For amperometric detectors the efficiency is somewhat less (often 1–10%). In both cases current is measured, and the instantaneous current is given by

$$\begin{aligned} i_t &= \frac{dQ}{dt} = nF\frac{dN}{dt} \\ &= (9.65 \times 10^4) \times (\text{equivalents converted per second}) \end{aligned} \tag{8}$$

A current of 10^{-9} A (easily measured in practice) therefore corresponds to a conversion rate of 10^{-14} eq/s. The conversion efficiency in cells designed for routine use is difficult to predict mathematically. Since the LC is operated at a constant flow rate, the fact that the conversion efficiency varies with flow rate is of little practical concern.

3.3.2. *Amperometry vs. Coulometry*

Amperometric detectors presently have the advantage of being more sensitive and less complex than coulometric detectors. But how can the sensitivity be better if less material is converted? This is often true in practice, because of the special geometry required to achieve complete conversion in a flowing stream. As the electrode surface area is increased to improve efficiency, each added increment of surface area contributes proportionately less to the total amount converted, but approximately equally to the background current from solvent, electrolyte, and electrode breakdown. Coulometric cells tend to be somewhat awkward in design. They involve electrodes made from beds of conducting particles or fibers to increase the area-to-volume ratio and thus the conversion efficiency. In general, coulometric cells are usually less satisfactory for routine use by chromatographers.

Perhaps more important is the fact that the advantage of coulometry for titrations and the like is not often realized in practice for chromatography detection where sample preparation and injection are normally the primary sources of error and relative measurements (internal and/or external standards) predominate. In any case, coulometric detection will only provide an absolute measurement for completely resolved compounds. Since one often cannot afford to operate with baseline resolution between peaks (it takes too long and is unnecessary), the question of absolute coulometric detection is academic. Nevertheless, coulometric detectors for LC have been applied to some important problems, particularly by Dennis Johnson and his students at Iowa State[87–89] and more recently by Poppe and co-workers.[90,91]

3.3.3. *Cell Design*

LC amperometric detection has been most widely used for applications involving the trace analysis of organic components in complex biological media. The active region of most amperometric detectors consists either of a tubular electrode or a thin-layer cell. The latter has been more popular in practice due to the ease of achieving a small volume (less than 1 μl) and in accommodating a variety of electrode materials. Thin-layer cells have been constructed with fluid flow strictly parallel to an electrode imbedded in a rectangular channel[92] or by directing the stream perpendicular to the surface followed by radial dispersion.[93] A popular design of the former type is illustrated in Figure 28. In this case the electrode surface is part of the channel wall formed by sandwiching a fluorocarbon gasket (typically 50–125 μm) between two blocks machined from Plexiglas, Kel-F, or similar engineering plastics.[94] A variety of electrode materials can be used (platinum, gold, glassy carbon, etc.); however, carbon paste has been found to be most satisfactory for a number of important analytical applications involving the oxidation of phenols, aromatic amines, and heterocycles. Normally electrodes with diameters of 2–4 mm are used in these cells.

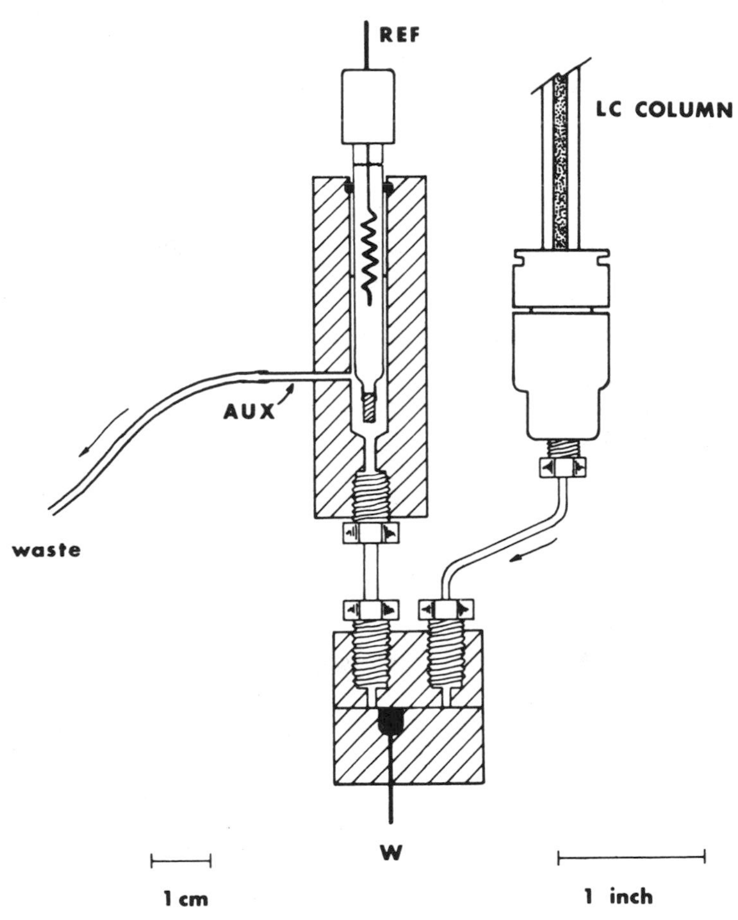

Figure 28. Thin-layer amperometric detector. (Reprinted from reference 114, courtesy of the American Chemical Society.)

3.3.4. Electrode Materials

Mercury has been the most widely used electrode material in electroanalytical chemistry. Its smooth, reproducible surface and high hydrogen overvoltage are very attractive characteristics. Unfortunately, in using mercury for LC/EC, a number of difficulties arise. First, at potentials greater than approx. +0.2 V (vs. the saturated calomel electrode), mercury is usually oxidized and thus the oxidation of most organic compounds is not possible using this electrode material. Furthermore, the mechanical design of a reliable low-dead-volume cell is quite difficult.

Platinum has been used to a limited extent, although it is not attractive for general organic use in aqueous solutions, since adsorption or filming of material at the electrode surface often makes frequent and careful cleaning necessary. These problems are less severe in nonaqueous solvents, where platinum is often an ideal choice.

The most commonly used electrode materials in LC/EC are carbon paste and glassy carbon. Carbon paste is made by mixing a fine graphite powder with mineral oil (or another dielectric material) to obtain a paste. This paste is then used to fill a small cavity facing into the flow stream. Carbon paste has a fair potential range which can be moderately adjusted by using different waxes or oils in the paste. In general, carbon paste is used for oxidizable species and easily reducible species. The major advantage of carbon paste is the relative freedom from surface filming. Usually an electrode will last for weeks without resurfacing. Glassy carbon is obtained as a solid rod which can be polished to a mirror finish at one end and inserted into the flow stream. Glassy carbon has a good useful potential range in aqueous solutions (+1.0 to −0.8 V) so that both oxidizable and reducible organic species can be electrolyzed with this system. The glassy carbon electrodes in current use are subject to adsorption problems, and may have to be cleaned and polished frequently. The material also has a reputation for not being reproducible from lot to lot. Hopefully these difficulties will be overcome soon.

Because of the low currents encountered in trace analysis work, it is possible to operate an LC/EC cell in the classical two-electrode mode. When three electrodes are used, the auxiliary electrode and reference electrode can be placed downstream as illustrated in Figure 28. This arrangement, while convenient, does result in nonlinear behavior when large samples are injected (typically >200 ng for compounds with small k') unless the ionic strength is very high or the gasket very thick. The nonlinearity arises from an ohmic drop along the thin film of solution, which at high currents causes the electrode potential to decrease. Positioning the auxiliary electrode across the flow stream from the working electrode reduces the uncompensated resistance to a negligible value even when low-ionic-strength (<10 mM) mobile phases are used. With a low uncompensated resistance, the interfacial electrode potential is not significantly influenced by sample concentration and a wide linear range is obtained. Another modification of the thin-layer detector cell involves the use of two identical counterposed working electrodes to enhance conversion efficiency. Electrodes are mounted in both the upper and lower blocks (see Figure 28) and electrically connected to a single current-to-voltage converter.

3.3.5. *Instrumentation*

Although constant-potential amperometry using conventional three-electrode circuitry is the only LC/EC approach which has been widely used in practice, it is, of course, possible to carry out more sophisticated experiments. There has been some interest in the possibility of programming the potential during a chromatographic run using a triangular wave in the manner of cyclic voltammetry. There are several reasons why this idea has little practical merit. Steady-state hydrodynamic electroanalysis has the advantage of being relatively free of double-layer charging effects, and there is no point in discarding this very desirable feature. Furthermore, Faradaic reorganization of electrode surface groups (particularly on carbon) is often a slow process which wastes charge and leads to very significant background currents when trace analysis is attempted with a ramp excitation. If "cyclic voltammetry" is possible in an electrochemical detector, then the concentration of analyte will be sufficiently great that a sample could just as easily be collected and studied under stationary solution conditions using various microcells.[95]

Potential pulse experiments are far more practical than ramp excitations due to the possibility of discriminating against background effects as is done in pulse voltammetry.[96] For example, repetitive waveforms using several pulses of different amplitude can be used with sample and hold circuitry to provide "simultaneous" multipotential chromatograms. It is also possible to selectively detect components eluted together by using small-amplitude differential-pulse techniques. Referring to Figure 29, B could

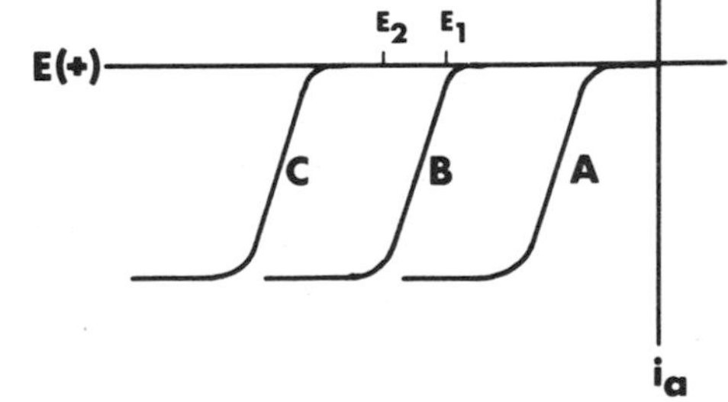

Figure 29. Hydrodynamic voltammograms for individual solutions of three oxidizable substances. (Reprinted from reference 114, courtesy of the American Chemical Society.)

be detected in the presence of C using a constant potential (E_2), however, B could not ordinarily be selectively detected in the company of species A. A pulse train may be applied with the potential stepped between E_1 and E_2 and the current sampled at the end of each half-cycle. If the *difference* in the sampled currents is plotted, the response to molecule A will subtract out and that to molecule B will be enhanced. An example of this approach has recently been published by Swartzfager.[97] For the reasons described above, pulse experiments tend to have significantly reduced sensitivity when compared to the constant-potential experiment. In most cases, simple DC amperometry will provide the best performance.

Another approach to obtaining improved selectivity is to operate two (or more) electrodes at different potentials at the same time, avoiding the problem with pulsed waveforms.[98] The several electrodes can be used independently or downstream electrodes can be used to detect the products from upstream electrodes. Although this concept works, it very likely will never see widespread application to practical chromatographic problems. As far as LC/EC is concerned, improved chromatography is always more desirable than adoption of a more complex scheme for operation of the detector.

3.4. Applications of Amperometric and Coulometric Detectors

3.4.1. Range of Applicability

In addition to being extremely sensitive, the electrochemical detector is quite selective in that only compounds electroactive at a given potential are detected. This feature can be viewed as either an advantage or disadvantage, depending on the analytical goal. For general use it is important to have a detector which is sensitive to a large number of compounds. On the other hand, when determining a trace component in a matrix as complex as body fluid or plant material, detector selectivity is a great advantage and can compensate for limited resolution by the chromatographic column. Since the majority of organic compounds are not electroactive in an easily accessible potential range, the electrochemical detector is relatively selective. Fortunately, a large number of extremely important compounds of biomedical and environmental interest are electroactive and therefore can be studied by LC/EC.

The electrochemical properties of a molecule are most often determined by the nature of the electroactive functional groups in the molecule. Electroactive molecules are either oxidizable, reducible, or both. One can anticipate with reasonable certainty that all phenols, aromatic amines, quinones, imines, and nitro compounds will be electroactive. Often the behavior of different molecules bearing the same electroactive function will be quite similar, and therefore voltammetric studies may not be necessary prior to setting up an electrochemical detector.

When determining whether or not a particular compound can be successfully analyzed by LC/EC, it is not sufficient to know that the compound can react electrochemically. The type of electrode surface, nature of the solvent, and relative ease of oxidation or reduction must be carefully considered, before one can be sure that such an analysis is feasible. Perhaps the most important consideration for trace analysis is the relative contribution which can be expected from the background current under a given set of conditions. In general, detection limits are lower for more easily oxidized (or reduced) substances, since these can be determined at potentials well inside the available potential window (see Figure 27).

There are special experimental problems associated with the LC/EC detection of trace amounts of reducible substances such as nitro compounds, nitrosamines, and imines. Reduction of dissolved oxygen, trace metal ions, and hydrogen ions can be difficult to eliminate. Although these problems (which lead to high background currents) can certainly be overcome, most practical applications which have been published involve oxidizable components. This trend may eventually be mitigated by recent improvements in electrode materials.

3.4.2. Recent Applications of LC/EC

The detector design schematically shown in Figure 28 was originally developed[92] in response to the need for a convenient method for determination of trace amounts of tyrosine metabolites in tissue samples and body fluids. The initial applications were directed toward measurement of catecholamines in small-animal brain tissue. Recently a number of improved methods have been developed for these molecules and their metabolites in brain[99] and urine.[100]

Figure 30 illustrates the power of amperometric detection for selective detection of easily oxidized components.[101] Homogentisic acid (2,5-dihydroxyphenylacetic acid, HGA) is an important compound in the primary catabolic pathway for tyrosine in man. In the LC/EC determination of HGA in serum (Figure 30), it is clearly advantageous to operate the detector at a low potential (+0.45 V) where possible interferences are less reactive. This degree of selectivity is rarely possible with a UV absorption detector since the electronic spectra of aromatic compounds are broad and are usually less sensitive to variation in substituents than is the oxidation potential.

Amperometric detection is broadly applicable in biomedical and phytochemical research, largely because of the predominance of aromatic

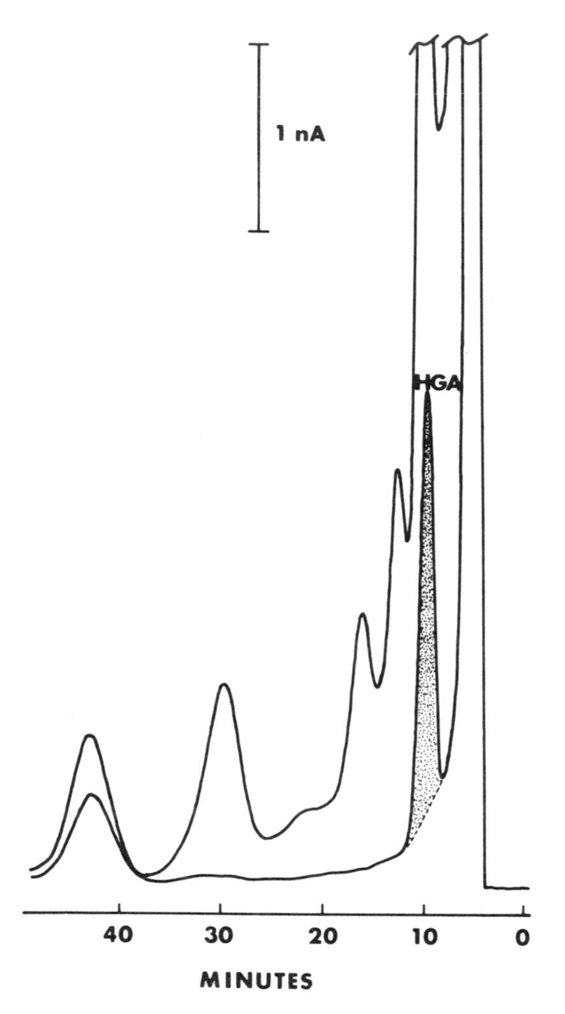

Figure 30. Determination of homogentisic acid in serum at 100 ng/ml, illustrating the influence of electrode potential. Two 7-ng samples were injected. The electrode potential was +0.75 V (upper trace) and +0.45 V (lower trace). Vydac SAX column (50 cm × 2.1 mm) with 0.5 M pH 4 acetate buffer as mobile phase (0.2 ml/min). (Reprinted from reference 101, courtesy of the American Chemical Society.)

hydroxylation in small-molecule metabolism. Many biochemicals, pharmaceuticals, food additives, pesticide residues, industrial antioxidants, plant phenolics, and so forth are also ideal candidates for LC/EC methods development. Simple aromatic amines and their derivatives (e.g., hydroxylamines, amides, quinoneimines, etc.) represent another large class of important molecules of scientific and commercial interest. Most of these are not well suited to GLC, but can often be handled with ease by the LC/EC technique. Many important compounds have been studied including ascorbic acid[102,103]; acetaminophen[104]; cysteine, glutathione, and penicillamine[105]; chlorogenic acid[106]; diethylstilbestrol[107]; benzidine[108]; folic acid, vitamin B_6, indoleacetic acid, kynurenic acid, and 5-hydroxytryptamine[109];

pentachlorophenol[110]; the parabens[110]; tetrahydroisoquinoline alkaloids[111]; and the phenothiazines.[90]

A frequently encountered problem in the use of LC for trace analysis is the positive identification of a trace constituent. Figure 31 illustrates a chromatogram of a human urine extract which was optimized for detection of 3-methoxy-4-hydroxyphenylethanolamine (normetanephrine, NM), metanephrine (M), and 3-methoxytyramine (3-MT). Tyramine (T) and serotonin (5-hydroxytryptamine, 5-HT) were also carried through the isolation procedure. Because most LC/EC determinations involve nanogram amounts of injected material, it is virtually impossible to obtain spectral information on collected fractions. It is therefore useful to carry out the measurement under at least two substantially different chromatographic conditions. Quantitative agreement between two such sets of data is excellent support for the absence of significant interfering components. In our laboratory, we also use TLC for this purpose. If a preliminary TLC step

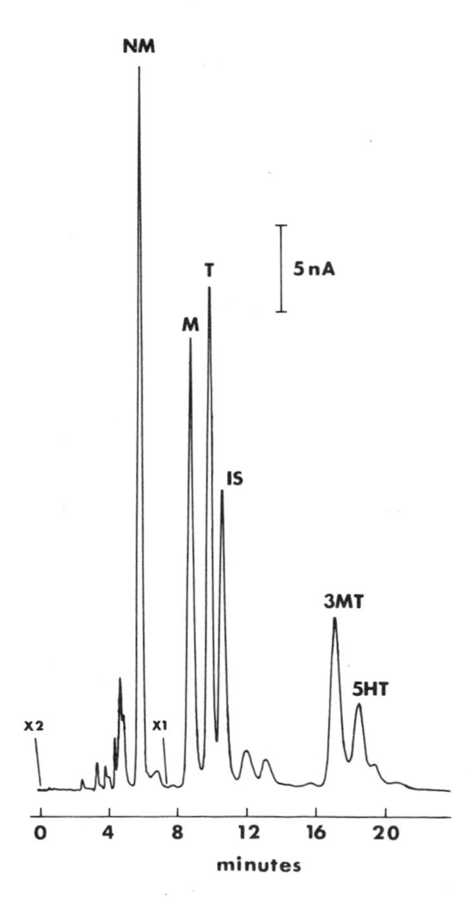

Figure 31. Chromatogram of an extract of a urine sample from a healthy individual. Operating conditions are as described in text. Concentrations are: NM (normetanephrine) = 320 μg/liter; M (metanephrine) = 170 μg/liter; 3-MT (3-methoxytyramine) = 160 μg/liter. (Reprinted from reference 100, courtesy of the American Association for Clinical Chemistry.)

(see Section 1.2.) does not change the results, it is likely that the peak(s) of interest are not representative of more than one substance.

Additional confirmation can be obtained by plotting hydrodynamic voltammograms for individual chromatographic zones. The curves are constructed by making repeated injections of a specimen extract at different potentials and plotting the peak current (i.e., peak height) vs. the applied potential. In order to facilitate comparison of the voltammogram with that for a standard, it is useful to ratio all the recorded currents to that measured at the most extreme potential. The resulting normalized current functions (ϕ) for standards and sample zones were plotted for all peaks identified in Figure 31. Excellent agreement was obtained, as illustrated in Figure 32 for tyramine.

3.4.3. *Reaction Detectors Based on LC/EC*

The limitations of amperometric and coulometric detection can be overcome in some cases by the use of post-column reactions.[112] A reagent may be mixed with the column effluent in order to convert the sample components into electroactive molecules. More commonly, a change in the reagent itself is measured. Sulfhydryl compounds are difficult to oxidize at graphite electrodes, but they reduce ferricyanide with ease. It is therefore possible to detect sulfhydryl compounds (e.g., penicillamine, cysteine, and

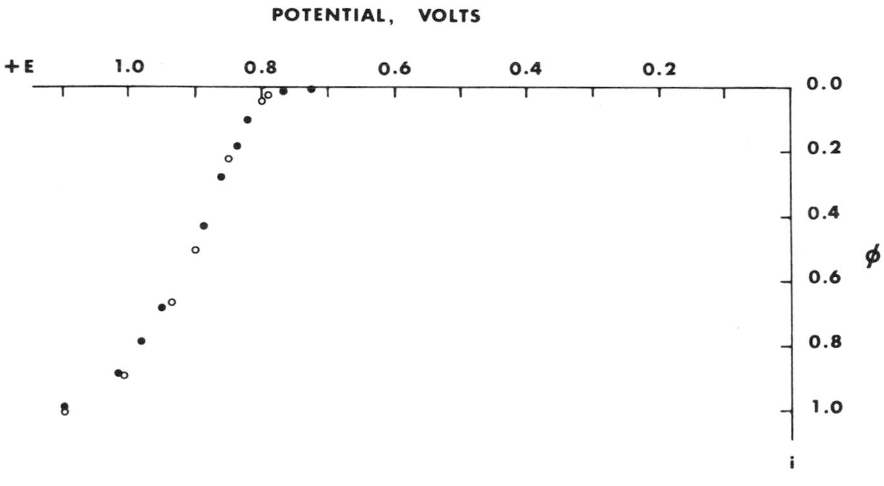

Figure 32. Hydrodynamic voltammograms of authentic tyramine standard (●) and urine extract (○). The normalized current function (ϕ) for each case (described in the text) is plotted versus the detector potential. (Reprinted from reference 100, courtesy of the American Association for Clinical Chemistry.)

glutathione) by reaction with ferricyanide and detection of the resulting ferrocyanide.[105]

$$\text{Reaction:} \quad RSH + Fe(CN)_6{}^{3-} \rightarrow RS\cdot + Fe(Cn)_6{}^{4-} + H^+$$

$$\text{Detection:} \quad Fe(Cn)_6{}^{4-} - 1\,e^- \rightarrow Fe(CN)_6{}^{3-}$$

This concept will work with a large number of compounds due to the disparity between thermodynamic E^0 values and half-wave potentials. Many organic compounds react slowly at electrode surfaces and therefore exhibit "irreversible" voltammetric waves at potentials ($E_{1/2}$) far in excess of the standard potential (E^0). Because of this situation, many seemingly ideal candidates for LC/EC cannot be detected with adequate selectivity in complex samples. Fortunately, the homogeneous redox reactions of many of these same compounds can be made to be quite rapid by using transition-metal oxidants. The reduced form of these reagents often oxidizes readily ("reversibly") at an electrode. The over-all result of this chemistry is to shift the required detector potential to a value where better selectivity and sensitivity can be anticipated.

Phenols and unsaturated compounds (e.g., prostaglandins) can be determined in a two-stage electrochemical reaction detector. The compounds are reacted with coulometrically generated bromine followed by amperometric detection of the bromine thus consumed. Perhaps the simplest "reaction detector" for LC/EC is one in which the reagent is mixed with the mobile phase for purposes of changing the pH and/or ionic strength to enhance sensitivity. It is also possible to use post-column mixing to permit the chromatography to be carried out with purely organic mobile phases of very low dielectric constant. A miscible solvent containing a supporting electrolyte is added just prior to the detector cell.[113] Research in this area is just beginning, with initially encouraging results. The additional complexity of post-column reactions always detracts somewhat from the convenience of direct measurement of mobile-phase constituents. On the other hand, the added selectivity and/or sensitivity can be important advantages for certain analytical problems.

Some of the material presented above has been briefly discussed in an earlier report.[114] A comprehensive review of the history, theory, and applications of electrochemical detectors will be published elsewhere.[115]

4. Vapor-Phase Detection Schemes

4.1. Introduction

The most extensive advances in chromatographic detectors have been made in gas–liquid chromatography. The development of such sensitive devices as the electron-capture detector and the chromatograph-coupled mass spectrometer have revolutionized the breadth of the gas chromatograph in analytical problem-solving. Accordingly, the versatility of gas-phase detection schemes in GLC has inspired a number of investigators to extend these techniques to liquid chromatography and eliminate some of the deficiencies in present LC detectors. Although some of the LC schemes mentioned below have no GLC counterpart, most will be recognized as simple modifications of existing GLC detectors. All involve the physical transformation of the liquid LC effluent to the gas phase prior to or at the time of detection.

Gas-phase LC detectors have not enjoyed widespread popularity. A few are commercially available, but most still remain within the haunts of academia. Although the reasons for this poor showing are varied and depend on the detector, general problem areas exist in all past approaches. Major obstacles include:

1. In total-volatilization schemes, the liquid effluent must undergo expansion to a volume 1000–10,000 times that originally occupied. During this process zones may become broadened or distorted. Alternate designs employ effluent "splitters" that funnel only a certain percentage (1–20%) of the liquid flow into the detector. Although large gaseous volumes are avoided, splitters obviously reduce sensitivity.
2. Nearly all gas-phase LC detection methods pass the totally vaporized effluent to the detector without attempting prior removal of mobile phase. The actual process of detection, therefore, is largely solvent-mediated and -controlled. In comparison to GLC, the higher relative proportions of solvent to solute molecules in LC particularly accentuate this effect. In LC/MS, for example, the mobile phase may act as the chemical ionization reagent and largely determines the mass spectrum. Gradient elution techniques are therefore difficult to implement. Another case in point is the LC electron-capture detector. The quenching effect of large populations of solvent molecules on the standing current is chiefly responsible for the 100-fold decrease in sensitivity of the device relative to its GLC cousin. It will be seen in other examples as well that the mobile phase is largely responsible for the disparities in performance between the LC and GLC versions.
3. Contrary to GLC, where monoatomic gases are often used for the mobile phase, in LC the differences in physical properties between analyte and mobile phase are often much less distinct. The business of devising suitable "separators" for analyte enrichment analogous to those in GC/MS consequently becomes quite complicated. "Moving-wire" approaches (see Section 4.2.4) claim complete removal of

mobile phase while leaving isolated solute behind but only provide workable results when the solute is much less volatile than the solvent. In cases where mobile phase and analyte are of comparable volatility, the selectivity of these separators for the latter is minimal.

4. Vapor-phase techniques work best with volatile, nonpolar mobile phases (such as hexane or chloroform). Polar solvents, particularly water or aqueous buffers, cannot be used in the majority of these approaches. A further problem lies in the inability of gas-phase detectors to cope with dissolved salts. Vaporization of buffers, for example, leaves behind salt deposits in the detector interface that eventually disable the entire system. These obstacles generally preclude their application to all ion-exchange and a significant portion of reverse-phase separations.
5. The costs of engineering and manufacture of these devices are fairly steep, with some units requiring an investment of $30,000 or more for detector alone. In spite of the cost, the mechanical complexity of many designs reduces their reliability for routine analysis.

4.2. Detector Designs: Principles and Applications

In spite of the above problems, a number of useful but limited LC detectors have emerged utilizing post-column volatilization. As with any new instrumental device, many reports are simply small modifications of established designs; it is our intent to discuss only those approaches that warrant mention for their versatility or usefulness.

4.2.1. Coulson Electrolytic Conductivity Detector

As an element-selective detector (primarily for N, S, and halogens), the Coulson electrolytic conductivity detector (CECD)[116] was first proposed in 1965 for use in gas–liquid chromatography and soon found extensive use in the GLC analysis of heteroatom-rich pesticides. However, in many situations, assay by GLC was not possible due to the nonvolatile or thermally labile character of these compounds. It appeared that liquid chromatography could be the solution in dealing with these substances if a similarly selective detector were available. A relatively simple interface[117] was designed to bridge the gap between the liquid chromatograph and a commercial GLC Coulson electrolytic conductivity detector, as illustrated in Figure 33. Liquid effluent (in this case, aqueous methanol) is quickly volatilized by passage through a 900°C furnace and then mixed with a concentric sheath of preheated hydrogen gas. The mixture is carried into a hot quartz pyrolysis tube farther down-line where the reductive nature of the gaseous hydrogen atmosphere converts S, N, and X (halogen) atoms

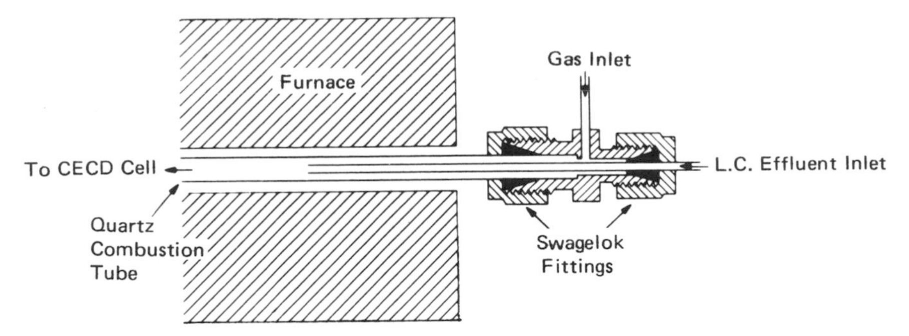

Figure 33. Details of interface to standard Coulson electrolytic conductivity detector. (Reprinted from reference 117, courtesy of the American Chemical Society.)

in the system to the corresponding gas-phase H_2S, NH_3, and HX. The gaseous reaction products are finally mixed with a previously deionized circulating water or water/alcohol solution and then passed into a standard CECD cell where changes in electrolytic conductivity due to ionized H_2S, NH_3, and HX are measured between platinum electrodes in an AC conductance bridge circuit. Since detection is a multistep process, it is not surprising to note that optimization strongly depends on such conditions as the nature of the pyrolysis gas, the furnace temperature, and the presence of a catalyst in the pyrolysis tube (necessary for production of ammonia from N-containing compounds). The system as reported was optimized for organochlorine compounds, but it is not unlikely that conditions for N-, S-, and P-containing compounds could also be devised. Simple modifications of the operating conditions extend its versatility. Malissa *et al.*,[118] for example, devised a similar approach for determining sulfur in organic solvents by nebulizing the LC effluent in an oxygen atmosphere to produce SO_2 and finally H_2SO_3.

For many pesticides and their natural conversion products, the scheme may be a viable alternative to mass-spectrometric and electron-capture methods. Detection limits of approx. 50 ng for heavily chlorinated lindane ($k' \sim 1.5$) and a linear range of about 10^5 have been reported.[117] The device has great mechanical simplicity and would not appear difficult to service. Nevertheless, direct volatilization approaches intrinsically display certain weaknesses. The massive quantities of mobile phase that must pass the orifice and become volatilized often fail to do so cleanly and eventually clog the inlet. This drawback usually stipulates limiting flow rates into the detector to 0.5 ml/min or less. For higher LC elution rates, effluent splitters would be necessary. As suggested by the authors, the more sensitive Hall electrolytic conductivity detector design[119] could alleviate many of these deficiencies, particularly in regard to sensitivity.

4.2.2. Photometric Vapor-Phase Techniques

Atomic Absorption. The analytical problems of organometallic chemistry, particularly those concerned with the role of chelation in biological chemistry, could benefit from the development of specific metal-ion detection schemes in LC. A fresh approach in LC detectors was taken by Jones and Manahan,[120–123] who, by direct connection of the column effluent tube to the nebulizer of a standard atomic absorption spectrophotometer (AA), effected a very simple interface. Mobile phase from the LC containing the free or chelated metal ions of interest is passed directly into the AA, readied at the spectral line of interest. The narrow bandwidths of atomic absorption transitions ensure adequate specificity for the particular metal ion desired. Extensive matrix cleanup procedures are usually not required. This detector would appear to be ideally suited to the selective determination of organometallics as well as those substances which can undergo "derivatization" via chelation. UV-absorbing solvents may be used without concern about the effects of high background, thereby adding to the variety of elution conditions possible for achieving a given separation. Spectrophotometric grade solvents and elaborate mobile-phase preparation are also not necessary.

The influence of the nebulization efficiency of the burner, the free atom population in the flame, and the variation in flame temperature with LC flow rate all interact to determine the maximum sensitivity attainable. Increasing mobile phase flow rates, for example, will quench the flame and produce a smaller free atom fraction, but will also deliver more analyte to the AA per unit time. In general, solutions containing metals in the 1–10 μg/ml range (or greater) may be analyzed. The utility of the system is demonstrated in Figure 34 with the detection of EDTA, NTA, EGTA, and DCTA copper chelates on an anion-exchange system. The authors estimated detection limits at ~13–450 ng injected, depending on the chelate's capacity factor and its behavior in the flame.

Flame Emission. With the advent of atomic absorption principles in LC detection, it was not surprising to see flame-emission techniques also being devised. Freed outlined a modular on-line approach[124] in which LC effluent was aspirated directly into the flame. The emission spectrum was optically monitored with a monochromator, photomultiplier tube, and associated amplifier electronics. The major advantages of this detector include virtual elimination of spectral overlap problems and less stringent conditions for chromatographic resolution of matrix constituents from the ions of interest. Mobile phase flow rates, however, must be closely matched to the normal aspiration rates of the burner; if too high, the flame is cooled with subsequent loss in sensitivity. Lower flow rates starve the flame and

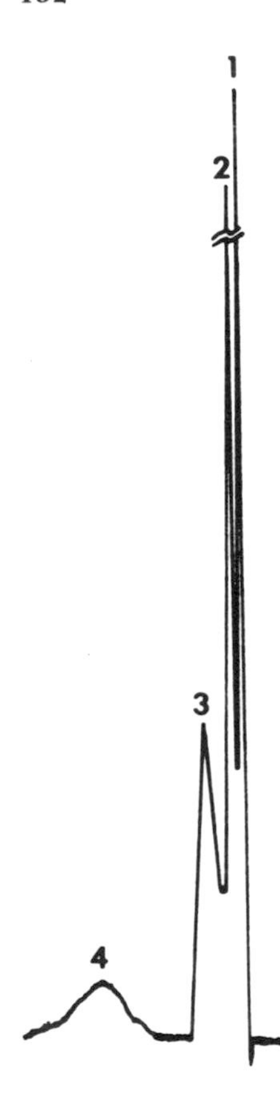

Figure 34. Separation of four chelating agents as Cu(II) complex: (1) $Cu_2(EGTA)$, (2) $Cu(NTA)^-$, (3) $Cu(EDTA)^{2-}$, (4) $Cu(CDTA)^{2-}$. Column: Aminex A-14 anion exchange resin, 5 cm × 2.1 mm (ID); detector: atomic absorption. (Reprinted from reference 121, courtesy of the American Chemical Society.)

likewise decrease performance. The system is, of course, necessarily sample-destructive.

Julin *et al.*[125] followed a similar approach in the development of a selective flame-emission detector for sulfur and phosphorus in LC effluent; their system, however, entailed the design of a special burner (see Figure 35) to handle flow rates from the LC of up to 2 ml/min. Effluent is nebulized and mixed with H_2 and N_2 to give a cool flame which may be selectively monitored by a simple filter monochromator and PMT. About one quarter of the nebulized effluent actually reaches the flame for emission. A cool flame was required to avoid breakdown of the emitters in this specific application. Phosphorus and sulfur were observed from the emissions of the HPO and S_2 species at 526 and 383 nm, respectively. Sensitivity for P was about 10^{-8} g of phosphorus/ml of effluent; sulfur measurements were an order of magnitude less sensitive.

Many of the previously mentioned limitations of aspirator interfaces apply here as well. Burner gas flow rates must be carefully regulated to avoid noise and baseline drift and maintain steady flame temperature. Organic solvents (even 1% aqueous methanol) and metal ions cause an unpredictable decrease in phosphorus emissions. The detector appears best suited for aqueous ion-exchange and reverse-phase separations.

Gas-Phase Chemiluminescence. Initially developed for use in GLC, the so-called "thermal energy analysis" detector described by Fine and coworkers[126] for *N*-nitroso compounds has been modified for use with LC effluent.[127] The selectivity of the device depends on the relatively weak nature of the *N*-nitroso bond and involves the following mechanism:

$$\mathrm{R{-}N{-}NO\ (gaseous)} \rightarrow \mathrm{RN\cdot} + \mathrm{NO\cdot}$$

$$\mathrm{NO\cdot} + \mathrm{O_3} \rightarrow \mathrm{NO_2^*} + \mathrm{O_2}$$

$$\mathrm{NO_2^*} \rightarrow \mathrm{NO_2} + h\nu$$

Under relatively weak pyrolysis conditions, the N–NO bond is split to liberate the nitrosyl radical which then reacts with ozone to produce electronically excited nitrogen dioxide. As the NO_2^* decays back to the ground

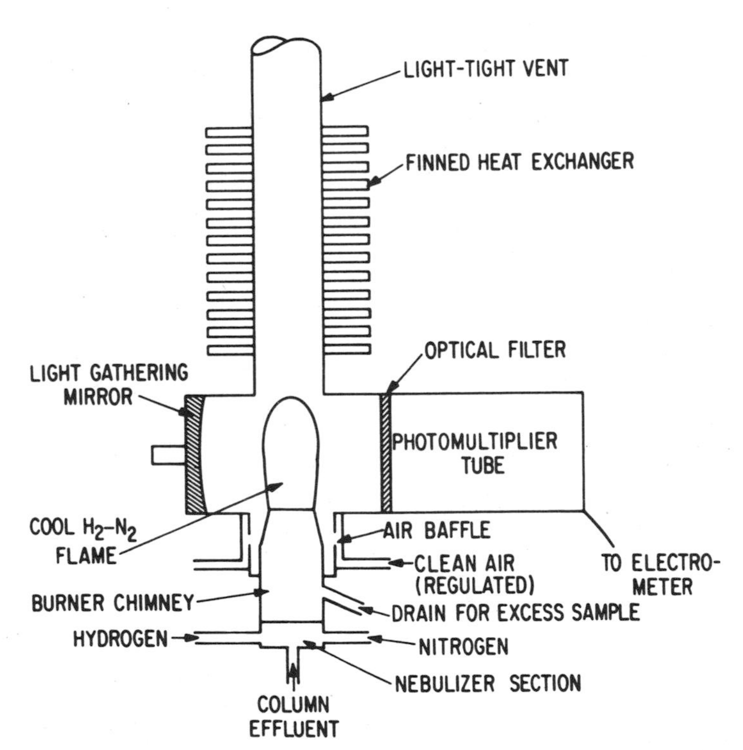

Figure 35. Flame emission burner and associated optics. (Reprinted from reference 125, courtesy of Elsevier Scientific Publishing Co.)

state, radiation in the 0.6- to 3.0-μm region is released. The mechanism is implemented as shown in Figure 36. Nozzle-atomized LC effluent enters as sample into a heated quartz pyrolysis tube packed with catalyst. The ruptured free radical is subsequently swept into an evacuated reaction chamber filled with ozone to yield chemiluminescent nitrogen dioxide. Since the emission intensity is directly related to the N—NO content, the response is nearly independent of the compound's over-all structure.

The LC/TEA system responds to *N*-nitroso compounds, organic nitrates and nitrites, and some inorganic nitrites. As the detector is not 100% selective to *N*-nitroso substances, users must be aware of these false positives and guard against them with sufficient chromatographic resolution. Some substances such as sodium nitrate must be avoided entirely, as they slowly decompose over several hours to disrupt baseline response. Relative to the GC detector, the LC system is about three orders of magnitude less sensitive, presumably due to the collision deactivation of NO_2^* by the large number of mobile-phase molecules. Detection limits are reported to be 20–50 ng of injected *N*-nitroso compound. At present, the greatest drawback to the detector is its complexity and cost (approx. $30,000). Its inherent selectivity, however, may allow a sizable savings in time otherwise spent in more extensive sample preparation.

A further application of LC/TEA extends the detector's range to include organic nitrogen compounds.[128] The only major modification involves replacement of the pyrolyzer with a combustion chamber in which

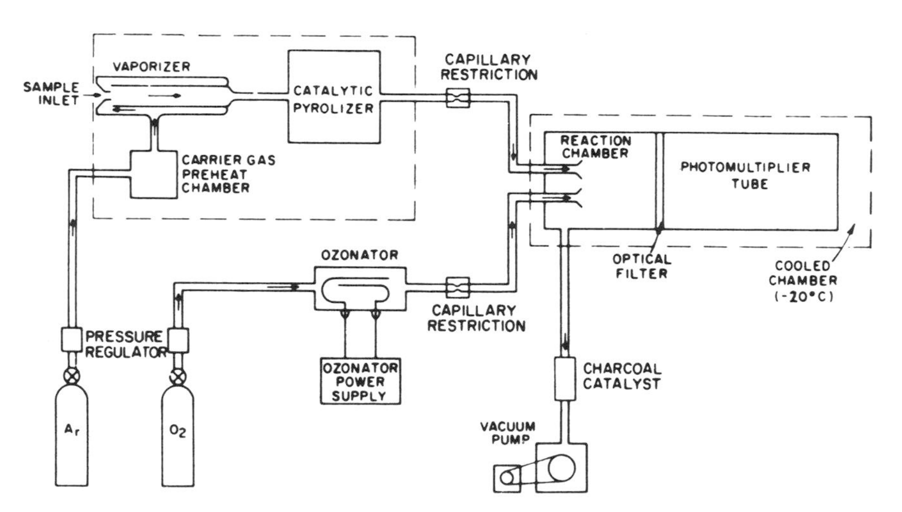

Figure 36. Schematic of the thermal energy analyzer. Atomized LC effluent enters at sample inlet on left. (Reprinted from reference 126, courtesy of the American Chemical Society.)

LC effluent is mixed with oxygen and burned at 1000°C to produce CO_2, H_2O, and NO. Major restrictions include detector flow rate limitations of 0.1–0.5 ml/min and the inability to use N-containing organic solvents for mobile phases. About 100 pg of organic nitrogen could be detected for weakly retained substances, at least two orders of magnitude more sensitive than the expected limits using an electrolytic conductivity cell optimized for total nitrogen.

4.2.3. *UV Photoionization*

An alternative detection approach suggested by Schmermund and Locke[129] utilizes selective ionization of vaporized LC effluent under vacuum UV radiation to produce a measurable ion current. The detector is represented in Figure 37. The totally vaporized mobile phase (transformed from the liquid state by an on-line low-dead-volume heater assembly) enters at the bottom and passes down a set of concentric electrodes mounted axially to the high-vacuum UV discharge lamp. Although this orientation maximized the effectiveness of the ionizing radiation, the process yielded an ionization efficiency of only 10^{-6}.

Only those substances with an ionization potential less than the energy of the ionizing UV radiation will give rise to a detectable increase in ion current. Luckily most solvents employed in LC mobile phases have ionization potentials greater than the energies emitted by the UV lamps and therefore pose no problem in causing significant background currents. Such an attribute can be highly desirable in gradient elution techniques.

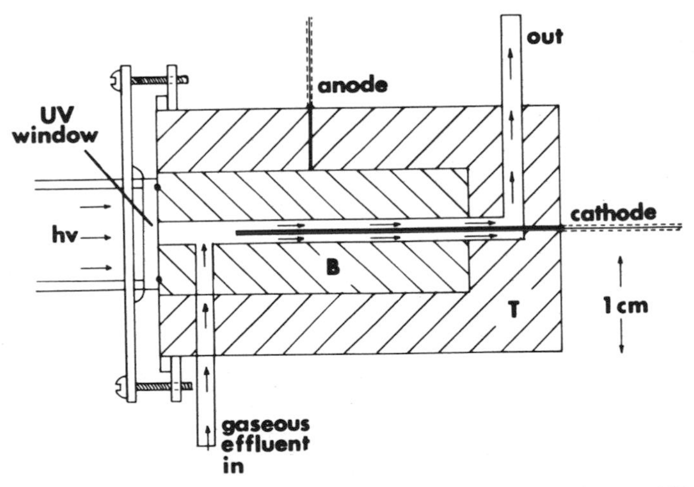

Figure 37. Diagram of photoionization detector. (Redrawn from reference 129, courtesy of Marcel Dekker, Inc.)

Minimum detectable concentrations in the effluent on the order of 10^{-9} g/ml and a linear range of 10^5 were reported. At present the design is still in its infancy; the potential benefits to analytical problem-solving are yet to be demonstrated.

4.2.4. *Flame Ionization*

A recurring problem in all of the vapor-phase approaches described so far is the deleterious presence of both solute and mobile-phase molecules during detection. In some cases the response from analyte can be severely attenuated due to the overwhelming proportion of solvent molecules present. In the chemiluminescent nitric oxide–ozone reaction employed in thermal energy analysis, for example, the presence of gaseous mobile-phase molecules can reduce the desired emission intensity. Obvious entanglements ensue if both solute and solvent yield a positive response; background effects, particularly with gradient elution techniques, become prohibitive.

The apparent solution to these problems would be the total fractionation of solute from the mobile phase. Without solvent to interfere, the solute could be chemically or physically manipulated in any manner so desired for detection. This attribute forms the central theme of the so-called "solute transport" class of detectors. In these systems, effluent from the LC is applied by drop, continuous stream, or spray to a moving transport carrier. Usually a belt, chain, or wire, the carrier conveys the sample to sequential stations where the mobile phase is evaporated—leaving only solute—and the residual analyte then detected farther down-line. Nearly all such systems employ flame-ionization sensors.

Most transport detectors have been based on the early prototypes developed by Haahti and co-workers.[130] Their important features are exemplified in Figure 38. In this particular design,[131] a stainless-steel wire serves as the transport medium for passage through different oven stages and a FID. As a preliminary step, the wire is heated in a nitrogen atmosphere to remove contaminants and reduce standing noise levels. Effluent is then coated onto the wire by placing its path in very close juxtaposition to the liquid exit. The thin film of solute and mobile phase on the transport passes into a heated evaporator. Since the mobile phase is more volatile than the solute in most separations, it can therefore be differentially removed in the evaporator prior to the wire's passage into the flame ionization detector unit. A tubular oven held at 600–700°C farther along pyrolyzes the remaining solute to fragmentation products which are subsequently swept by nitrogen into the flame-ionization detector. Several obstacles, however, precluded the widespread use of the detector. Mechanical difficulties in maintaining optimum operating conditions, noise problems due

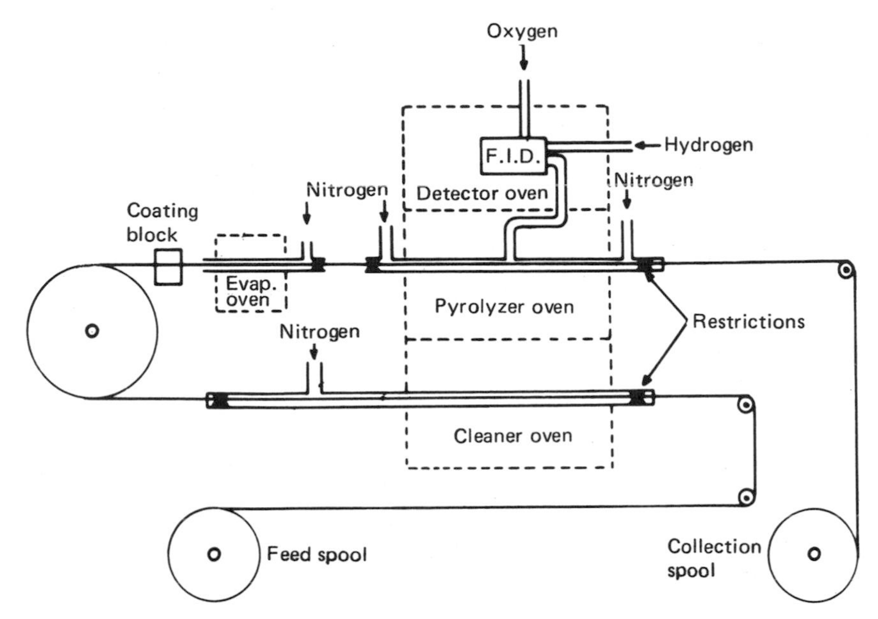

Figure 38. Normal wire transport flame ionization detector. LC effluent is applied to wire at coating block. (Reprinted from reference 132, courtesy of Preston Publications, Inc.)

to wire transport speed variations, and its failure to adequately detect highly oxygenated compounds led to its demise. Pyrolysis efficiencies were often quite low, usually amounting to only a few percent of the solute mass coating the wire. Maximum sensitivities of about 4 μg/ml *in the effluent* were possible. With respect to an injected sample, this sensitivity is quite poor.

An alternate design[132] replaces the pyrolysis oven with an oxygen-rich oxidation unit maintained at 800°C. The modified unit is shown in Figure 39. Instead of pyrolysis, solute is combusted to completion to yield carbon dioxide and water; the products are swept over a specially prepared nickel catalyst in the presence of hydrogen gas to form methane. The methane is finally detected by the FID. Since combustion affords more complete removal of solute from the wire than pyrolysis, the minimum detectable concentration in the effluent accordingly decreased to 1 μg/ml levels (again in the effluent). Another advantage is the predictable, proportional response of the system to organic carbon in the sample. Probably the greatest strength (or weakness, depending on viewpoint) is its applicability to nearly all classes of relatively nonvolatile compounds. For samples such as lipids, fats, carbohydrates, and polymers, which do not fluoresce or absorb UV–visible radiation, solute transport detectors may be competitive for the detection of microgram amounts.

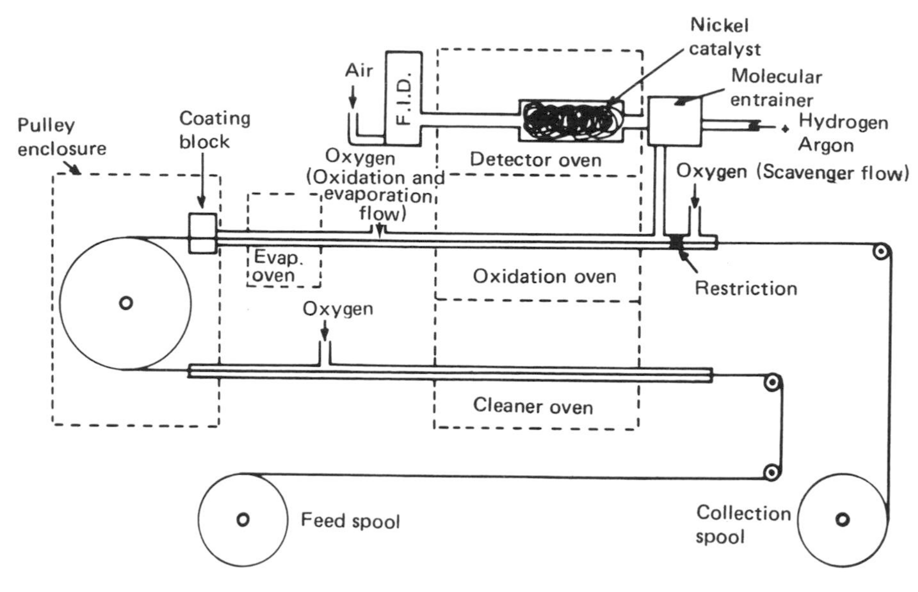

Figure 39. Modified flame ionization detector. (Reprinted from reference 132, courtesy of Preston Publications, Inc.)

A number of modifications for improving the performance of the flame-ionization detector have been proposed, usually as a direct outgrowth of some mechanical or physical problem in keeping the device operable. A major problem centers on the uniform application of effluent on the transport carrier. The physical separation of sample into irregular "pools" during coating and the coalescence of wet retreating solute as the sample approaches the hot evaporator all contribute to its random noise. The latter effect is thought to be responsible for the "spiking" phenomenon which shows up as a series of sharp spikes superimposed on the over-all elution profile. Dispersion of LC effluent as a "semidry" atomized spray reduced coalescence phenomena to some extent,[133] but does not appear applicable to all mobile phases. Transport systems are also plagued by low coating efficiencies. Investigators have proposed other solute carriers, namely chains,[134] porous-coated wires,[135] rotating disks,[136,137] and helical encapsulated cable.[138,139] At present, the latter approach appears most successful, both in eliminating spiking and in allowing higher sample-coating capacity on the wire. This modification permitted 25 ng of triolein to be detected in an actual chromatographic separation.[140] It has been suggested that the problems of uneven coating and poor coating efficiencies with wire transport detectors might be solved by using an electrostatic spray-painting technique, whereby a potential difference would be applied between the effluent nozzle and the moving wire.[141] Effluent would be attracted to the wire uniformly by electrostatics.

The choice of a particular mobile phase for the chromatography may be contradictory to the requirements for solute and solvent volatility in these detectors. If a separation necessitates using a polar solvent or ionic buffer, for example, it may not be possible to leave the solute intact on the carrier during the more vigorous conditions required for the evaporation of that mobile phase. Buffer salts are troublesome since they remain on the wire throughout the process.

4.2.5. *Electron Capture*

For the analysis of nonvolatile pesticides and their ubiquitous residues, the applicability of an electron-capture detector to liquid chromatography could be a productive development. To date, two designs have been proposed. The first involves a moving-wire approach[142] as described in the preceding section and therefore will not be discussed here. The other design,[143] shown in Figure 40, avoids a transport system entirely and rather employs total vaporization of the effluent in heated transfer tubing. As vaporization occurs, the expansion forces the gaseous mobile phase into a small chamber fitted with a ^{63}Ni β-emitter. The minimum detectable concentration for aldrin is approx. 1×10^{-10} g/ml of effluent. When compared on an equivalent basis to GLC designs, the performance of the LC electron-capture scheme is approximately two orders of magnitude less sensitive, presumably due to the overwhelming presence of mobile-phase molecules in reducing the probability of e^- capture by solute. Flow rates of up to 3.5 ml/min, however, can be tolerated in the detector if extreme sensitivity is not required. The greatest restriction of the electron-capture detector is the limited choice of mobile phases. Obviously, such solvents as chloroform and carbon tetrachloride are ineligible. Some polar solvents may be used, but only in dilute (1–5%) solutions of saturated hydrocarbons. Most of the separations demonstrated so far have required a solvent such as *n*-hexane as the mobile phase. A major advantage of the electron-capture detector is its selectivity for compounds that contain halogen and nitro functional groups. This feature is exemplified in Figure 41, where the UV and electron-capture traces for the separation of nanogram amounts of two nitrated toluene isomers are presented.

4.2.6. *Liquid Chromatography/Mass Spectrometry*

For trace organic analysis, the universal liquid chromatography detector remains a Pandora's box. Chromatographers, in their desire to design a detector that can sense *any* component, often fail to realize that it will

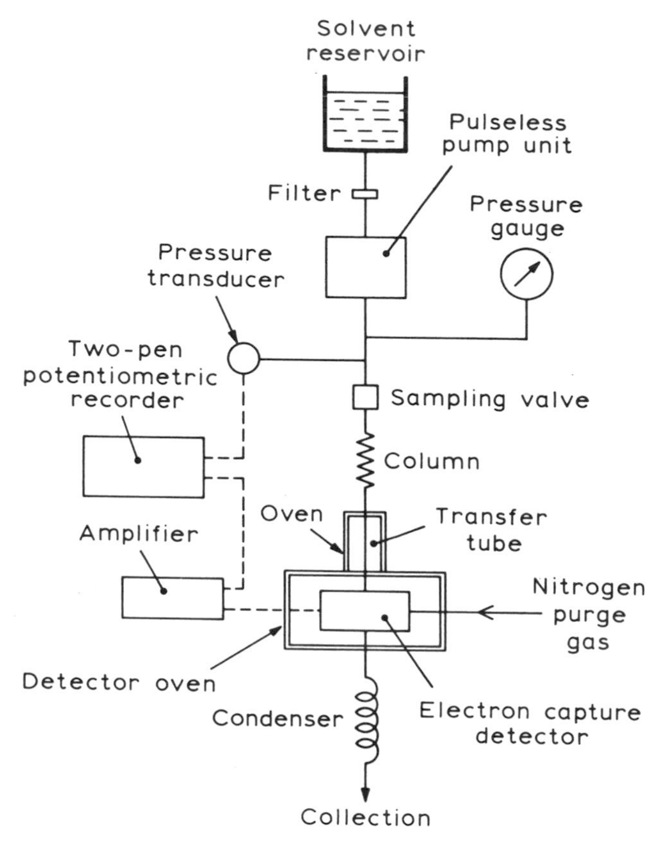

Figure 40. Schematic diagram of liquid chromatograph interfaced to electron capture detector. (Reprinted from reference 143, courtesy of Preston Publications, Inc.)

usually detect *all* substances as well. More often than not, the analytical sample is sufficiently complex that even the most efficient chromatographic column is incapable of resolving all the components present. With a universal detector, the peaks of interest—perhaps a drug and its metabolites—may be so obscured by the background of the sample matrix that even qualitative information could be highly suspect. The analyst must then reconsider his approach and search for more selective methods of detection.

Even in specific techniques, the issue of unequivocal peak identification remains a major obstacle. Often peak identities are based only on the capacity factor which is determined by injecting a known standard. Obvious difficulties arise in scaling up this limited identification method to complex mixtures of unknown composition.

The ideal LC detector would have subnanogram sensitivity, a *tunable selectivity* for all molecules, and a wide linear dynamic range. In addition, the detection should be accompanied by positive identification of the sample component. This system might eventually be derived from a combination of liquid chromatography plus mass spectrometry (LC/MS). LC would provide optimum separation and MS would provide sensitivity and specific structural information.

The principal problem in coupling an LC to a mass spectrometer occurs at the proposed interface, since there is apparently no simple way of modifying the LC effluent stream for introduction into the low-pressure ion source of the mass spectrometer. The problem is analogous to that faced earlier in the development of gas chromatograph–mass spectrometers (GC/MS). In GC/MS a liquid sample is volatilized in the injection port prior to chromatography. The purpose of the interface is to enrich the solute relative to the mobile phase during an over-all pressure reduction of a continuously gaseous sample. With LC/MS, a phase change must occur

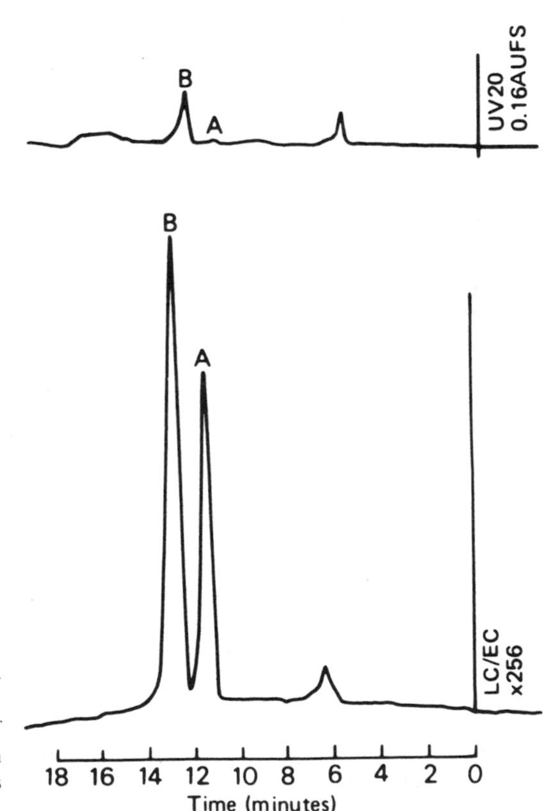

Figure 41. Separation of nitro compounds on adsorbent Partisil 10 column. Top trace: UV detector, 0.16 AUFS; bottom trace: electron-capture detection of same sample. (Reprinted from reference 144, courtesy of Philips Electronic Instruments.)

at the interface to the mass spectrometer. The excess of vaporized mobile phase is a greater problem if solute enrichment is to be achieved, since LC solvents are of higher molecular weight and therefore harder to separate by diffusional processes (vs. GC carrier gases such as H_2 or He).

Most papers in the field of LC/MS are concerned with the solution of these two problems in the design and operation of a suitable interface. In all of the interfaces described below, it should be remembered that the primary problem is in achieving enrichment of solute during an over-all phase change from liquid to gas.

Four main types of LC/MS interfaces have been developed: (1) flash evaporation of solvent on a solid probe followed by periodic introduction of the probe into the ion source of the MS, either manually[(145)] or mechanically[(146)]; (2) a continually moving wire or belt[(147,148)] upon which the eluate is coated, the solvent evaporated, and the resulting solute passed directly through the ion source, on the carrier; (3) direct introduction of the eluate into a chemical ionization mass spectrometer via a capillary tube[(149,150)]; and (4) atmospheric-pressure ionization,[(151,152)] whereby the LC eluate is vaporized by preheated carrier gas and passed into an ionizing source external to the low-pressure region of the MS where ^{63}Ni foil or a corona discharge causes ionization. The solid-probe approach in on-line LC/MS has largely fallen from use and will not be discussed further in this review.

Moving Wire. The LC/MS moving-wire (or belt) interface basically shares the same transport system as the flame ionization detector discussed in Section 4.2.4. The most significant advantage of these systems is their freedom from mobile-phase effects in MS. Solvent-mediated fragmentation processes are no longer a factor. The analyst can employ electron impact or chemical ionization techniques at his discretion. The earliest design along these lines was offered by Scott and co-workers,[(147)] who adapted a commercial FID transport system for travel through the mass spectrometer ion source. A small current passed through the transport wire is responsible for the thermal desorption of the solute from the wire into the ion source. The performance of the device, however, prohibits its use in trace organic analysis. Most of the separations reported using the detector required samples on the order of 1 mg!

The Finnigan LC/MS interface[(148)] is probably the best variation on the moving-wire theme. The major improvements, illustrated in Figure 42, include an infrared reflector for accommodating increased solvent flow rates and a flash vaporizer. A cleaning element after the ion source reconditions the metal belt prior to another round of coating on the next revolution. Depending on the polarity of the mobile phase, flow rates from 0.2 to 2.0 ml/min can be tolerated. Sensitivity under actual chromatographic conditions varies with the background spectrum. Solvent residues remain-

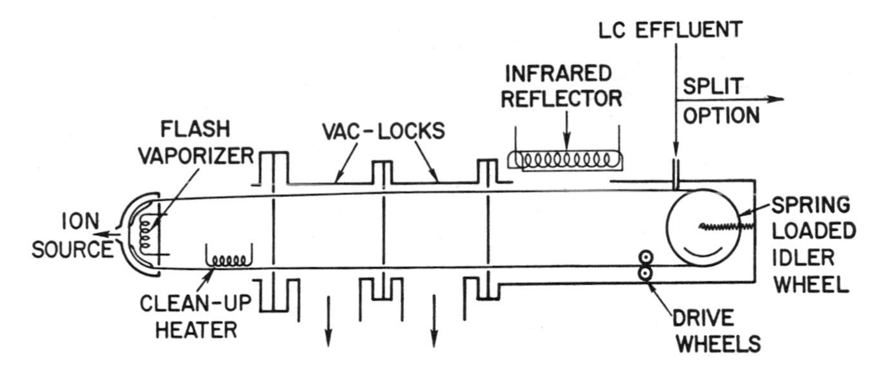

Figure 42. Finnigan LC/MS interface. Moving belt serves as transport carrier for solute. (Reprinted from reference 148, courtesy of Finnigan Corp.)

ing on the carrier may increase noise in the 100–300 atomic-mass-unit region to the extent that detection limits may rise as high as several hundred nanograms for a full scan. At higher mass values, this background problem is less severe and detection limits for injections of 50 ng of solute or less can be achieved. With selected ion monitoring, subnanogram sensitivity is attainable.

Direct Inlet Sampling. McLafferty *et al.*[149,150] have developed systems representative of the third class of interface. Direct introduction of the LC effluent via a capillary tube into a chemical ionization (CI) mass spectrometer allows solution to continuously enter the source at about 1% of the normal effluent flow rate. The detection limits for the 1% split ratio are about 1–10 ng of injected solute, and 10–100 pg have been detected on direct injection. As Figure 43 shows, effluent is partitioned between the LC detector and the MS capillary inlet by a needle valve. The Teflon capillary structure is surrounded by vacuum, and the capillary is constricted at the CI source to give suitable flow. The whole structure fits into the opening made for the solid sample probe. The entire system has been coupled to a laboratory minicomputer.

Obvious advantages to this design are the lower detection limits, the nondestructive nature of the system, and the relative ease of adapting the interface to commercial mass spectrometers. The disadvantages in this design are physical rather than mechanical. Compounds with insufficient vapor pressure for the chemical ionization technique are not detectable. Also, flow rates greater than 10 μl/min can cause high-voltage breakdown in the ion source. In general, this system cannot accept buffers used for ion-exchange or reverse-phase chromatography except at very low ionic strengths. Finally, the mass spectrum is heavily dependent on the solvent employed, since the solvent acts as the agent for chemical ionization. This is not a major disadvantage, as such effects can usually be predicted using

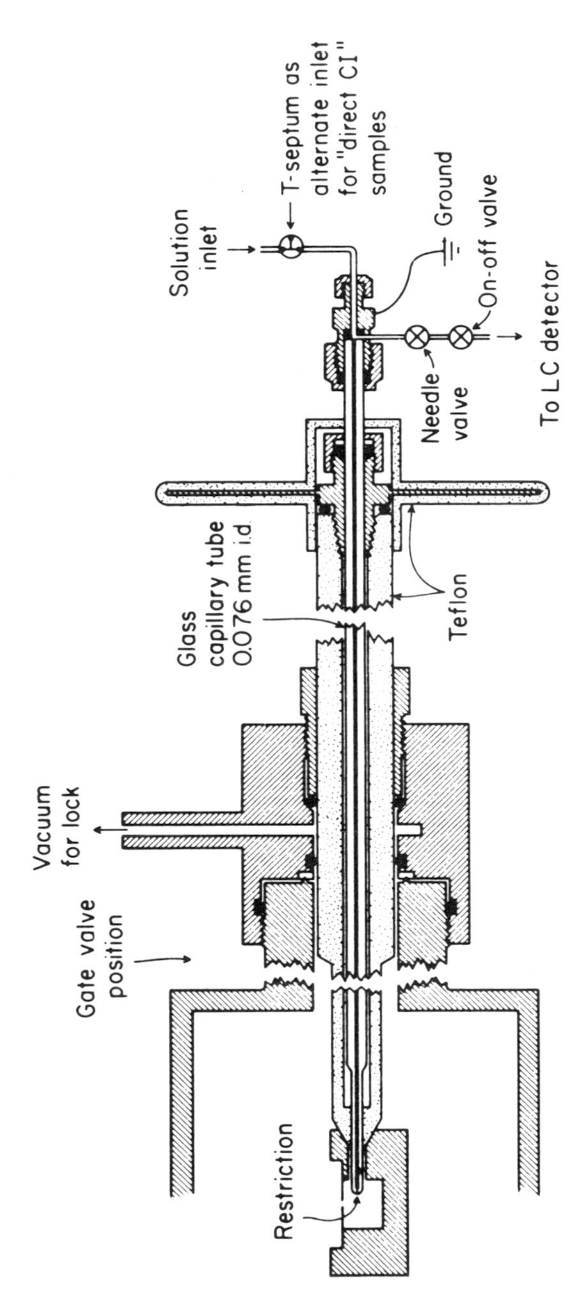

Figure 43. McLafferty inlet probe for continuous introduction of LC effluent into mass spectrometer. Constructed of Teflon, the assembly is designed for insertion into the vacuum lock made for the solids introduction probe. (Reprinted from reference 149, courtesy of Heyden & Son, Ltd.)

model compounds. However, in cases when gradient elution is desired, the composition of the solvent (and therefore the chemical ionization reagent gas) is continually in a state of flux during separation. As a result the spectrum is changing so that the predominant quasi-molecular ion might, for example, change from MH^+ to $M—H^+$. Chemical ionization also, by nature of the technique, yields little structural information; electron impact would be the desired mode for any determination of structure.

Atmospheric-Pressure Ionization (API). Work in this regard has been carried out by the Hornings and co-workers.[151,152] In spite of complications and difficulties to be discussed later, this type of system has been proved to be the most sensitive under optimum conditions.

The basic schematic of the API-LC/MS is given in Figure 44 and one version of the API source itself in Figure 45. Eluate is passed through a UV detector and then fed into a heated (275°C) vaporizing chamber, while being intermixed with preheated carrier gas (usually N_2). This aids in vaporization and also serves for CI reactions in the API source. The vapors are then passed into a 0.025-cm^3 chamber where the β-emitting ^{63}Ni foil is placed. The latter source extends the linear dynamic range of the detector into the microgram region. Subsequent electron–molecule collisions produce the ions, which then diffuse through a 25-μm pinhole to the mass spectrometer focusing elements. It is important to remember that only after the ions pass through the aperture do they enter a low-pressure region.

The complicated aspect of the system is the interpretation of the

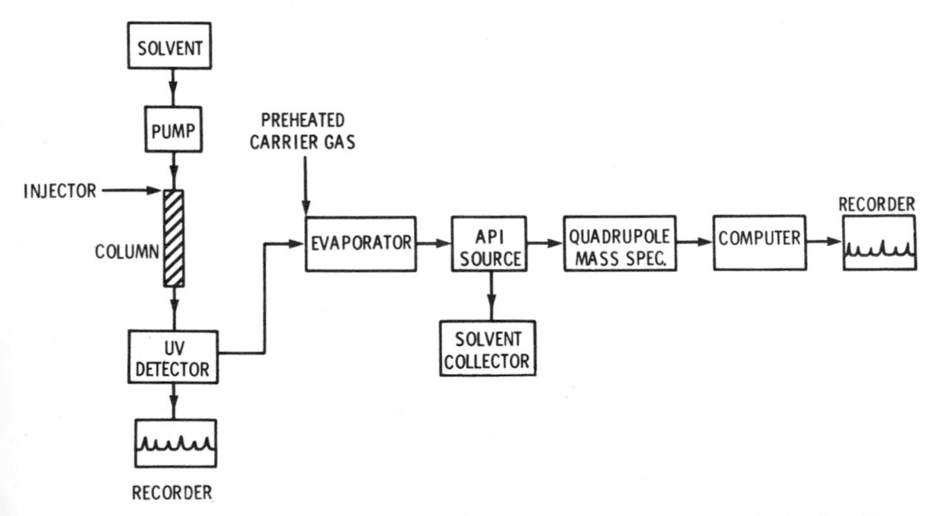

Figure 44. Block diagram of LC/MS system using atmospheric pressure ionization. (Reprinted from reference 151, courtesy of Elsevier Scientific Publishing Co.)

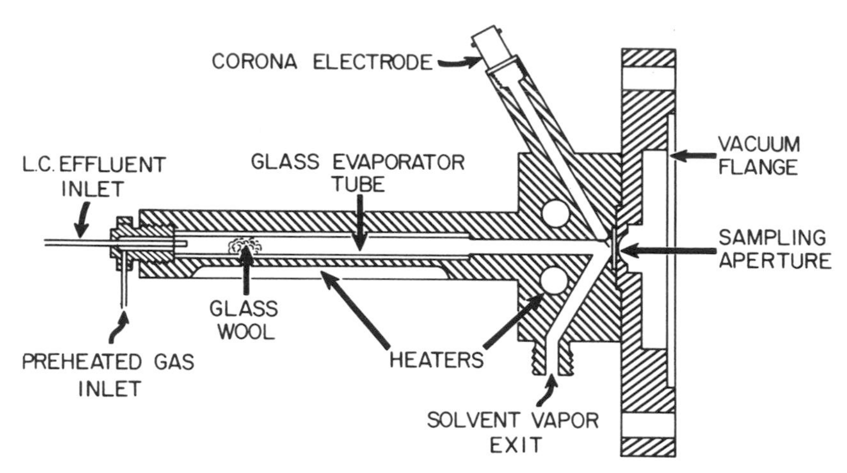

Figure 45. Diagram of corona discharge atmospheric-pressure ionization source. The electrode is positioned in close proximity to the sampling aperture leading to the low-pressure region of the MS. (Reprinted from reference 152, courtesy of the American Chemical Society.)

spectra produced. Due to the possible variety of conditions inside the source, a complex series of ion–molecule reactions can occur for a relatively simple situation. Also, the particular ion output for the same chemical system is a sensitive function of such factors as the ionization chamber temperature, various component concentrations, and the chemical properties of the material in the gas phase.

Overall, the advantage of API is its great sensitivity. Sample size can be limited to the picogram level in an ion-monitoring mode. Disadvantages include the greater variability of operational conditions and the greater number of background ions present. This solvent mediation could pose a large problem for widescale use of the system, since establishment of a known and stable background would be impossible when using gradient elution techniques. Another problem is matching solvents with the necessary characteristics for resolution of a particular mixture on the LC with those required for solvent-mediated ionization. The over-all efficiency of the sytem is drastically reduced by such factors as incomplete ionization and effusion. The many complicating factors in the sensitive API system are being studied by several groups.

The major drawback to API-LC/MS is the requirement that the compound be volatile at relatively high temperatures and 1 atmosphere pressure. Since most biologically important molecules have high molecular weights and low volatility, this limitation—common to some degree in the other interfaces as well—severely undercuts their worth as LC detectors. An extreme case in point is the obviously doomed membrane separator design.[153] Oddly enough, therefore, those molecules which comprise LC's forte are not easily determined with these systems. At present, barring some major development in ion-source design or sample volatilization, it appears unlikely that LC/MS will attain the versatility and sensitivity now enjoyed by GC/MS.

5. Miscellaneous Detectors

5.1. Bulk-Solution Property Detectors

The search for universality of detection often leads to attempts to utilize measurement of some physical property of the solution which undergoes a change in the presence of the analyte. Unlike selective detection schemes, such as the UV and electrochemical detectors, which respond to an analyte property ideally immeasurable in its absence, bulk-property detectors must transduce minute changes in a quantity which is large relative to the change. This fact is the most serious, fundamental limitation inherent in the two detection schemes described in this subsection. The problem is comparable to arriving at the weight of a passenger by weighing the Boeing 747 before and after he has boarded.

Bulk-property methods generally attain lower absolute sensitivity, but perform satisfactorily among a wider class of molecular types. Such is the variety of analytical problems to which liquid chromatography is being applied that this fact is alternately construed as a significant advantage and a major drawback. Nonselective detectors have enjoyed great popularity in gas chromatography. Their application to LC has been thwarted by the fact that solutes in a homogeneous liquid phase generally have a smaller relative perturbation on its physical properties than a comparable amount present in a gas. Some detection schemes alleviate this limitation by volatilizing the column effluent prior to detection, but this may defeat one of LC's chief advantages over GC, namely its ability to separate labile and nonvolatile components. In addition, the vaporized mobile phase often causes problems, as noted in Section 4.

The most popular bulk property detector is the differential refractometer, described in brief below.

5.1.1. *Refractive-Index Detector*

The popularity of the differential refractometer is second only to the UV detector. There are many instances in which it is by far the best alternative currently available. The measured quantity is a change in the refractive index of the mobile phase, defined as

$$n = \frac{\sin \phi_A}{\sin \phi_M} \tag{9}$$

The quantities ϕ_A and ϕ_M are the angles of incidence and departure of a beam crossing the interface between air and the liquid phase. The refractive index of a liquid is particularly sensitive to the presence of solutes, the basis for its frequent use in determining liquid purity.

Two distinctly different LC detector designs are available commercially. One is based on the deflection of a penetrant beam, and one measures total reflectance, a function of n. A diagram of one deflection-type detector is shown in Figure 46. The deflection of a beam passing twice through a tandem sample reference flow cell can be measured by focusing it on a photoelectric position sensor. Figure 47 depicts a reflectance, or Fresnel, design. Although the optics are slightly simpler, two prism cells are required to cover the full range of n. Only the ratio of incident to reflected intensity is required and, since it senses only the minuscule volume of mobile phase at the prism surface, very small dead-volume cells are achieved.

In practice, the performance of these two designs is essentially the same. The limiting factor for both is the control of temperature. Thermal coefficients of refractive index are extremely high, as much as 10^{-4} °C/n. A high premium must be placed on precise thermostatting, often with cumbersome immersion baths. Heat conduction through the arriving effluent is a major problem. Relief is offered by long runs of narrow-bore heat-exchanging tubing prior to the cell, but these invariably introduce extra dead volume and back pressure. Measures more extensive than placing the column in a water jacket are not often considered. Pre-equilibration of both eluant and sample would be beneficial.

This detection mode offers a versatile, convenient, and reliable alternative if the demands for trace levels are not too encumbering. Under optimum conditions, quantitation down to 50 ng injected is possible.[154] Refractive-index detectors are not currently in a state of development, and a significant improvement in sensitivity does not appear to be forthcoming.

5.1.2. Dielectric-Constant Detector

Two parallel conducting planes of common area A separated by a thickness d of nonconductive (dielective) matter constitute a capacitor of capacitance

$$C = \epsilon \frac{A}{d} \epsilon_0 \tag{10}$$

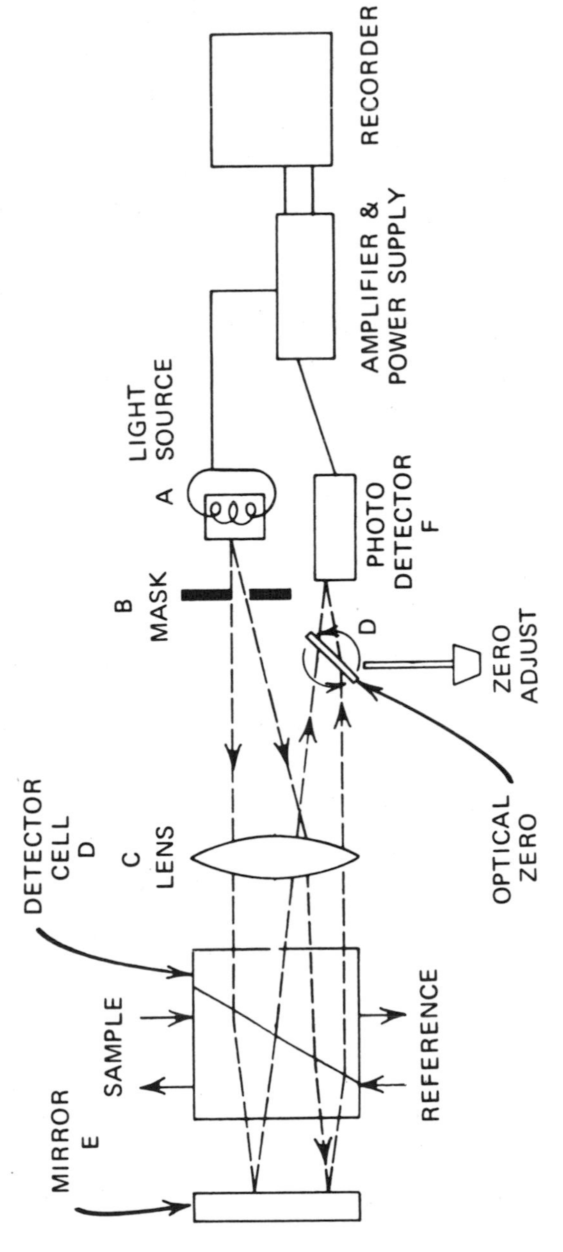

Figure 46. Beam-deflection refractometer LC detector. (Courtesy of Waters Associates.)

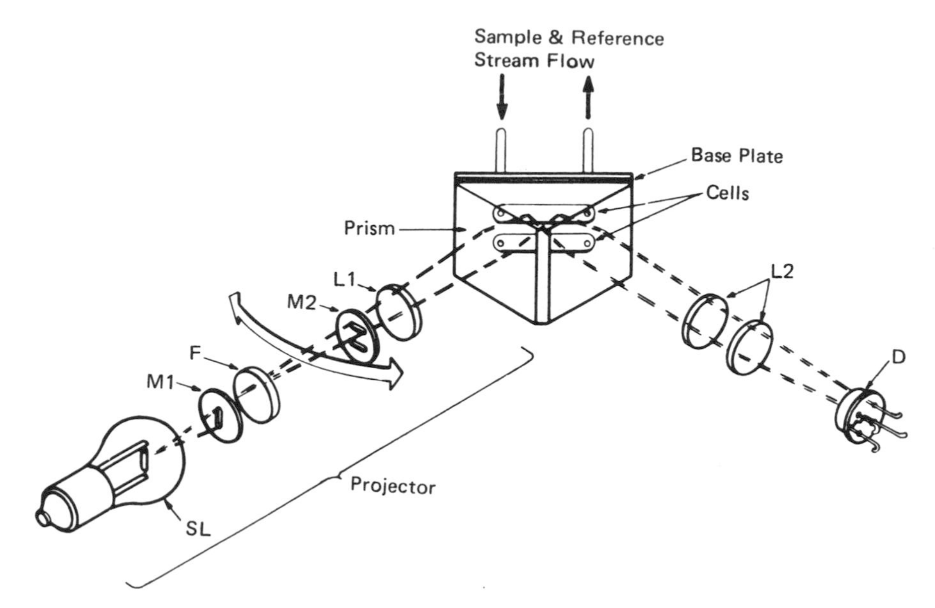

Figure 47. Fresnel refractometer LC detector. L designates a lens, M a mask, and D the detector photocell. (Courtesy of DuPont Instrument Products Division.)

ϵ_0 is a universal constant and ϵ is the dielectric constant, a fundamental property of the dielectric material. The dielectric constant of a liquid can be altered by the presence of a solute, forming the basis for the use of a flow-through capacitor as an LC detector. In dilute binary solutions, dielectric constants are to a good approximation additive, weighted by volume fractions:

$$\epsilon_T = V_x \epsilon_x + (1 - V_x)\epsilon_m \tag{11}$$

where V_x = volume fraction of minor component; ϵ_x = dielectric constant of minor component; and ϵ_m = dielectric constant of pure mobile phase.

It should be noted that a slightly more complicated expression is more satisfactory when polar solvents are used.[155] The change in capacitance is given by the difference of the cell capacitance with only mobile phase, C_c, and with added solute C_s,

$$\Delta C = (C_s - C_c) = (\epsilon_T - \epsilon_m)C_c = (\epsilon_x - \epsilon_m)C_c V_x \tag{12}$$

In principle, adequate sensitivity in this mode is possible only if the sample dielectric constant differs significantly from that of the mobile phase, perhaps by more than three units for trace levels.[156] The capacitance change should be linear with volume fraction and thus with concentration at a fixed temperature. This has been shown to be the case for at least one detector design in one of the most complete evaluations yet reported.[156]

The underlying principles of dielectric-constant measurement is somewhat more complex than equation (12) would suggest. When an oscillating electric field is applied to a capacitance cell, as is always the case in LC detection, the dielectric constant can be separated into two orthogonal components, ϵ and ϵ'. The actual measured property is the complex permittivity

$$\epsilon^* = \epsilon + j\epsilon' \tag{13}$$

a function of applied frequency. The ϵ component is identified with conventional bulk polarization opposing the applied field. The term ϵ' is the loss component, due to dissipation of energy by dielectric absorption. The latter is in phase with the applied electric field and is more familiar as the basis for the operation of the microwave oven. Liquid chromatography detectors employing a capacitance cell are sometimes called permittivity detectors,[155] although this designation is only correct when ϵ' is appreciable at the operating frequency. The frequency can, in principle, be extended upwards into the near infrared, where the dipoles can no longer follow the applied field by rotation and vibration. Here the loss term ϵ' vanishes, but measurement using a capacitor becomes impractical. The Maxwell relationship shows the connection between dielectric constant and refractive index:

$$\epsilon = n^2 \tag{14}$$

In a sense, a dielectric constant measurement is a very-low-frequency optical refractive-index measurement, and it is reasonable to expect comparable performance as LC detectors. However, this is strictly true only for nonpolar, homogeneous gas-phase atoms. Local field effects in liquids, hydrodynamic anomalies in flow cells, temperature sensitivity, electronic excitation, permanent dipole moments, inherent polarizability, and solute interactions make theoretical comparison intractable.

The equivalent circuit for a generalized cell is shown in Figure 48. To the cell capacitance C_c must be added residual capacitance C_r due to surroundings and wiring not influenced by liquid flow. All capacitors exhibit residual series resistance and inductance R_r and L_r from the plates, leads, and connections. The term L_r becomes important only in the microwave region, establishing an upper usable frequency limit well above that commonly used for detectors. Proper design can minimize R_r. The term R_g represents all forms of electronic conduction across the capacitor and is absent only if a vacuum is used as the dielectric. Dielectric "leakage" can be a serious problem in flow capacitors if ionic solutes or buffers are used. Most designs strive to minimize conduction (maximize R_g). The

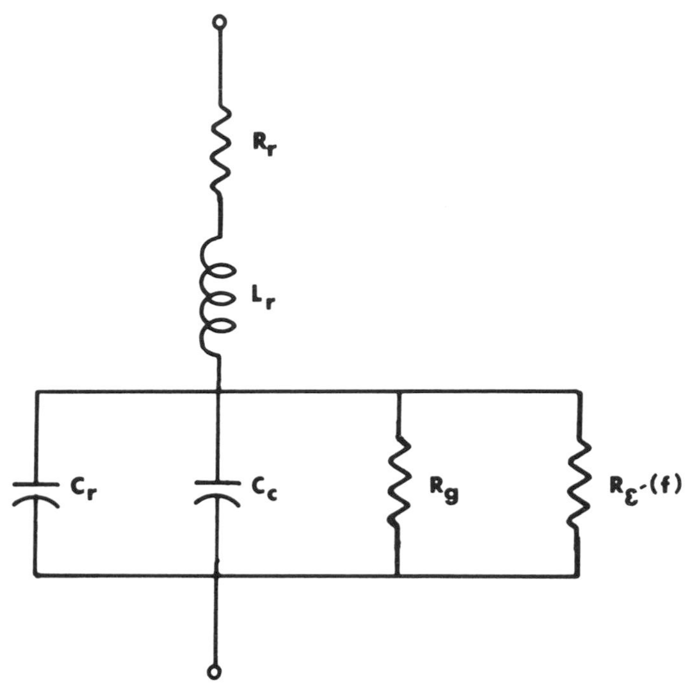

Figure 48. Electrical equivalent circuit of a dielectric constant cell.

$R_{\epsilon'}$ component represents dielectric loss. Strictly speaking, all these parameters are functions of temperature, the precise control of which is an important consideration in practical cell design.[157]

The simplest means of detecting a change in capacitance is to use the flow cell as a variable reactance in a resonant circuit. The frequency of a tuned oscillator then becomes a function of cell capacitance and can be measured or converted to a convenient form. In practice, greater precision and stability is afforded by a heterodyne method,[158] wherein the sample cell oscillator is mixed with a stable, fixed-frequency oscillator. The filtered mixer output is a signal whose frequency is the difference of the sample and reference oscillator frequency. The chromatogram may be taken directly from a frequency-to-voltage converter. A detector employing this circuit has recently been evaluated.[159] Employing a thin-layer, direct-contact flow cell, this arrangement appears to compare favorably with the refractive-index detector. A minimum detectable increment of 1×10^{-4} $\Delta\epsilon$ ($\Delta f - 25$ Hz) afforded a detection limit of 1–50 ng injected for several simple organic solutes. A strong response dependence on the nature of the solute was observed, evidence that the nonspecific nature of this mode has been overemphasized. Dielectric loss and conductance tend to dampen oscillation and sharply degrade performance, rendering this mode unsuitable for eluants of high dielectric constant and for ionic solutes.

A conventional AC bridge circuit can measure capacitance changes, but the need to maintain proper phasing with a variable capacitance in the reference arm proves impractical. A related circuit, an operational amplifier difference differentiator devised by Erbelding,[160] gave satisfactory performance, but offers no relief for the problems of the heterodyne mode. The tubular brass flow cell of this author has excessive dead volume (700 μl) and contributes to a flow-dependent response not usually observed in thin-layer geometrics.

Poppe and Kuysten[155] presented an inductively coupled 1-MHz bridge for use with their two-compartment thin-layer cell. A full assessment of this arrangement with respect to sensitivity and noise shows the former to be adequate, but the latter, disappointing. The source of this background noise appears to be external to the electronics. Despite careful thermostatting, pulseless pumping, and a reference cell, this unit achieves a lower limit of detection only about a factor of three better than a commercial refractive-index detector under the most favorable circumstances.

Detectors measuring only capacitance, such as the AC bridge and heterodyne circuits, usually suffer ill effects from R_r, C_r, R_g, and $R_{\epsilon'}$. However, as pointed out by Haderka,[161] it is possible to design detectors which respond only to the in-phase components or to the combined effects of both phases. This author[161] has devised a quadrature circuit capable of extracting two chromatograms from one detector, one reactive and one resistive. The possibility of operating a capacitance detector in a quasi-selective mode is intriguing, but no practical evaluation of a working model has been forthcoming.

Klatt[162] has implemented the most advanced electronic scheme yet applied to a working model. A block diagram of his phase-locked loop detector is shown in Figure 49. The detector, a concentric tubular geometry of established merit,[162] shown in Figure 50, forms half of a parallel LC network. The cell admittance contributes to that of the over-all network in additive fashion. From AC circuit theory, it is a simple matter to arrive at the phase angle and frequency across R in Figure 49:

$$\phi = \tan^{-1}\left[\frac{(R/\omega L)^* (1 - \omega^2/\omega_0^2)}{1 + R/R_s}\right] \tag{15}$$

$$R_s = (R_g^{-1} + R_{\epsilon'}^{-1})^{-1}$$

$$\omega_0 = (LC_c)^{-1/2} \tag{16}$$

The resonant angular frequency is a function of the cell capacitance. Since $\phi = 0$ only when $\omega = \omega_0$, the condition of resonance, a control signal which is a function of all the variable cell parameters can be derived. The

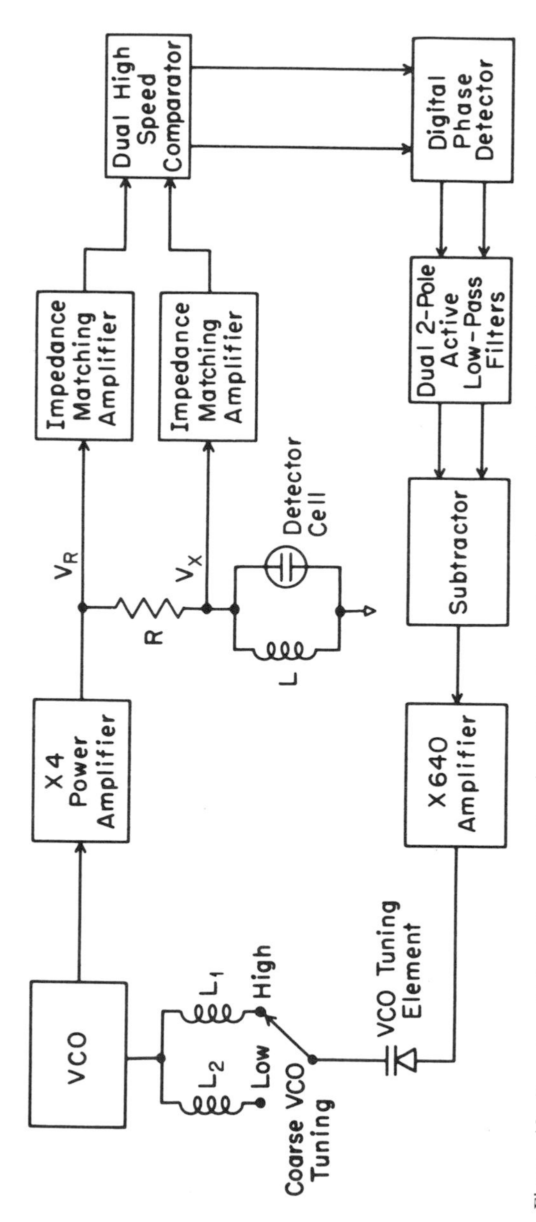

Figure 49. Functional block diagram of a phase-locked loop circuit for use with the dielectric constant detector shown in Figure 50. (Reprinted from reference 162, courtesy of the American Chemical Society.)

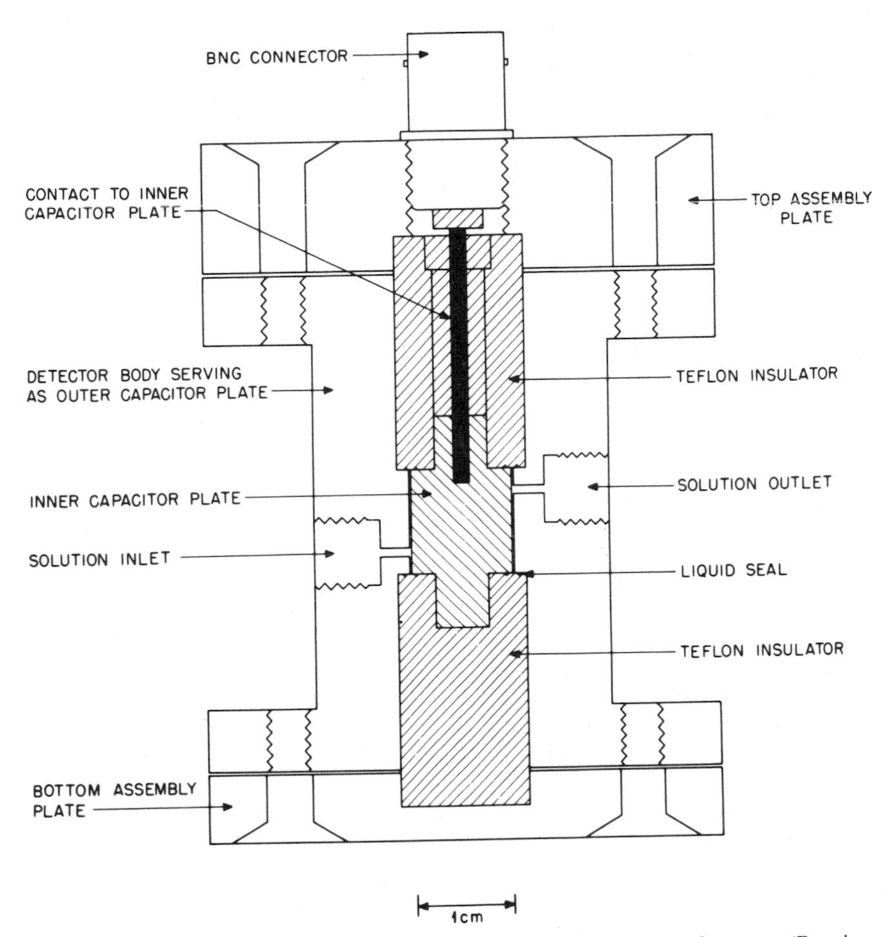

Figure 50. Construction of a tubular flow cell dielectric constant detector. (Reprinted from reference 162, courtesy of the American Chemical Society.)

off-resonance signal is fed back to a voltage-controlled oscillator, forcing it to adjust the resonant frequency as the cell environment changes. In this way, the circuit tracks the natural resonance across a broad range of frequencies with one coarse tuning adjustment, from 0.6 to 5 MHz in this model. The upper limit of eluant dielectric constant is consequently extended to the point that, for the first time, largely aqueous mobile phases are possible. Digitally sampling the oscillator frequency shift provides a chromatogram showing positive deflections for both increasing and decreasing frequencies.

Noise limits the minimal detection increment to about 5×10^{-4} $\Delta\epsilon$ units, translating to around 50 ng injected for favorable samples. More extensive evaluation of this and other designs will be needed before their applicability to a wide base of chromatographic systems is firmly established. An experimental performance comparison[163] of the capacitance detector of Vespalec and Hána,[156] later improved by Vespalec,[164] to two commercial LC refractometers concluded that the former is competitive but somewhat underperforms the popular refractive-index units in sensitivity, noise, and detection limit. One distinct advantage of capacitance detectors seems to be a linear range wider by almost one order of magnitude. This advantage may be offset by the higher noise level, a recurrent problem in this mode of detection. This may reflect the natural $1/f$ dependence of noise power. The advantages of reasonably simple, noncritical design and the prospect for further reduction of instrumental noise should make pursuit along these lines attractive to the frustrated chromatographer in search of a reliable, inexpensive, nonspecific detector usable to the 100 ppm level.

5.2. Microadsorption Detector

Fundamental to chromatography in all its forms is the transition of a solute component from one phase to another, accompanied by a change in its free energy. In the particular instance of adsorption liquid chromatography, a molecule in the mobile phase must sacrifice free energy in the form of heat in order to adsorb to the lower entropy environment on the surface of the stationary phase. The passage of a zone of concentration through an adsorption column will always be accompanied by a zone of elevated temperature, if there is any interaction with the stationary phase. If there is not, the component will elute in the void volume and no useful chromatography will have taken place. A probe sensitive to very small temperature differentials would then seem to offer the ideal nonspecific LC detector.

Claxton[165] was first to apply this principle to liquid chromatography, as a detector for hydrocarbons eluting from silica gel, but the first design to achieve promising sensitivity was that of Hupe and Bayer.[166] A dual-chamber PTFE flow cell with embedded thermistors serves as a differential flow microcalorimeter. The upstream chamber contains only a thermistor bead projecting into the effluent stream. The second chamber is packed with an adsorbent material, usually the same used in the column. A conventional DC Wheatstone bridge circuit produces a difference output, which is amplified and recorded. At the sensitivity attained by these authors, the output is clearly not a conventional chromatogram, but rather one resembling the first derivative thereof.

Hupe and Bayer[166] report a detection limit below 50 ng injected for rather polar alcohols and ketones eluting from silica gel, but a figure as much as an order of magnitude higher is more realistic. The range of linear response was severely restricted, from about 0.5 to 20 μg, probably due to saturation of available sites on the silica. Furthermore, peak shape was found to be a function of solvent polarity, flow, and gradient elution rate, indicative of unsuitability for use with mobile-phase programming. This conclusion was further substantiated by Munk and Raval,[167] who carried out a rather detailed design study with emphasis on baseline instability and flow-rate sensitivity.

The most puzzling and detrimental detector characteristic is the asymmetry of the two halves of the peak, the desorption curve being normally broader and lower in amplitude. The degree of asymmetry changes markedly with elution volume, sample load, flow rate, temperature, and cell design. The output is not simply a derivative chromatogram, and integration yields a curve with drastic distortion and baseline shifts. Scott[168] has presented a theoretical study of the influence of many parameters on the peak shape. Using the plate theory and a minimum of assumptions, he has arrived at a theoretical response function which includes the influence of the thermodynamic properties of the cell, adsorbant, and mobile phase, as well as flow rate, cell geometry, and adsorbant capacity factor.

Mismatch between the plate height of the sensor cell and the column was blamed for most of the distortion. Only if the sensor cell and column are identical in packing geometry will the temperature profile approach the first derivative of concentration. One possibility is to place the sensor at the very bottom of the column itself, rather than in an external cell. Although this would eliminate column/detector mismatch, Scott shows that distortion would still arise from the convection of heat into the detection zone from the region above, unless perfect radial heat loss is achieved or the capacity factor is low. He concludes, "Bearing in mind other limitations of the detector, such as its inability to cope with gradient elution and temperature program methods, and that the adsorbent has to be changed from time to time, the practical difficulties of construction make the heat of adsorption detector not a viable system for detection in LC."[168]

Since this resounding assertion, the popularity of this detector, which was scarcely ever encouraging, has become virtually nil. The only commercial unit, sold by Varian, is no longer available. Because of the difficulty of interpretation of closely spaced peaks, the need to match carefully each cell to a specific column, the facile saturation of the adsorbant, and the very stringent thermostatting requirements, only a very few have ever reported application of the microadsorption detector to an actual detection problem.[169,170] Although excellent temperature resolution was demonstrated by McNair and Stafford,[171] suggesting that this mode can be made very sensitive, even these most recent workers appear to concede that the host of other experimental problems make the prospect for improvement and widespread acceptance very remote.

5.3. Radioactivity Detector

On-line monitoring of radiolabeled compounds in the effluent of an LC is quite attractive. Progress in biochemistry and related fields of study is heavily dependent on the ability to follow a labeled compound through a series of metabolic changes. In tracing the course of a tagged drug, for example, the simultaneous separation and measurement of radioactive metabolites would be extremely valuable in assessing the compound's bioavailability. Radiochemical detection would appear to be potentially useful as a general-purpose LC detector. Many classes of compounds could be made detectable through the use of radiolabeled derivatives, while a high degree of specificity could be maintained by the selection of the proper derivatizing agent.

Although a number of radiochemical detectors for LC have been designed, none are entirely satisfactory. This is due in part to the incompatibility of flow systems with counting techniques. The precision of a radiochemical measurement is dependent on the number of counts detected and is determined by the counting time. The precision of the method is proportional to the square root of the number of counts. In a batch mode, measurement of low activities with satisfactory precision is accomplished by appropriately increasing the counting time. In LC the counting time cannot be freely adjusted without adversely affecting the resolution. In a flow-through detector the counting time is increased by increasing the cell volume, and thereby the sample residence time in the cell. Large cell volumes may cause significant band broadening. Therefore one has a trade-off between counting precision and resolution. A mathematical treatment of the important parameters in radiochemical flow detectors is presented by Sieswerda.[172]

One of the earlier LC/radiochemical systems was constructed by placing a scintillation counting cell in the flow stream of an amino acid analyzer.[173] A cylindrical quartz cell packed with a solid anthracene fluorophore was constructed to fit in the chamber of a scintillation counter. Using this approach, counting efficiencies of 38% for ^{14}C and 0.9% for tritium were obtained. Labeled amino acids with activities of about 3000 dpm could be measured with 1–3% precision. The counting cell had a dead volume of over 1 ml. Although adequate for use with classical resins, the large dead volume would be disastrous when used with modern high-efficiency columns.

One of the most recent cell designs was developed by de Belleroce *et al.*[174] The 3-mm ID path of the cell is packed with 100-μm plastic phosphor beads. Using this design a counting efficiency of 30% was obtained for ^{14}C. The authors do not present a detailed characterization of the cell, so it is difficult to assess its performance in detail. Nevertheless, it proved to be satisfactory for study of the metabolism of the neurotransmitter dopamine in tissue extracts. An 8-μm polystyrene-based ion-exchange resin was used to resolve a number of amino acids, dopamine, and dopamine metabolites. The effluent from the column was simultaneously monitored by UV, *o*-phthalaldehyde/fluorescence, and by scintillation counting. Peak broadening due to the radiochemical cell was obvious, although this was of no consequence in the study.

6. Appendix 1

A derivation of the output of a low-pass filter for a peak input is given below. A Gaussian function is assumed to adequately represent an undistorted chromatographic peak arriving from an ideal detector. A simplified electrical equivalent of a filter is an RC network, shown in Figure 51A. This may be thought to represent the over-all time constant of the amplification electronics, such as the simple active filter in Figure 51B.

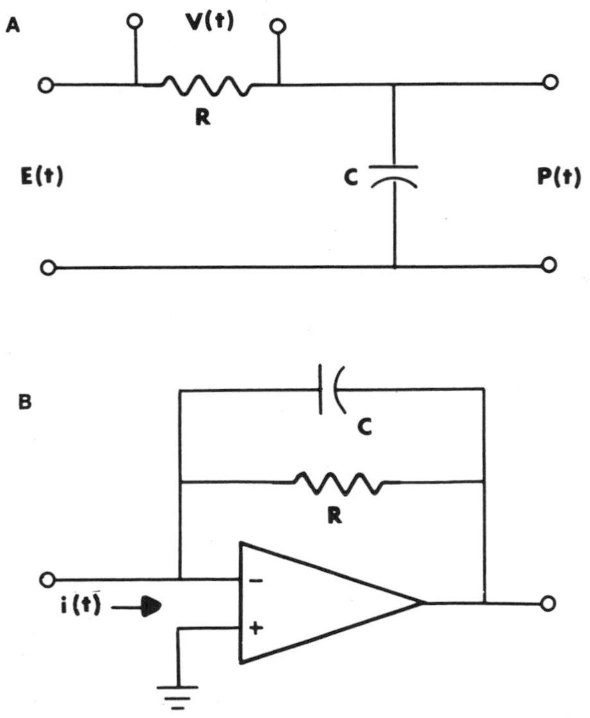

Figure 51. (A) RC circuit model for instrumental time constant. (B) Typical amplifier circuit for a current source transducer. The operational amplifier is considered free of phase shift and uniform in gain at the frequencies of interest.

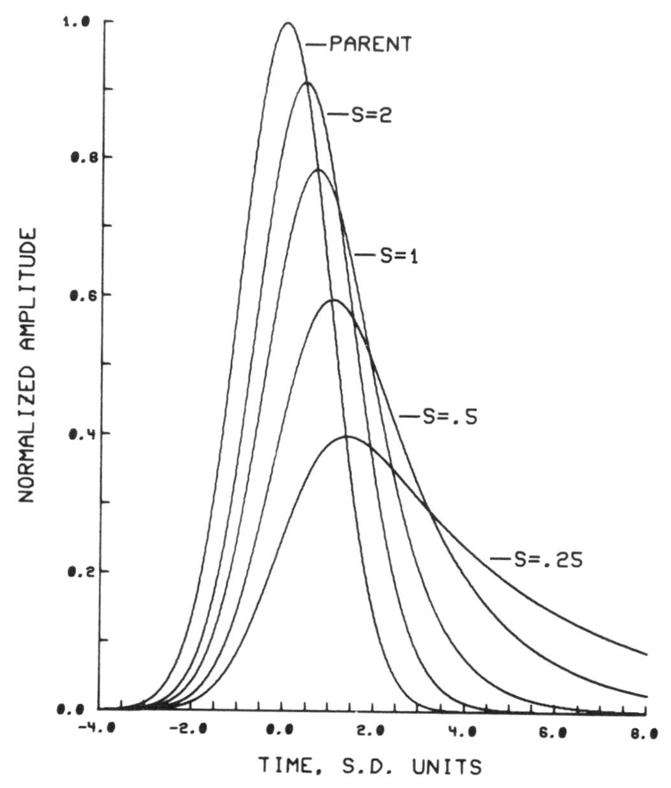

Figure 52. Distortion of a Gaussian peak due to a single time constant τ, shown for several values of $S = \sigma/\tau$.

The voltage V across the resistor in Figure 51A can be expressed from Ohm's law as

$$V(t) = i(t)R \tag{17}$$

The time-dependent current across the capacitor can also be expressed in terms of the instantaneous voltage $P(t)$ as

$$i(t) = C\frac{dP(t)}{dt} \tag{18}$$

The total potential impressed across the RC network is divided between the two components such that

$$E(t) = V(t) + P(t) = i(t)R + P(t) \tag{19}$$

Therefore

$$E(t) = \tau\frac{dP(t)}{dt} + P(t) \tag{20}$$

where $\tau = RC$ and $P(t)$ is a generalized response function. Equation (12) is the fundamental restraining relationship of an RC network. It dictates that any response function $P(t)$ must be a solution to this linear, first-order differential equation. The appropriate conditions are that

$$P(t) = 0, \; t \leq 0 \tag{21}$$

$$P(t) \rightarrow E(t), \; t \rightarrow \infty$$

The input function $F(t)$ is Gaussian in the present analysis. To simplify the symbology, we choose to center the peak at the origin $t = 0$ rather than at $t = t_r$, its retention time. Rearranging equation (20) and substituting the appropriate functional expression gives

$$\frac{dP(t)}{dt} + \tau^{-1}P(t) = \frac{A}{\tau}\exp\left[-\frac{1}{2}\left(\frac{t}{\sigma}\right)^2\right] \tag{22}$$

where A is a constant. Multiplying all terms by an integrating factor, $\exp(\int \tau^{-1}\, dt) = \exp(t/\tau)$, gives

$$\frac{dP(t)}{dt}\exp(t/\tau) + \tau^{-1}\exp(t/\tau)P(t) = \frac{A}{\tau}\exp\left[-\frac{1}{2}\left(\frac{t}{\sigma}\right)^2\right]\exp(t/\tau) \tag{23}$$

Using a familiar property of derivatives of products in reverse on the left-hand side, we have

$$\frac{d}{dt}[P(t)\exp(t/\tau)] = \frac{A}{\tau}\exp\left[-\frac{1}{2}\left(\frac{t}{\sigma}\right)^2\right]\exp(t/\tau) \tag{24}$$

The response function is now readily found by integration of both sides. the lower limit of integration is $t = -t_r$, since the peak must have begun at the injection point. Since the detection is accomplished far removed from the starting point relative to the peak width, no error is accrued by extending the limit to $-\infty$. Using primes to denote the dummy variables, then integrating and rearranging, we have

$$P(t) = \frac{A}{\tau}\exp\left(-\frac{t}{\tau}\right)\int_{-\infty}^{t}\exp\left[-\frac{1}{2}\left(\frac{t'}{\sigma}\right)^2 + \frac{t'}{\tau}\right]dt' \tag{25}$$

Now let $T = t/\sigma$, and $S = \sigma/\tau$, hence $TS = t/\tau$, so

$$P(T) = AS\exp(-\,TS)\int_{-\infty}^{T}\exp(-\frac{1}{2}\,T'^2 + T'S)\,dT' \tag{26}$$

The function described by equation (26) is an exact solution for equation (20), but it may be arranged to a more useful form by completing the square in the exponential term, regrouping, and substituting $X = (T - S)/\sqrt{2}$.

$$P(T) = \sqrt{2}\, AS \exp\left(\frac{S^2}{2} - TS\right) \int_{-\infty}^{X} \exp(-X'^2)\, dX' \tag{27}$$

This result is easily couched in terms of the well-known normalized error function erf(X), using the fact that

$$\int_{-\infty}^{X} \exp(-X'^2) dX' = \frac{\sqrt{\pi}}{2} [1 + \text{erf}(X)] \tag{28}$$

Substituting (28) into (27) gives a readily computed solution

$$P(T) = \frac{\sqrt{2\pi}}{2} AS \exp\left(\frac{S^2}{2} - TS\right) [1 + \text{erf}(X)] \tag{29}$$

This is equivalent to the equation given by McWilliam and Bolton.[6] Equation (29) is plotted in Figure 52 for several values of S, along with the undistorted parent Gaussian. Note that the peak width, shape, and maximum value are altered by the filter, and that the position of the maximum shifts to longer times.

Returning to equation (25), the exponential term may be moved inside the integral, since it does not involve t', to give

$$P(t) = \frac{A}{\tau} \int_{-\infty}^{t} \exp\left[-\frac{1}{2}\left(\frac{t'}{\sigma}\right)^2\right] \exp\left(\frac{t'}{\tau} - \frac{t}{\tau}\right) dt' \tag{30}$$

Calling $H(t) = 1/\tau \exp\left(-t/\tau\right)$, this becomes

$$P(t) = \int_{-\infty}^{t} E(t')H(t - t')dt' = E * H \tag{31}$$

Equation (31) is in the form of a very important mathematical operation called the convolution integral.[175] The generality of this form is such that we could have written it directly by recognizing that $H(t)$ is the time response of an RC network to an infinitely-short-duration, infinitely-high-amplitude impulse of unit area [the $\delta(t)$ function].

In more physical terms, equation (31) describes the process of decomposing the excitation function into infinitely many segments of width dt', assessing the network response to each segment individually, then summing the results together to obtain the net response function. This decomposition/reconstitution process is valid only when the system concerned is linear; this, in essence, defines the criterion of linearity. Proper amplifier design strives to maintain linearity, but possible sources of nonlinear behavior exist, notably amplifier saturation and chart-recorder slewing.

Integrating equation (20) gives

$$\int_0^t E(t')dt' = \tau P(t) + \int_0^t P(t')\,dt' \tag{32}$$

The total area under the response function is then

$$\lim_{t\to\infty} \int_0^t P(t')dt' = \lim_{t\to\infty} \int_0^t E(t')dt' - \tau \lim_{t\to\infty} P(t) \tag{33}$$

From equation (29) or (21), $P(t)$ may be seen to approach zero at large t. We have, then, the rather unsurprising result that convoluting a peak with an exponential does not affect the total area. This may also be considered a natural consequence of the law of conservation of matter, since capacitances and dead volumes can only redistribute, not consume or produce.

Equation (20) is the key to another interesting property of this convolution operation. Rearrangement gives

$$\frac{dP(t)}{dt} = \frac{1}{\tau}[E(t) - P(t)] = 0 \tag{34}$$

at the output peak maximum. Thus the output reaches its maximum value as it crosses the trace of the input peak, since $P(t) = E(t)$ at that point.

Locating the position of the peak maximum requires differentiation of equation (29) to give

$$\sqrt{\frac{2}{\pi}}\exp(-X^2) - S[1 + \mathrm{erf}(X)] = 0 \tag{35}$$

Each value of S leads to a value of X which is a solution of equation (28), and hence a unique T_{max}. Once T_{max} is found, the peak height is readily computed from equation (34). There is no analytical solution to equation (35), but a successive approximation computer algorithm quickly yields the desired results, which are plotted in Figure 2. The width at half-maximum is often used to characterize a chromatographic peak. Since the maximum amplitude has been found, the half-width can be computed by a searching algorithm.

The peak moments have been proposed as a more appropriate, although less accessible, means of characterization.[176] The defining expression for the nth moment M_n is given in Appendix 2, equation (48). For the functions under consideration, the first and second moments are

$$M_1[E(t)] = t_r \ (= 0 \text{ in the above}) \tag{36}$$

$$M_1[H(t)] = \tau \tag{37}$$

$$M_2[E(t - t_r)] = \sigma^2 \tag{38}$$

$$M_2[H(t - \tau)] = \tau \tag{39}$$

$$M_1[P(t)] = t_r + \tau \tag{40}$$

$$M_2[P(t)] = \sigma^2 + \tau^2 \tag{41}$$

The last two relations above suggest a means for extracting σ and τ from the moments. A more reliable method has been proposed.[8]

7. Appendix 2

The concept of convolution introduced in Appendix 1 is here applied to show the effect of a finite detection zone on a Gaussian concentration profile. The result is shown to apply without modification to the case of an injected plug of finite width introduced into an idealized column.

Although the detector observes the column effluent passing across a region of space, it is more convenient to consider the time width τ of this zone as given by

$$\tau = V/F \tag{42}$$

where V = volume of detection zone; and F = volume flow rate of mobile phase. τ is the transit time of an analyte molecule across the detection zone, or, alternately, the time the injected zone takes to enter the column. The situation from the point of view of the detector is considered first.

The response of any linear system to any input may be determined from its response to a normalized impulse input $\delta(t)$. Such an "infinitely sharp" peak entering the detection zone will cause an instantaneous jump in output. The output remains constant for τ seconds as the impulse travels across the zone, vanishing instantaneously. The over-all output describes a rectangle function, given in origin-centered form as

$$H(t) = \begin{cases} 1, & -\tau/2 < t < \tau/2 \\ 0, & \text{elsewhere} \end{cases} \tag{43}$$

The response to a Gaussian input is then given by the convolution integral (equation 31, Appendix 1). Substituting the functional expression of a Gaussian in the time domain and expressing $H(t)$ by adjustment of the integration limits gives

$$P(t) = \frac{1}{\tau}\int_{t-\tau/2}^{t+\tau/2} \exp\left[-\frac{1}{2}\left(\frac{t'}{\sigma}\right)^2\right] dt' \qquad (44)$$

Let $z' = t'/\sqrt{2}\sigma$, thus

$$P(t) = \frac{\sqrt{2}\sigma}{\tau}\int_{(t-\tau/2)/\sqrt{2}\sigma}^{(t+\tau/2)/\sqrt{2}\sigma} \exp(-z'^2)\, dz' \qquad (45)$$

$$P(t) = \frac{\sqrt{2\pi}}{2} S\left[\operatorname{erf}\left(z + \frac{\sqrt{2}}{4S}\right) - \operatorname{erf}\left(z - \frac{\sqrt{2}}{4S}\right)\right] \qquad (46)$$

where $S = \sigma/\tau$ and $P(t)$ is normalized to unit amplitude. A plot of this function for several S values is shown in Figure 53. The width of the detection zone corresponding to S is depicted by a rectangle wave tangent to the maximum of each curve.

Differentiation of equation (46) leads to the simple result that the peak maximum always occurs at $t = Z = 0$. Thus the maximum amplitude is

$$P_{\max}(S) = \frac{\sqrt{2\pi}}{2} S\left[\operatorname{erf}\left(\frac{\sqrt{2}}{4S}\right) - \operatorname{erf}\left(-\frac{\sqrt{2}}{4S}\right)\right] \qquad (47)$$

This equation is used to generate the S values in Table 1. Note that to achieve reproduction faithful to better than 1% in amplitude, the detector zone should be less than half the true peak standard deviation.

There is no simple way to compensate for increased peak variance. It may be shown[175] that a peak derived from the convolution of two functions will have a second moment equal to the sum of the second moments of the two component functions, provided these moments exist. The nth moment of a time function is defined by

$$M_n[f(t)] = \int t^n f(t)\,dt \Big/ \int f(t)\, dt \qquad (48)$$

When

$$f(t) = \exp\left[-\frac{1}{2}\left(\frac{t}{\sigma}\right)^2\right], \qquad M_2[f(t)] = \sigma^2$$

the variance about the mean. In the present case

$$M_2[H(t)] = \int_{-\tau/2}^{\tau/2} t^2 dt \Big/ \int_{-\tau/2}^{\tau/2} dt = \frac{1}{\tau}\left(\frac{1}{3} t^3 \Big|_{-\tau/2}^{\tau/2}\right) = \frac{\tau^2}{12} = \sigma_z^2 \qquad (49)$$

From equation (3) it follows that the relative error in peak variance

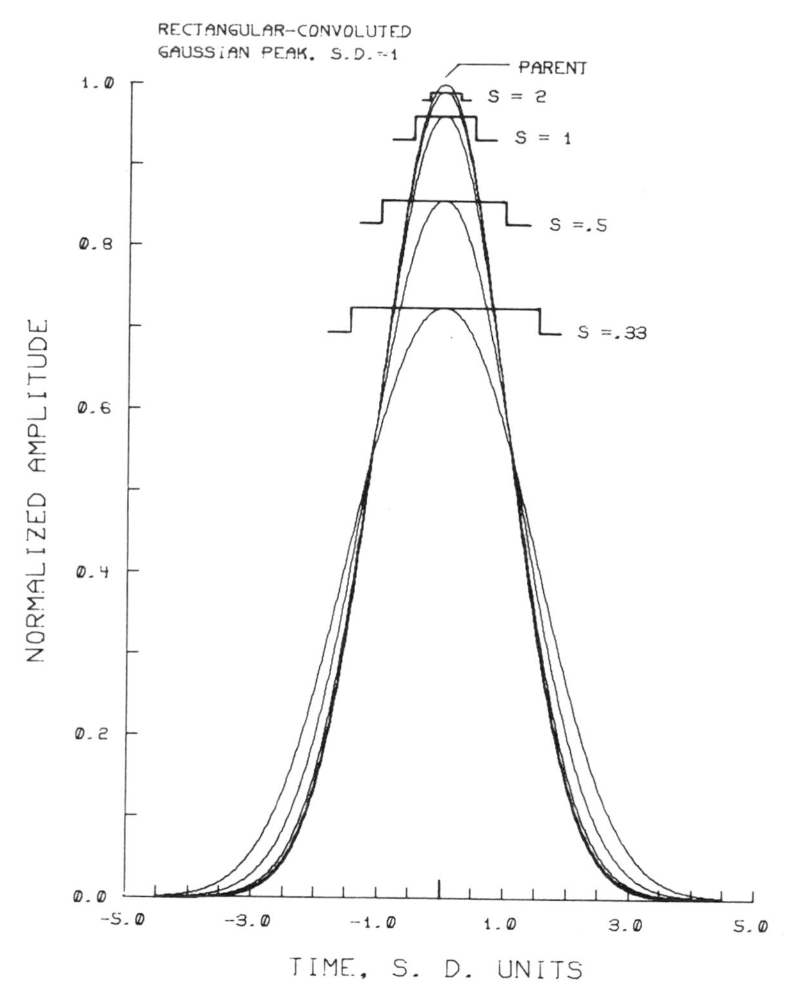

Figure 53. Symmetric broadening of a unit amplitude, unit standard deviation Gaussian due to a finite detection or injection zone width. The rectangle function tangent to each trace depicts the width of the zone convoluted with the parent to obtain that result.

due to the detector zone width is

$$\frac{\Delta\sigma^2}{\sigma^2} = \frac{\sigma_z^2}{\sigma^2} = \frac{1}{12}\left(\frac{\tau}{\sigma}\right)^2 \tag{50}$$

This equation supplies the rest of the S values in Table 1. Note that since the error in variance is inversely proportional to the square of the undistorted variance, sharp peaks will be severely flattened.

An important property of the convolution integral is commutativity

$$\int_{-\infty}^{t} E(t')H(t - t')dt' = \int_{-\infty}^{t} E(t - t')H(t')dt' \tag{51}$$

or, in simplified notation,

$$E * H = H * E \tag{52}$$

Hence it is valid to assume a rectangle function as the input from a loop injector and the impulse response of an ideal column as a true Gaussian. The operation and results of convoluting a finite-width injected zone with an idealized column response is mathematically identical to the above, except that τ is taken as the time required for the plug to penetrate the top of the column. Sternberg[12] has carried out the same calculations for this and other probable injection functions.

8. References

1. J. J. Kirkland, Preferred experimental conditions for trace analysis by modern liquid chromatography, *Analyst, 99,* 859–885 (1974).
2. C. N. Reilley, G. P. Hildebrand, and J. W. Ashley, Jr., Gas chromatographic response as a function of sample input profile, *Anal. Chem., 34,* 1198–1212 (1962).
3. P. T. Kissinger, C. S. Bruntlett, G. C. Davis, L. J. Felice, R. M. Riggin, and R. E. Shoup, Recent developments in the clinical assessment of aromatic metabolism by high-performance reverse-phase chromatography with amperometric detection, *Clin. Chem., 23,* 1449–1455 (1977).
4. R. M. Riggin and P. T. Kissinger, Determination of catecholamines in urine by reverse-phase liquid chromatography with electrochemical detection, *Anal. Chem., 49,* 2109–2111 (1977).
5. I. G. McWilliam and H. C. Bolton, Instrumental peak distortion. I. Relaxation time effects, *Anal. Chem., 41,* 1755–1762 (1969).
6. I. G. McWilliam and H. C. Bolton, Instrumental peak distortion. III. The analysis of overlapping curves, *Anal. Chem., 43,* 883–889 (1971).
7. H. M. Gladney, B. F. Dowden, and J. D. Swalen, Computer assisted gas–liquid chromatography, *Anal. Chem., 41,* 883–888 (1969).
8. W. W. Yau, Characterizing skewed chromatographic band broadening, *Anal. Chem., 49,* 395–403 (1977).
9. R. E. Pauls and L. B. Rogers, Band broadening studies using parameters for an exponentially modified Gaussian, *Anal. Chem., 49,* 625–628 (1977).
10. E. Grushka, Characterization of exponentially modified Gaussian peaks in chromatography, *Anal. Chem., 44,* 1733–1738 (1972).
11. B. L. Karger, M. Martin, and G. Guiochon, Role of column parameters and injection volume on detection limits in liquid chromatography, *Anal. Chem., 46,* 1640–1647 (1974).
12. J. C. Sternberg, in: *Advances in Chromatography* (J. C. Giddings and R. A. Keller, eds.), Vol. 2, pp. 205–270, Marcel Dekker, New York (1966).
13. P. R. Griffiths, Recent applications of FT-IR spectrometry in chemical and environmental analysis, *Appl. Spectrosc., 31,* 497–505 (1977).

14. J. H. Ross and M. E. Casto, A method for high-temperature exclusion chromatography of polyethylenes, *J. Polym. Sci. C, 21,* 143–152 (1968).
15. R. Yost, J. Stoveken, and W. MacLean, Positive peak identification in LC using absorbance ratioing with a variable-wavelength spectrophotometric detector, *J. Chromatogr., 134,* 73–82 (1977).
16. R. N. Smith and M. Zetlein, Use of dual-wavelength detection in high-pressure liquid chromatography for the quantitative determination of unresolved or partially resolved compounds, *J. Chromatogr., 130,* 314–317 (1977).
17. R. G. Berg, C. Y. Ko, J. M. Clemons, and H. M. McNair, Characterization of unresolved components in high pressure liquid chromatography, *Anal. Chem., 47,* 2480–2482 (1975).
18. J. N. Little and G. J. Fallick, New considerations in detector–application relationships, *J. Chromatogr., 112,* 389–397 (1975).
19. J. M. Essigman and N. Catsimpoolas, Simple derivative mode detector for LC, *J. Chromatogr., 103,* 7–13 (1975).
20. A. Bylina, D. Sybilska, Z. R. Grabowski, and J. Koszewski, Rapid scanning spectrophotometry as a new system in chromatography, *J. Chromatogr., 83,* 357–362 (1973).
21. M. S. Denton, T. P. DeAngelis, A. M. Yacynych, W. R. Heineman, and T. W. Gilbert, Oscillating mirror rapid scanning ultraviolet–visible spectrometer as a detector for liquid chromatography, *Anal. Chem., 48,* 20–24 (1976).
22. A. McDowell and H. L. Pardue, Application of a vidicon tube as a multiwavelength detector for liquid chromatography, *Anal. Chem., 48,* 1815–1817 (1976).
23. K. M. Aldous and J. S. Garden, The use of a linear photodiode array in a multichannel detector for liquid chromatography, Pittsburgh Conference on Analytical Chemistry and Applied Spectroscopy (March 1975), Paper #434.
24. R. E. Dessy, W. G. Nunn, C. A. Titus, and W. R. Reynolds, Linear photodiode array spectrometers as detector systems in automated liquid chromatographs, *J. Chromatogr. Sci., 14,* 195–200 (1976).
25. M. J. Milano, S. Lam, and E. Grushka, Rapid scanning diode array as a multiwavelength detector in liquid chromatography. *J. Chromatogr., 125,* 315–326 (1976).
26. M. J. Milano and E. Grushka, Diode array detector in liquid chromatography, II. Enhanced sensitivity via first derivative *(dA/dλ)* chromatograms, *J. Chromatogr., 133,* 352–354 (1977).
27. Varian varichrom product literature, SEP-2005.
28. C. Bollet and M. Caude, Séparation par chromatographie en phase liquide rapide des dérivés phénylthiohydantoine des amino-acides rencontreés lors de la dégradation d'Edman, *J. Chromatogr., 121,* 323–328 (1976).
29. C. L. Zimmerman, E. Appella, and J. J. Pisano, Advances in the analysis of amino acid phenylthiohydantoins by high performance liquid chromatography, *Anal. Biochem., 75,* 77–85 (1976).
30. I. Kato, W. J. Kohr, and M. Laskowski, Limited proteolyses of ovomucoids caused by staphylococcal proteinase, Federation of American Societies for Experimental Biology Abstract #2592 (1977).
31. J. F. Lawrence and R. W. Frei, *Chemical Derivatization in Liquid Chromatography,* Elsevier, Amsterdam (1976).
32. N. G. Anderson, R. H. Stevens, and J. W. Holleman, Analytical techniques for cell fractions. X. High-pressure ninhydrin reaction system, *Anal. Biochem., 26,* 104–117 (1968).
33. P. J. Lamothe and P. G. McCormick, Role of hydrindantin in the determination of amino acids using ninhydrin, *Anal. Chem., 45,* 1906–1911 (1973).
34. R. A. Henry, J. A. Schmit, and J. F. Dieckman, The analysis of steroids and deri-

vatized steroids by high speed liquid chromatography, *J. Chromatogr. Sci., 9,* 513–520 (1971).
35. F. A. Fitzpatrick, S. Siggia, and J. Dingman, Sr., High speed liquid chromatography of derivatized urinary 17-keto steroids, *Anal. Chem., 44,* 2211–2216 (1972).
36. M. A. Carey and H. E. Persinger, Liquid chromatographic determination of traces of aliphatic carbonyl compounds and glycols as derivatives that contain the dinitrophenyl group. *J. Chromatogr. Sci., 10,* 537–543 (1972).
37. L. J. Papa and L. P. Turner, Chromatographic determination of carbonyl compounds as their 2,4-dinitrophenylhydrazones II. High pressure liquid chromatography, *J. Chromatogr. Sci., 10,* 747–750 (1972).
38. I. R. Politzer, G. W. Griffin, B. J. Dowty, and J. L. Laseter, Enhancement of ultraviolet detectability of fatty acids for purposes of liquid chromatographic-mass spectrometric analysis, *Anal. Lett., 6,* 539–546 (1973).
39. D. R. Knapp and S. Krueger, Use of *O-p*-nitrobenzyl-*N,N'*-diisopropylisourea as a chromogenic reagent for liquid chromatographic analysis of carboxylic acids, *Anal. Lett., 8,* 603–610 (1975).
40. H. D. Durst, M. Milano, E. J. Kikta, Jr., S. A. Connelly, and E. Grushka, Phenacyl esters of fatty acids via crown ether catalysis for enhanced ultraviolet detection in liquid chromatography, *Anal. Chem., 47,* 1797–1801 (1975).
41. E. Grushka, H. D. Durst, and E. J. Kikta, Jr., Liquid chromatographic separation and detection of nanogram quantities of biologically important dicarboxylic acids, *J. Chromatogr., 112,* 673–678 (1975).
42. R. F. Borch, Separation of long chain fatty acids as phenacyl esters by high pressure liquid chromatography, *Anal. Chem., 47,* 2437–2439 (1975).
43. N. E. Hoffman and J. C. Liao, High pressure liquid chromatography of *p*-methoxyanilides of fatty acids, *Anal. Chem., 48,* 1104–1106, (1976).
44. W. Morozowich and S. L. Douglas, Resolution of prostaglandin *p*-nitrophenacyl esters by liquid chromatography and conditions for rapid, quantitative *p*-nitrophenacylation, *Prostaglandins, 10,* 19–40 (1975).
45. M. V. Merritt and G. E. Bronson, High-performance liquid chromatography of *p*-nitrophenacyl esters of selected prostaglandins on silver ion-loaded microparticualte cation-exchange resin, *Anal. Biochem., 80,* 392–400 (1977).
46. W. Morozowich and S. L. Douglas, Detection of prostaglandins by HPLC after conversion to *p*-(9-anthroyloxy)phenacyl esters, *Anal. Chem.,* submitted.
47. L. H. Thacker, A miniature flow fluorometer for liquid chromatography, *J. Chromatogr., 73,* 117–123 (1972).
48. J. C. Steichen, A dual-purpose absorbance fluorescence detector for high-pressure liquid chromatography, *J. Chromatogr., 104,* 39–45 (1975).
49. F. Martin, J. Maine, C. C. Sweeley, and J. F. Holland, A novel fluorescence detector for high-performance liquid chromatography, *Clin. Chem., 22,* 1434–1437 (1976).
50. W. Slavin, A. T. Williams, and R. F. Adams, A fluorescence detector for HPLC, *J. Chromatogr., 134,* 121–130 (1977).
51. G. J. Diebold and R. N. Zare, Laser fluorimetry: Subpicogram detection of aflatoxins using high-pressure liquid chromatography, *Science, 196,* 1439–1441 (1977).
52. M. J. Sepaniak and E. S. Yeung, Laser two-photon excited fluorescence detection for high pressure liquid chromatography, *Anal. Chem., 49,* 1554–1556 (1977).
53. M. J. Wirth and F. E. Lytle, Two-photon excited molecular fluorescence in optically dense media, *Anal. Chem., 49,* 2054–2057 (1977).
54. H. Hatano, Y. Yamamoto, M. Saito, E. Mochida, and S. Watanabe, A high speed liquid chromatograph with a flow-spectrofluorimetric detector and the ultramicrodetermination of aromatic compounds, *J. Chromatogr., 83,* 373–380 (1973).

55. E. D. Pellizari and C. M. Sparacino, Scanning fluorescence spectrometry combined with ultraviolet detection of high pressure liquid chromatographic effluents, *Anal. Chem., 45,* 378–381 (1973).
56. J. R. Jadamec, W. A. Saner, and Y. Talmi, Optical multichannel analyzer for characterization of fluorescent liquid chromatographic petroleum fractions, *Anal. Chem., 49,* 1316–1321 (1977).
57. I. M. Warner, J. B. Callis, E. R. Davidson, and G. D. Christian, Multicomponent analysis in clinical chemistry by use of rapid scanning fluorescence spectroscopy, *Clin. Chem., 22,* 1483–1492 (1976).
58. J. F. McKay and D. R. Latham, Fluorescence spectrometry in the characterization of high-boiling petroleum distillates, *Anal. Chem., 44,* 2132–2137 (1972).
59. M. A. Fox and S. W. Staley, Determination of polycyclic aromatic hydrocarbons in atmospheric particulate matter by high pressure liquid chromatography coupled with fluorescence techniques, *Anal. Chem., 48,* 992–998 (1976).
60. J. A. Robertson, W. A. Pons, Jr., and L. A. Goldblatt, Preparation of aflatoxins and determination of their ultraviolet and fluorescent characteristics, *J. Agric. Food Chem., 15,* 798–801 (1967).
61. J. Chelkowski, Spectral behavior of aflatoxins in different solvents, *Photochem. Photobiol., 20,* 279–280 (1974).
62. W. Przybylski, Formation of aflatoxin derivatives on thin layer chromatographic plates, *J. Assoc.Off. Anal. Chem., 58,* 163–164 (1975).
63. W. A. Pons, Jr., Resolution of aflatoxins B_1, B_2, G_1 and G_2 by high-pressure liquid chromatography, *J. Assoc. Off. Anal. Chem., 59,* 101–105 (1976).
64. R. C. Garner, Aflatoxin separation by high-pressure liquid chromatography, *J. Chromatogr., 103,* 186–188 (1975).
65. L. M. Seitz, Comparison of methods for aflatoxin analysis by high pressure liquid chromatography, *J. Chromatogr., 104,* 81–89 (1975).
66. D. M. Takahashi, Reversed-phase high-performance liquid chromatographic analytical system for aflatoxins in wines with fluorescence detection, *J. Chromatogr., 131,* 147–156 (1977).
67. S. Udenfriend, S. Stein, P. Bohlen, W. Dairman, W. Leimgruber, and M. Weigele, Fluorescamine: A reagent for assay of amino acids, peptides, proteins and primary amines in the picomole range, *Science, 178,* 871–872 (1972).
68. S. DeBernardo, M. Weigele, V. Toome, K. Manhart, and W. Leimgruber, Studies on the kinetics of reaction and hydrolysis of fluorescamine, *Arch. Biochem. Biophys., 163,* 400–403 (1974).
69. R. W. Frei, L. Michel, and W. Santi, Post-column fluorescence derivatization of peptides: Problems and potential in high-performance liquid chromatography, *J. Chromatogr., 126,* 665–677 (1976).
70. J. A. F. de Silva and N. Strojny, Spectrofluorometric determination of pharmaceuticals containing aromatic or aliphatic primary amino groups as their fluorescamine (fluram) derivatives, *Anal. Chem., 47,* 714–718 (1975).
71. K. Samejima, Separation of fluorescamine derivatives of aliphatic diamines and polyamines by high-speed liquid chromatography, *J. Chromatogr., 96,* 250–254 (1974).
72. K. Imai, Fluorometric assay of dopamine, norepinephrine and their 3-*O*-methyl metabolites by using fluorescamine, *J. Chromatogr., 105,* 135–140 (1975).
73. M. Roth, Fluorescence reaction for amino acids, *Anal. Chem., 43,* 880–882 (1971).
74. S. S. Simons, Jr., and D. F. Johnson, The structure of the fluorescent adduct formed in the reaction of *o*-phthalaldehyde and thiols with amines, *J. Am. Chem. Soc., 98,* 7098–7099 (1976).

75. J. L. Meek, Application of inexpensive equipment for high pressure liquid chromatography to assays for taurine, γ-amino butyric acid and 5-hydroxytryptophan, *Anal. Chem., 48,* 375–379 (1976).
76. E. Bayer, E. Grom, B. Kaltenegger, and R. Uhmann, Separation of amino acids by high performance liquid chromatography, *Anal. Chem., 48,* 1106–1109 (1976).
77. R. W. Frei, W. Santi, and M. Thomas, Liquid chromatography of dansyl derivatives of some alkaloids and the application to the analysis of pharmaceuticals, *J. Chromatogr., 116,* 365–377 (1976).
78. R. W. Frei, J. F. Lawrence, J. Hope, and R. M. Cassidy, Analysis of carbamate insecticides by fluorogenic labeling and high-speed liquid chromatography, *J. Chromatogr. Sci., 12,* 40–44 (1974).
79. W. Dunges, High pressure liquid chromatographic analysis of barbituates in the picomole range by fluorometry of their DANS-derivatives, *J. Chromatogr. Sci., 12,* 655–657 (1974).
80. N. E. Newton, K. Ohno, and M. M. Abdel-Monem, Determination of diamines and polyamines in tissues by high-pressure liquid chromatography, *J. Chromatogr., 124,* 277–285 (1976).
81. S. Katz, W. W. Pitt, Jr., and G. Jones, Jr., Sensitive fluorescence monitoring of aromatic acids after anion-exchange chromatography of body fluids, *Clin. Chem., 19,* 817–820 (1973).
82. S. Katz, W. W. Pitt, Jr., and J. E. Mrochek, Comparative serum and urine analyses by dual detector anion-exchange chromatography, *J. Chromatogr., 104,* 303–310 (1975).
83. A. W. Wolkoff and R. H. Larose, A highly sensitive technique for the LC analysis of phenols and other environmental pollutants, *J. Chromatogr., 99,* 731–743 (1974).
84. P. A. Asmus, J. W. Jorgenson, and M. Novotny, Fluorescence enhancement, new selective detection principle for liquid chromatography, *J. Chromatogr., 126,* 317–325 (1976).
85. H. Small, T. S. Stevens, and W. C. Bauman, Novel ion exchange method using conductimetric detection, *Anal. Chem., 47,* 1801–1809 (1975).
86. C. Anderson, Ion chromatography: A new technique for clinical chemistry, *Clin. Chem., 22,* 1424–1426 (1976).
87. D. C. Johnson and J. Larochelle, Forced-flow liquid chromatography with a coulometric detector, *Talanta, 20,* 959–971 (1973).
88. R. J. Davenport and D. C. Johnson, Determination of nitrate and nitrite by forced-flow liquid chromatography with electrochemical detection, *Anal. Chem., 46,* 1971–1978 (1974).
89. L. R. Taylor and D. C. Johnson, Determination of antimony using forced-flow liquid chromatography with a coulometric detector, *Anal. Chem., 46,* 262–266 (1974).
90. U. R. Tjaden, J. Lankelma, H. Poppe, and G. Munsze, Anodic coulometric detection with a glassy carbon electrode in combination with reversed-phase high-performance liquid chromatography, *J. Chromatogr., 125,* 275–286 (1976).
91. J. Lankelma and H. Poppe, Design and characterization of a coulometric detector with glassy carbon electrode for high-performance liquid chromatography, *J. Chromatogr., 125,* 375–378 (1976).
92. P. T. Kissinger, C. J. Refshauge, R. Dreiling, L. Blank, R. Freeman, and R. N. Adams, An electrochemical detector for liquid chromatography with picogram sensitivity, *Anal. Lett., 6,* 465–477 (1973).
93. B. Fleet and C. J. Little, Design and evaluation of electrochemical detectors for HPLC, *J. Chromatogr. Sci., 12,* 747–752 (1974).

94. R. E. Shoup and P. T. Kissinger, A versatile thin-layer detector cell for high performance liquid chromatography, *Chem. Instrum., 7,* 171–177 (1976).
95. M. Karolczak, R. Dreiling, R. N. Adams, L. J. Felice, and P. T. Kissinger, Electrochemical techniques for study of phenolic natural products and drugs in microliter volumes, *Anal. Lett., 9,* 783–793 (1976).
96. S. C. Rifkin and D. H. Evans, Analytical evaluation of differential pulse voltammetry at stationary electrodes using computer-based instrumentation, *Anal. Chem., 48,* 2174–2180 (1976).
97. D. G. Swartzfager, Amperometric and differential pulse voltammetric detection in high performance liquid chromatography, *Anal. Chem., 48,* 2189–2192 (1976).
98. C. L. Blank, Dual electrochemical detector for liquid chromatography, *J. Chromatogr., 117,* 35–46 (1976).
99. R. Keller, A. Oke, I. Mefford, and R. N. Adams, Liquid chromatographic analysis of catecholamines. Routine assay for regional brain mapping, *Life Sci., 19,* 995–1004 (1976).
100. R. E. Shoup and P. T. Kissinger, Determination of urinary normetanephrine, metanephrine, and 3-methoxytyramine utilizing liquid chromatography with amperometric detection, *Clin. Chem., 23,* 1268–1274 (1977).
101. P. H. Zoutendam, C. S. Bruntlett, and P. T. Kissinger, Determination of homogentisic acid in serum and urine by liquid chromatography with amperometric detection, *Anal. Chem., 48,* 2200–2202 (1976).
102. K. V. Thrivikraman, C. Refshauge, and R. N. Adams, Liquid chromatographic analysis of nanogram quantities of ascorbate in brain tissue, *Life Sci., 15,* 1335–1338 (1974).
103. L. A. Pachla and P. T. Kissinger, Determination of ascorbic acid in body fluids, foodstuffs, and pharmaceuticals by liquid chromatography with electrochemical detection, *Anal. Chem., 48,* 364–367 (1976).
104. R. M. Riggin, A. L. Schmidt, and P. T. Kissinger, Determination of acetaminophen in pharmaceutical preparations and body fluids by high performance liquid chromatography with electrochemical detection, *J. Pharm. Sci, 64,* 680–683 (1975).
105. L. A. Pachla and P. T. Kissinger, Oxidative reaction detector for liquid chromatography using thin-layer amperometric detection, manuscript in preparation.
106. L. J. Felice, W.. P. King, and P. T. Kissinger, A new liquid chromatographic approach to plant phenolics. Application to the determination of chlorogenic acid in sunflower meal, *J. Agric. Food Chem., 24,* 380–382 (1976).
107. T. M. Kenyhercz and P. T. Kissinger, Determination of diethylstilbestrol residues by reverse-phase liquid chromatography with amperometric detection, *J. Anal. Toxicol.,* in press.
108. I. Mefford, R. W. Keller, R. N. Adams, L. A. Sternson, and M. S. Yllo, Liquid chromatographic determination of picomole quantities of aromatic amine carcinogens, *Anal. Chem., 49,* 683 (1977).
109. D. R. Koch and L. A. Pachla, unpublished results.
110. W. P. King, K. J. Thengumthyil, and P. T. Kissinger, unpublished results.
111. R. M. Riggin, M. J. McCarthy, and P. T. Kissinger, Identification of salsolinol as a major dopamine metabolite in the banana, *J. Agric. Food Chem., 24,* 189–191 (1976).
112. Y. Takata and G. Muto, Flow coulometric detector for liquid chromatography, *Anal. Chem., 45,* 1864–1868 (1974).
113. M. Lemar and M. Porthault, Amperometric detection in high performance liquid chromatography in the case of nonconducting eluants, *J. Chromatogr., 130,* 373 (1977).

114. P. T. Kissinger, Amperometric and coulometric detectors for high-performance liquid chromatography, *Anal. Chem., 49,* 447A–456A (1977).
115. P. T. Kissinger, Electrochemical detectors for liquid chromatography, *Adv. Chromatogr.,* manuscript in preparation.
116. D. M. Coulson, Electrolytic conductivity detector for gas chromatography, *J. Gas Chromatogr., 3,* 134–137 (1965).
117. J. W. Dolan and J. N. Seiber, Chlorine-selective detection for liquid chromatography with a Coulson electrolytic conductivity detector, *Anal. Chem., 49,* 326–331 (1977).
118. H. Malissa, J. Rendl, and W. Buchberger, Ein schwefelselektiven Detektor für die Flüssigkeitschromatographie auf konduktometrischer Basis, *Anal. Chim. Acta, 90,* 137–141 (1977).
119. R. C. Hall, A highly sensitive and selective microelectrolytic conductivity detector for gas chromatography, *J. Chromatogr. Sci., 12,* 152–160 (1974).
120. D. R. Jones IV and S. E. Manahan, Atomic absorption detector for chromium organometallic compounds separated by HSLC, *Anal. Lett., 8,* 569–574 (1975).
121. D. R. Jones IV and S. E. Manahan, Aqueous phase high speed liquid chromatographic separation and atomic absorption detection of amino carboxylic acid-copper chelates, *Anal. Chem., 48,* 502–505 (1976).
122. D. R. Jones IV, H. C. Tung, and S. E. Manahan, Mobile phase effects on atomic absorption detectors for high speed liquid chromatography, *Anal. Chem., 48,* 7–10, (1976).
123. D. R. Jones IV and S. E. Manahan, Detection limits for flame spectrophotometric monitoring of high speed liquid chromatographic effluents, *Anal. Chem., 48,* 1897–1899 (1976).
124. D. J. Freed, Flame photometric detector for liquid chromatography, *Anal. Chem., 47,* 186–187 (1975).
125. B. G. Julin, H. W. Vanderborn, and J. J. Kirkland, Selective flame emission detection of phosphorous and sulfur in high-performance liquid chromatography, *J. Chromatogr., 112,* 443–453 (1975).
126. D. H. Fine, F. Rufeh, D. Lieb, and D. P. Rounbehler, Description of the thermal energy analyzer (TEA) for trace determination of volatile and nonvolatile *N*-nitroso compounds, *Anal. Chem., 47,* 1188–1191 (1975).
127. P. E. Oettinger, F. Huffman, D. H. Fine, and D. Lieb, Liquid chromatograph detector for trace analysis of non-volatile *N*-nitroso compounds, *Anal. Lett., 8,* 411–414 (1975).
128. D. H. Fine, An organic nitrogen specific detector for HPLC, *Anal. Lett., 10,* 305–307 (1977).
129. J. T. Schmermund and D. C. Locke, A universal photoionization detector for liquid chromatography, *Anal. Lett., 8,* 611–625 (1975).
130. E. Haahti and T. Nikkari, Continuous detection of fractions in effluents of silicic acid chromatography, *Acta Chem. Scand., 17,* 2565–2568 (1973).
131. R. J. Maggs, Commercial detector for monitoring the eluent from liquid chromatographic columns, *Chromatographia, 1,* 43–48 (1968).
132. R. P. W. Scott and J. G. Lawrence, An improved moving wire liquid chromatography detector, *J. Chromatogr. Sci., 8,* 65–71 (1970).
133. J. H. van Dijk, Sensitivity improvement of a moving wire liquid chromatography detector, *J. Chromatogr. Sci., 10,* 31–34 (1972).
134. R. H. Stevens, Noise reduction in flame ionization type LC monitors: Development of an improved method for sample transport, *J. Gas Chromatogr., 6,* 375–383 (1968).
135. V. Pretorius and J. F. J. van Rensburg, Improvements to the wire solute transport detector, *J. Chromatogr. Sci., 11,* 355–357 (1973).

136. H. Dubský, A disc detector for liquid chromatography, *J. Chromatogr., 71,* 395–403 (1972).
137. J. J. Szakasits and R. E. Robinson, Disk conveyor flame ionization detector for liquid chromatography, *Anal. Chem., 46,* 1648–1652 (1974).
138. A. Stolywho, O. S. Privett, and W. L. Erdahl, An improved FID and associated transport system for LC, *J. Chromatogr. Sci., 11,* 263–267 (1973).
139. O. S. Privett and W. L. Erdahl, in: *Anal. Lipids Lipoproteins 1975* (E. G. Perkins, ed.), pp. 123–137, American Oil Chemists' Society, Champaign, Illinois (1975).
140. O. S. Privett, personal communication (actual chromatographic conditions unknown).
141. L. S. Snyder, personal communication.
142. R. J. Maggs, Use of the electron capture detector as a monitor for liquid chromatograph columns, *Column, 2*(4), 5–7 (1968).
143. F. W. Willmott and R. J. Dolphin, A novel combination of liquid chromatography and electron capture detection in the analysis of pesticides, *J. Chromatogr. Sci., 12,* 695–700 (1974).
144. The application of an electron capture detector to liquid chromatography, Liquid Chromatography Application 13, Philips Electronic Instruments, Mount Vernon, New York.
145. H.-R. Schulten and H. D. Beckey, Potentiality of the coupling of column liquid chromatography and field desorption mass spectrometry, *J. Chromatogr., 83,* 315–320 (1973).
146. R. E. Lovins, S. R. Ellis, G. D. Tolbert, and C. R. McKinney, Liquid chromatography–mass spectrometry. Coupling of a liquid chromatograph to a mass spectrometer, *Anal. Chem., 45,* 1553–1556 (1973).
147. R. P. W. Scott, C. G. Scott, M. Munroe, and J. Hess, Jr., Interface for on-line liquid chromatography–mass spectroscopy analysis, *J. Chromatogr., 99,* 395–405 (1974).
148. W. H. McFadden, H. L. Schwartz, D. C. Bradford, and L.H. Wright, Applications of combined liquid chromatography/mass spectrometry, Pittsburgh Conference on Analytical Chemistry and Applied Spectroscopy, Cleveland, Ohio, February 28–March 4 (1977).
149. P. Arpino, M. A. Baldwin, and F. W. McLafferty, Liquid chromatography–mass spectrometry II—continuous monitoring, *Biomed. Mass Spectrom., 1,* 80–82 (1974).
150. F. W. McLafferty, R. Knutti, R. Venkataraghavan, P. J. Arpino, and B. G. Dawkins, Continuous mass spectrometric monitoring of a liquid chromatograph with subnanogram sensitivity using an on-line computer, *Anal. Chem., 47,* 1503–1505 (1975).
151. E. C. Horning, D. I. Carroll, I. Dzidic, K. D. Haegele, M. G. Horning, and R. N. Stillwell, Liquid chromatograph–mass spectrometer–computer analytical systems. A continuous-flow system based on atmospheric pressure ionization mass spectrometry, *J. Chromatogr., 99,* 13–21 (1974).
152. D. I. Carroll, I. Dzidic, R. N. Stillwell, K. D. Haegele, and E. C. Horning, Atmospheric pressure ionization mass spectrometry: Corona discharge ion source for use in liquid chromatograph–mass spectrometer–computer analytical system, *Anal. Chem., 47,* 2369–2372 (1975).
153. P. R. Jones and S. K. Yang, A liquid chromatograph/mass spectrometer interface, *Anal. Chem., 47,* 1000–1003 (1975).
154. L. R. Snyder and J. J. Kirkland, *Introduction To Modern Liquid Chromatography,* pp. 149–153, Wiley, New York, (1974).
155. H. Poppe and J. Kuysten, Construction and evaluation of a thermostatted permittivity detector for high performance column liquid chromatography, *J. Chromatogr., 132,* 369–378 (1977).
156. R. Vespalec and K. Hána, Performance of the capacitance detector for liquid chromatography, *J. Chromatogr., 65,* 53–69 (1972).

157. S. Haderka, Permittivity and conductivity detectors for liquid chromatography, *J. Chromatogr., 91,* 167–179 (1974).
158. P. H. Monaghan, P. B. Moseley, T. S. Burkhalter, and O. A. Nance, Detection of chromatographic zones by means of high frequency oscillators, *Anal. Chem., 24,* 193–195 (1952).
159. N. Watanabe, M. Azuma, and E. Niki, Study of a dielectric constant detector for high-speed liquid chromatography, *Bunseki Kagaku, 26,* 295–300 (1977).
160. W. F. Erbelding, Dielectric constant detector for liquid chromatography, *Anal. Chem., 47,* 1983–1987 (1975).
161. S. Haderka, The prospects of selective detection by capacitance detectors in liquid chromatography, *J. Chromatogr., 57,* 181–191 (1971).
162. L. N. Klatt, Universal detector for liquid chromatography based upon dielectric constant, *Anal. Chem., 48,* 1845–1850 (1976).
163. M. Krejčí and N. Pospíšilova, Experimental comparisons of some detectors used in high-efficiency liquid chromatography, *J. Chromatogr., 73,* 105–115 (1972).
164. R. Vespalec, Improvement of the performance of the capacitance detector for liquid chromatography, *J. Chromatogr., 108,* 243–254 (1975).
165. G. C. Claxton, Detector for liquid chromatography, *J. Chromatogr., 2,* 136–139 (1959).
166. K.-P. Hupe and E. Bayer, A micro adsorption detector for general use in liquid chromatography, *J. Gas Chromatogr., 5,* 197–201 (1967).
167. M. N. Munk and D. N. Raval, Flow sensitivity of micro adsorption detector, *J. Chromatogr. Sci., 7,* 48–55 (1969).
168. R. P. W. Scott, A theoretical treatment of the heat-of-adsorption detector, *J. Chromatogr. Sci., 11,* 349–357 (1973).
169. H. P. Warren and D. P. McKay, The response of the micro adsorption detector to inorganic cations, *J. Chromatogr. Sci., 13,* 117–122 (1975).
170. M. N. Munk, Some practical aspects of the micro adsorption detector, *Am. J. Clin. Pathol., 53,* 719–730 (1970).
171. H. M. McNair and D. T. Stafford, Micro-adsorption detector. I. Principles of operation and mechanisms of response, *J. Chromatogr. 133,* 31–36 (1977).
172. G. B. Sieswerda, H. Poppe, and J. F. K. Huber, Flow versus batch detection of radioactivity in column liquid chromatography, *Anal. Chim. Acta, 78,* 343–358 (1975).
173. K. A. Piez, Continuous scintillation counting of carbon-14 and tritium in effluent of the automatic amino acid analyzer, *Anal. Biochem., 4,* 444–458 (1962).
174. J. de Belleroche, C. R. Dykes, and A. J. Thomas, The automated separation and analysis of dopamine, its amino acid precursors and metabolites, and the application of the method to the measurement of specific radioactivities of dopamine in striatial symaptosomes, *Anal. Biochem., 71,* 193–203 (1976).
175. R. N. Bracewell, *The Fourier Transform and its Applications,* pp. 24–48, McGraw-Hill, New York (1965).
176. E. Grushka, Characterization of exponentially modified Gaussian peaks in chromatography, *Anal. Chem., 44,* 1733–1738 (1972).

4

The Radioimmunoassay of Enzymes

J. Landon, J. A. Carney, and D. J. Langley

1. Introduction

Many sensitive and elegant assays are available which determine enzyme levels in terms of their catalytic activity. The application of these assays to the measurement, in biological fluids, of a wide range of enzymes has proved of immense clinical value. It is appropriate, therefore, that clinical enzymology and its exponents now occupy a valued role in most large pathology departments. Nonetheless, the time may have come to question the present emphasis on the measurement of enzymes by means of their catalytic effects. The continued acceptance of such assays may reflect their relative simplicity rather than excluding the possibility that other analytical techniques may provide equally relevant clinical information.

The large majority of serum enzymes routinely measured in clinical chemistry do not normally affect their biological role in the serum and are in fact derived from the various tissues. As such they can be considered waste products en route for removal from the body. The purpose of assaying each of these enzymes is to monitor the condition of the tissues of origin, e.g., serum creatine kinase increases dramatically at the time of myocardial infarction. Enzymes are actually proteins which possess readily measured biological activity. However, their biological activity is not necessarily directly proportional to the concentration of that protein. For instance, proteases such as trypsin are partially inactivated in serum by the presence of specific circulating inhibitors.

Most of the normally assayed enzymes are tissue markers. As in the case of proteins measured in monitoring tumors or fetoplacental function,

J. Landon • Department of Chemical Pathology, St. Bartholomew's Hospital, London, EC 1, England **J. A. Carney and D. J. Langley** • Technicon Methods and Standards Laboratory, London, EC 1, England

it is possibly more important to know the concentration of serum enzymes rather than their catalytic capacity. The advantages and disadvantages of radioimmunoassay (RIA) of enzymes are summarized in Table 1.

The present paper reviews the recent introduction of RIA for the measurement of enzyme levels in terms of their concentration (Table 2). It argues that while catalytic assays will continue to occupy the predominant role in some situations and that a combination of the two will provide new information in others, measurement of concentration may gradually replace conventional assays for some enzymes in biological fluids. Although the technique of measuring enzymes by RIA is new, the immunochemistry of many enzymes has been investigated in detail (viz., reviews by Cinader[1] and Arnon).[2] Indeed, an antiserum to amylase was raised more than 50 years ago.[3]

2. Categories of Analytical Techniques

All analytical procedures can be subdivided into one of three categories (Figure 1). The complexity of enzyme molecules precludes their assay by chemical or physical techniques, which are based on specific structural characteristics of the compounds to be assayed. Measurement of the catalytic activity of an enzyme is a biological assay, based on determining the rate at which it converts a substrate to its reaction products. Such an assay

Table 1. Relative Advantages and Disadvantages of RIA and Activity Assay of Enzymes

Advantages of RIA
1. Parameter measured—The absolute amount of an enzyme present in serum is measured by RIA, whereas activity measurements quantitate the catalytic activity of the enzyme as expressed in serum.
2. Independence of circulating inhibitors—Certain enzymes are partially inactivated in serum by the presence of inhibiting proteins, e.g., trypsin, antitrypsin. Measurements made by RIA are independent of the presence of such specific inhibitors.
3. Practicability—Far fewer factors influence measurements made by RIA (e.g., temperature, presence of cofactors, metal ions).
4. Sensitivity—For enzymes with a low turnover number, an RIA will be more sensitive.
5. Specificity—Unlike most activity assays, RIA can allow measurements to be made of specific isoenzymes.

Disadvantages of RIA
1. Practicability—Radioimmunoassays can require long periods of incubation, especially where second antibody precipitation is employed.
2. Specificity—It may be difficult or impossible to obtain an antiserum specific for one particular isoenzyme. Furthermore the enzyme used as immunogen, standard, and label must be of human origin and is therefore not readily available.

Table 2. Some Available Radioimmunoassays for Enzymes

Enzyme	Species	Reference
Plasminogen, plasmin	Human	56
Trypsin, trypsinogen, chymotrypsin chymotrypsinogen, carboxypeptidase A, amylase, lipase	Bovine, porcine, rat, human	28, 29, 38–41
Cl esterase	Human	42, 43
Carbonic anhydrase I and II	Human, monkey	52, 53
Pepsinogen	Human	54
Fructose-1,6-diphosphatase	Rabbit	15, 21, 26
A-chymotrypsin	Bovine	55
Group 1 pepsinogens	Human	31, 44
Collagenase	Human, rat	56, 57
Placental alkaline phosphatase	Human	58, 59, 60
Dopamine β-hydroxylase	Bovine, human	45–47, 61–63
Erythrocyte catalase	Human	64
Elastase	Canine	14, 30
Creatine kinase	Human	35
Carboxypeptidase B	Human	33
Prostatic acid phosphatase	Human	65, 66
Pancreatic amylase	Porcine	13, 48
Hexosaminidases	Human	20
Prolyl hydroxylases	Human, chick	49

assesses the functional integrity of the active site of the enzyme which, in turn, depends on the conformation (tertiary structure) of the appropriate part of the molecule and may require the presence of activators, such as divalent metals. Bioassays are usually unaffected by the presence of proenzymes or the products of enzyme degradation. However, they will usually determine isoenzymes from other tissues, the equivalent enzyme from other species and, indeed, related enzymes which may differ considerably in part of their primary structure.

Radioimmunoassays, which are the subject of this review, are examples of binding assays used to measure the concentration of a ligand (in this instance an enzyme). They employ a specific antibody, as the binding protein, and isotopically labeled enzyme, as a tracer, to assess the distribution of total enzyme between the antibody-bound and free fractions. It is the antibody which determines, in large part, both the sensitivity and specificity of the assay, with the latter depending upon the antigenic determinant—a sequence of some three to five amino acids in the enzyme molecule to which the antibody binds. RIA will usually determine levels of proenzymes and enzyme breakdown products containing this antigenic determinant as well as the enzyme. They are usually unaffected by isoenzymes, functionally related enzymes, or the equivalent enzyme from other

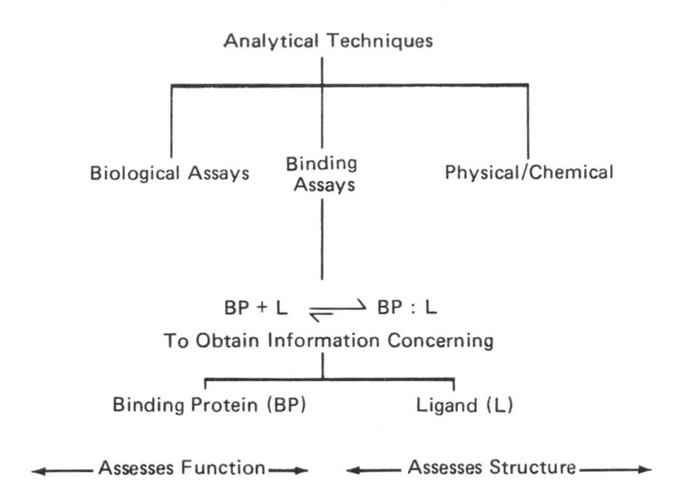

Figure 1. Categories of analytical techniques currently employed in clinical chemistry.

species. The results of a bioassay may correlate with those obtained by RIA; however, this is not always the case, and the finding of different values by a structural (binding assay) and a functional (bioassay) technique does not invalidate either.

Certain additional points require note. First, each antiserum is unique with regard to its titer, affinity, and antigenic determinant(s). Second, immunoassays can be used to determine the rate at which an enzyme is either degrading its substrate or producing a reaction product (Table 3). Such assays determine biological activity, and RIA is used only for endpoint detection. Finally, enzymes have frequently been the subject of immunological studies because they possess catalytic activity which can be, and often is, altered by their combination with antibodies.

Specific antisera frequently inhibit enzyme activity probably because the enzyme–antibody combination sterically hinders access of the substrate to the active site. Thus the larger the substrate, the more likely is such inhibition, although its degree must also depend on the position of the epitope relative to the active site. McGeachin and Reynolds[4] noted a 93% inhibition of hog pancreatic α-amylase caused by its antiserum at 1/20 dilution, whereas a later study on the human enzyme observed the same effect at a 1/5000 dilution of antiserum.[5] The antibodies in the latter case would appear to have been directed at a determinant more closely related to the active site, although the relative size of the substrates employed may also have affected the results.

Such inhibition studies have recently been found to be diagnostically useful in the measurement of the creatine kinase isoenzyme MB, which is released into the circulation only during myocardial infarction. Jockers-Wretou and Pfleiderer[6] have raised antibodies to the brain-specific BB isozyme, as well as the muscle-specific MM isoenzyme and used them in enzyme inhibition assays of the heart-specific MB.

Antibody-induced enzyme inhibition allows studies to be made on molecular evolution. The relationship between phylogenically homologous enzymes involves chemical and functional similarity and also conformational homology. Immunological cross-reaction of the antisera is corroborating evidence in this area, as evidenced, for example, by studies with trypsin[7] and amylase.[8]

In certain cases enzymes can be activated by their combination with antiserum, while the antisera raised to enzymes such as penicillinase have been found to contain populations of both inhibiting and stimulating antibodies.[9]

3. Requirements for an Enzyme Radioimmunoassay

3.1. Antigen

A few milligrams of highly purified enzyme are required for use as the standard, for labeling, and for immunization. Purification seldom occasions difficulties because of the rigid conformation and relatively high concentration of most enzymes in certain tissues, as well as the applicability of affinity chromatography. Whereas some enzymes, such as amylase, maintain their catalytic activity during isolation, others, such as the brain isoenzyme of creatine kinase (CK), are partially inactivated.[10] This does not preclude their use as immunogens, although, by preference, one should attempt to maintain enzymic activity and thereby guarantee the integrity of the total tertiary structure. Globular proteins, which include enzymes, possess mostly conformational antigenic determinants.[7] This has been well demonstrated by Maron and his colleagues,[11] using fragments of egg-white lysozyme. The complete unwinding of the polypeptide chain of

Table 3. Use of RIA in Biological Assays for Enzymes

Determine rate of production of a reaction product
for plasma renin: based on RIA of angiotensin I or angiotensin II
for tissue prostaglandin synthetase: based on RIA of PGE produced from arachidonic acid[24]
Determine rate of degradation of a substrate
for plasma angiotensinase: based on RIA of angiotension I[50]
for plasma vasopressinase: based on RIA of vasopressin (Rosenbloom *et al.*, 1975)
Combine problems of biological assays and of the RIA used for endpoint detection

ribonuclease is also accompanied by a complete loss of both enzymic activity *and* the ability to cross-react with antibodies to the native enzyme.[12] Nonetheless, in most cases the state of the active site is probably unimportant when immunizing because: (1) The active-site antigenic determinant(s) constitutes only a small proportion of the many potential immunogenic sites within the protein molecule. (2) The active site is seldom exposed on the surface of the molecule in the native protein. (3) Conservation of catalytic activity throughout evolution may require the conservation of the structure of the active site of any enzyme,[1] thereby making it a poor immunogen.

The enzyme must be of appropriate origin with regard to both tissue and species, and there is presently a paucity of pure human enzymes. Endocrinologists brought up on the dictate that "one must compare like with like" are bemused by the many species and sites of origin of the enzyme preparations currently used for standardization, which can greatly affect the results obtained. For example, studies in this laboratory have shown that antisera to human pancreatic α-amylase do not cross-react with the corresponding commercially available hog enzyme.[8] This finding is supported by the evidence of Ryan and his colleagues,[13] using an antiserum to the hog amylase. Furthermore, Carballo and Troiano[14] have demonstrated almost complete lack of cross-reactivity between hog and human elastase.

3.2. Antiserum

As emphasized above, the most important requirement is a high-titer, high-affinity antiserum which is specific to the enzyme being assayed. There is seldom difficulty in producing such antisera, since enzymes are very immunogenic as a result of their rigid conformation, large molecular weight, and the structural differences which exist with similar enzymes in the animals being immunized. Thus, all four rabbits immunized with human pancreatic α-amylase produced antisera of acceptable titer and affinity within a few weeks (Figure 2). However, large numbers of animals may be required to obtain one antiserum specific for a single isoenzyme.

3.3. Labeled Antigen

Isotopically labeled enzyme is essential for use as the tracer. Most groups have attempted to radioiodinate with ^{125}I using the chloramine-T method and have experienced difficulties in obtaining a high-specific-activity product with some enzymes. In this laboratory it was found that porcine pancreatic α-amylase readily accepted an ^{125}I label but that there was a very poor incorporation of iodine into the corresponding human

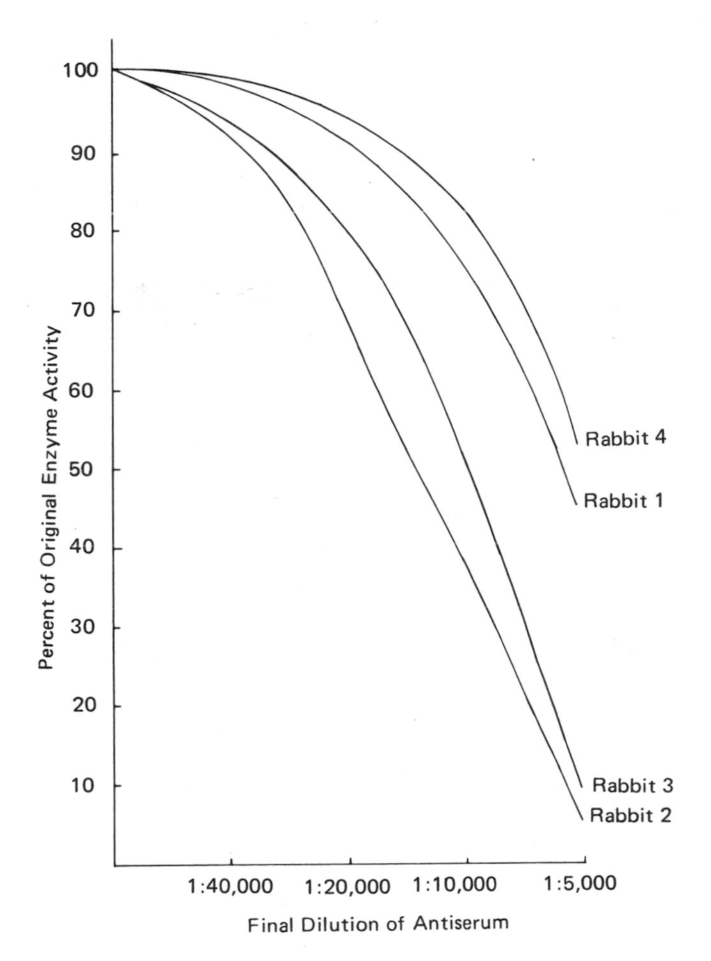

Figure 2. Inhibition of serum amylases by four antisera raised against human pancreatic isoamylases.

enzyme. However, we have found relatively high concentrations of circulating human serum amylase in normal subjects (650 ± 150 ng/ml) which increase to as high as 35 μg/ml in acute pancreatitis. For this reason large amounts of labeled amylase could be employed, and its low specific activity did not adversely affect the sensitivity of clinical measurements.

Radioiodination is often accompanied by a rapid loss of catalytic activity.[15] Taniguchi and his colleagues[16] found that the chloramine-T procedures for human carbonic anhydrase β gave rise to a complete loss of antigenicity. However, this difficulty was overcome by obtaining a monospecific antiserum and labeling the antibodies. This method would be of use in any assay where iodination damaged the enzyme either antigenically or catalytically. In our experience either chloramine-T or lactoperoxidase labeling of brain CK totally destroys enzymic activity, probably by the oxidation of thiol groups necessary for the maintenance of the tertiary structure.

3.4. Separation Procedure

It is necessary to employ a suitable method to separate the antibody bound and free fractions. In many RIAs the antigen measured has a much lower molecular weight than its antibody, e.g., T_4, ACTH, and steroids. In such cases the antibody–antigen complex is a much larger molecule than the free antigen, and a wide range of techniques can be employed for separation of free and bound fractions. These techniques include chromatoelectrophoresis, gel filtration, charcoal adsorption, as well as precipitation with neutral salts or organic solvents.[17] However, enzymes, being proteins, have relatively large molecular weights, and the above methods may not be able to completely differentiate between the free and antibody-bound protein. For example, the use of sodium sulfate to separate free and bound fructose-1,6-diphosphatase resulted in the nonspecific precipitation of about 20% of free enzyme.[15]

The only two methods which are universally applicable in the RIA of enzymes are: (1) fractional precipitation by use of a second antibody and (2) use of antibodies immobilized on a solid phase.

4. Differences in the Use of Enzyme Assays in Biochemistry and Clinical Chemistry

The biochemist has applied enzyme assays predominantly to tissue homogenates in studies of metabolic pathways, rate-limiting steps, and enzyme kinetics. It is essential, in such situations, to determine biological activity. When biochemists moved into the field of clinical chemistry, they continued to employ the same analytical approach. There is, however, a fundamental difference in that the assays were being applied predominantly to blood (and urine) rather than tissue homogenates (Table 4). Such an approach was appropriate for those enzymes that effect their role in the biological fluid being assayed—for example, in assays of circulating renin or the enzymes involved in clotting. It was not essential, however, for those enzymes which do not effect their role in the biological fluid, and more than 99% of the enzyme assays performed in most routine clinical chemistry departments fall into this category. In these situations the enzyme is being employed as a tissue marker, and it is irrelevant whether or not it is biologically active.

Table 4. Applications of Enzyme Assays

Biochemistry: predominantly applied to tissue homogenates to study metabolic pathways, rate-limiting steps, enzyme kinetics, etc.
Clinical chemistry: predominantly applied to biological fluids and can be further subdivided into:
Enzymes that effect their role in the biological fluid (i.e., renin, enzymes involved in clotting)
Enzymes that do not effect their role in the biological fluid (i.e., all enzymes determined in urine and most in blood)

Additional information may be obtained if an immunoassay is employed to determine enzyme concentration as well as a bioassay to measure catalytic activity. Thus in plasma the ratio of active plasmin to the inactive plasminogen is about 1:50. Plasmin activity can be affected directly by circulating levels of agents such as vitamin E[18] actually inhibiting the enzyme or by agents altering the rate of conversion of plasminogen to plasmin. RIA for both precursor and enzyme would produce much more evidence in situations such as the increased fibrinolysis observed following venous occlusion. In a disease associated with the genetic absence of activity of a particular enzyme, this combination would enable differentiation between (1) failure to synthesize the enzyme, or (2) synthesis of an abnormal form of the enzyme which lacks catalytic activity. For instance in the Lesch–Nyhan syndrome, immunoreactive hypoxanthinylguanine phosphoribosyl transferase is present in the tissues of affected individuals, but its enzymic activity is totally lacking.[19] For the determination of the majority of enzymes which are being used as tissue markers (including, for example, CK in the diagnosis of myocardial infarction or acid phosphatase in a patient with a prostatic neoplasm), either an immunoassay or bioassay could be employed once the relationship between serum concentration and activity has been established.

5. Advantages of Enzyme Measurement by Concentration Rather Than by Catalytic Activity

Once it is appreciated that the majority of enzyme determinations performed in clinical chemistry are employing the enzyme as a tissue marker and that, therefore, either type of analytical technique may be appropriate, it remains to determine which is preferable. This decision should be based on such standard criteria as sensitivity, specificity, precision, practicability, and diagnostic performance. Thus, while RIAs are extremely sensitive and specific, they may take longer to perform and seem more complex than many conventional enzyme assays.

5.1. Sensitivity

RIA is more sensitive than bioassay for those enzymes which have a low catalytic number. Examples are given in Table 5. This is of practical importance with regard to the assay of hexoaminidases A and B in the diagnosis of Tay–Sachs disease.[20] Thus the improved sensitivity of the RIA enables diagnosis at the time of amniocentesis rather than having to culture the amniotic fluid cells for some weeks in order to attain levels that can be assayed by the conventional technique.

5.2. Specificity

Kolb and Grodsky[21] showed that their RIA for fructose diphosphatase was much more specific than bioassay, both with regard to the tissue and species of origin or the enzyme. This is not always the case since, in our experience, an assay for human pancreatic α-amylase also determined human salivary amylase[5] and pancreatic amylase from several other species, although not that of porcine origin, which is the standard usually employed in clinical chemistry.[8] Each antiserum is unique, and Table 6, based on the data of Ryan and his colleagues,[13] shows that their particular antiserum could differentiate porcine pancreatic α-amylase from salivary and other amylases—as demonstrated by the undetectable levels found following pancreatectomy. However, it is unlikely that one could differentiate between all human amylases because, although there are at least two genetic loci for amylase expression, there is only one gene product.[22] The differences seen in circulating isoenzymes result from modifications of the carbohydrate moiety which occur after the protein has been synthesized.

Table 7 summarizes examples from the literature where RIA has been employed to differentiate various isoenzymes. Many of the isoenzymes can

Table 5. Examples of Enzymes for Which RIA is More Sensitive than Bioassay

Author	Enzyme	Comment
Rabiner *et al.*[51]	Plasminogen	Requires activation prior to bioassay
Temler and Felber[29]	Trypsin, chymotrypsin	100,000× more sensitive
Bauer *et al.*[56]	Human skin collagenase	200–400× more sensitive
Eisen *et al.*[57]	Rat skin collagenase	1000× more sensitive
Tuderman *et al.*[49]	Prolyl hydroxylases	
Geiger *et al.*[20]	Hexosaminidases A and B	Improved sensitivity of great value in diagnosis of Tay–Sachs disease

Table 6. Effect of Pancreatectomy on Serum Pancreatic Amylase Levels in the Pig[13]

	Serum amylase	
Day	RIA (μg/ml)	Bioassay (units/100 ml)
0	2.4	3100
1	1.5	1680
2	0	1253
3	0	1204
4	0	1210
5	0	1214
6	0	1222
7	0	1212

also be differentiated by conventional techniques, but this involves an initial and time-consuming separation step, the assay of samples with and without prior heating, or the use of special substrates, etc. There are now grounds to suggest that RIA may prove more specific and, in some circumstances, simpler than the currently employed procedures.

Not all isoenzymes could be resolved immunologically. Although immunological differentiation has been demonstrated for the isoenzymes of CK, alkaline phosphatase, and GOT, the same cannot be said of serum LDH; for this reason one might not choose to set up an RIA for the measurement of serum LDH. The five isoenzymes are different tetrameric associations of two basic subunits, and the degree of cross-reactivity of these isozymes with antisera to each of the subunits is demonstrated for chicken in Table 8. In the case of LDH there already exists a rapid and specific enzymic method for measuring the heart-specific isoenzyme (H_4),[23] although this to some extent also measures M_1H_3.

Table 7. Examples of Use of RIA for Isoenzymes

Author	Isoenzymes
Headings and Tashian[52]	Carbonic anhydrase I or II
Samloff[44]	Group I pepsinogens
Jacoby and Bagshawe[59]; Iino *et al.*[58]	Placental alkaline phosphatase
Nicholson and O'Sullivan[35]	MM isoenzyme of CPK
Cooper and Foti[65]	Prostatic acid phosphatase
Geokas *et al.*[33]	Pancreatic carboxypeptidase B type II
Geiger *et al.*[20]	Hexosaminidase A or B

Table 8. Precipitation of Various Forms of Chicken H LDH with a Limiting Amount of Antibody

	Precipitation (%)	
	Anti-M	Anti-H
M_4	100	0
M_3H_1	65	17
M_2H_2	32	58
M_1H_3	8	95
H_4	0	100

5.3. Practicality

Many factors (Table 9) influence measurement of catalytic concentrations of enzymes. Very different amounts of substrate are reacted at 25°, as compared with 30° or 37° (which comprise some of the recommended temperatures). Indeed, an instrument which fluctuates from its preset temperature by one degree may lead to an error of some 7.5%. Other variables include the identity and concentration of the substrate, coenzymes (such as NAD and pyridoxal phosphate), and activators (such as the divalent metals, magnesium and zinc). The hydrogen ion concentration and ionic strength of the reaction mixture markedly influence results, as may the presence of enzyme inhibitors in the sample (drugs, anticoagulant, urea, oxalate, specific inhibitors), in reagents including water, on the glassware, or in the reaction mixture (for example, the products of the reaction). The concentration of enzymes employed in indicator and/or auxillary reactions for coupled systems must be optimal. Finally the assay should be based on the initial linear first-order reaction rate (which necessitates sophisticated equipment to enable continuous or multipoint monitoring), and care must be taken with choice of cuvettes or the use of hemolyzed, pigment-containing or lipemic samples which may affect endpoint detec-

Table 9. Factors Influencing Measurement of the Catalytic Activity of Enzymes

Factors
Standard
Temperature
Hydrogen ion concentration and ionic strength of reaction mixture
Identity and concentration of substrate, coenzymes, activators, and inhibitors
Concentration of enzymes employed in indicator and/or auxillary reactions when coupled systems are employed
Factors (such as hemolysis or lipemia) effecting endpoint detection
Equipment

tion. It is hardly surprising, therefore, that difficulties are experienced in ensuring reproducibility of results "within" and "between" laboratories and in recommending internationally accepted methods.

Antibody–antigen reactions appear to be more robust than enzyme–substrate interactions. Thus acid-denatured bovine liver esterase reacts equally with antiserum to the native enzyme,[24] while antibodies to acetylcholinesterase not only protect it from heat denaturation but also, by combining with the previously heat-denatured enzyme, can reactivate its catalytic capability.[25] Kolb[26] found that fructose-1,6-diphosphatase denatured by urea or guanidine hydrochloride had zero enzymic activity but measured by RIA gave standard curves identical with the native enzyme.

Immunoassays appear able to determine enzyme levels in the presence of inhibitors. Thus, while it has been stated that "naturally occurring inhibitors in serum do not seem to be present in concentrations high enough to have a significant influence on the *in vitro* assay of enzymes,"[27] there is now sufficient evidence in the literature to refute this statement.

Among the most elegant demonstrations of the effects of inhibitors are those by Temler and Felber.[28] In one study (Table 10) they obtained similar porcine trypsin levels in buffer, using the two analytical techniques, whereas there was a complete loss of biological activity only following the addition of plasma. Temler and Felber[29] also showed (Figure 3) that addition of Trasylol or DFP impaired the catalytic activity of trypsin without significantly affecting its immunological activity. A further example of such a dissociation was provided by Carballo and his colleagues,[30] who demonstrated that the induction of acute pancreatitis in dogs resulted in a rapid and marked rise in plasma elastase levels, as determined by RIA, but an immediate fall, as determined by bioassay. They concluded that this discrepancy was probably due to activation or augmentation of circulating enzyme inhibitors. Indeed, one use of a combination of bioassay and immunoassay is to accurately determine the levels of some inhibitors and, thereby, assess whether they play a role in disease.

Table 10. Porcine Trypsin Activity and Concentration Measured in Both Buffer and Plasma[28]

In buffer		In 1/5 dilution of plasma	
RIA (μg/ml)	Bioassay	RIA (μg/ml)	Bioassay
20	20	20	0
10	10	10	0
5	5	5	0
2	2	2	0

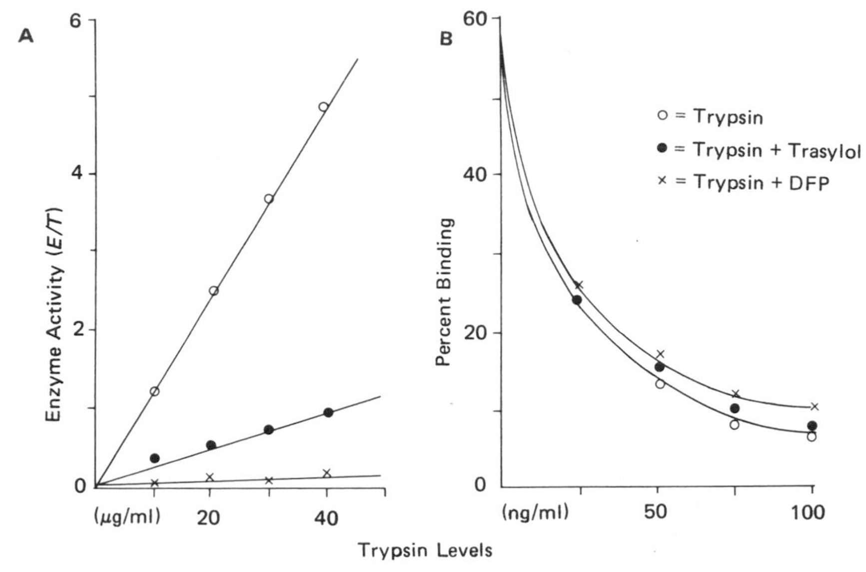

Figure 3. Comparison of bioassay and RIA of trypsin in the presence of enzyme inhibitors: (A) bioassay; (B) RIA.

In contrast, a direct linear relationship has been demonstrated, in our laboratory, between serum amylase concentration and activity, even during an attack of acute pancreatitis (Figure 4). This suggests the absence of any inhibitors or inactive enzyme fragments, and measurement of serum amylase activity provides, therefore, a direct assessment of concentration. Nonetheless, in view of the unreliable nature of many of the activity assays in routine clinical use for amylase, it is possible that RIA may become the method of choice for this determination also.

6. Examples of the Advantages of Enzyme Measurement by Concentration

6.1. Serum Levels of Proteolytic Enzymes

Until recently it was uncertain whether the apparent absence of certain proteolytic enzymes (e.g., pancreatic trypsin) in serum samples reflected their failure to enter the circulation or the presence of inhibitors which prevented their determination by catalytic assays. It is now known that the latter is correct, and the development of RIA to determine serum levels of such enzymes opens up new and exciting possibilities in, for example, gastroenterology.

This is illustrated by the work of Samloff and Liebman,[31] who developed and applied a specific RIA for the group I pepsinogens. These are limited to the chief and mucus neck cells in the oxyntic-gland mucosa of the stomach, while group II pepsinogens are also present in the pyloric gland area and in Brunner's glands. Thus serum levels of the former would be expected to correlate more closely with oxyntic gland function,[32] and this assay enabled excellent differentiation between control subjects and those with pernicious anemia or following total gastrectomy on the one hand and patients with Zollinger–Ellison syndrome on the other. More recently it has been found that pepsinogen 1 levels are elevated in patients with unoperated duodenal ulcer. There is a positive correlation between serum concentration and gastric peak acid output, and both are elevated in patients with a tendency to recurrent ulceration.[32] All of these findings may have diagnostic significance. A radioimmunoassay for human pancreatic carboxypeptidase B, by Geokas and his colleagues,[33] differentiated between patients with acute pancreatitis and control subjects. Of more

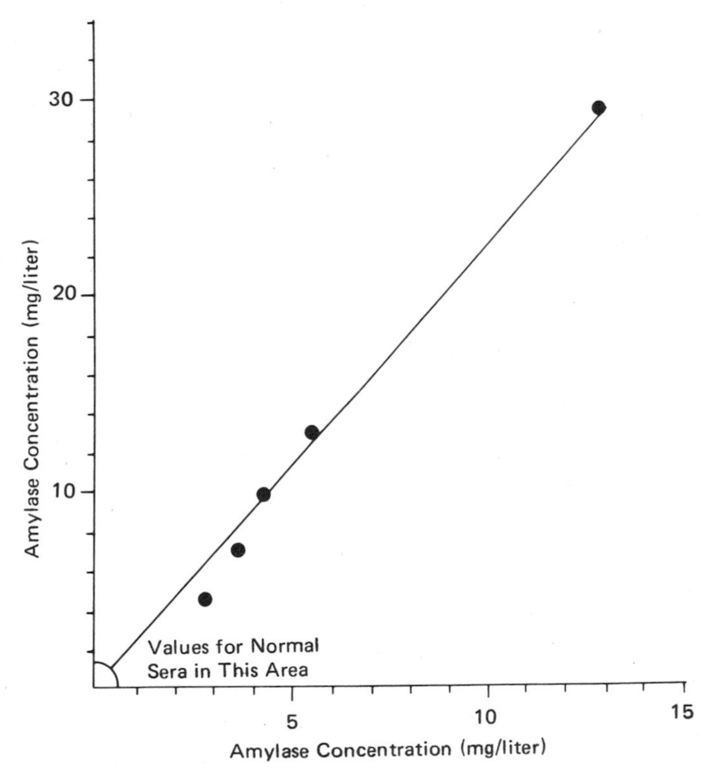

Figure 4. Correlation of activity and concentration for human serum amylase. (From reference 5.)

importance, the assay could be employed as the basis of a dynamic test of pancreatic function (Figure 5) to determine the serum carboxypeptidase response to the injection of secretin.

For obvious reasons most proteases (e.g., pepsin, trypsin, chymotrypsin, carboxypeptidase, thrombin, and plasmin) circulate or are stored mainly in the inactive proenzyme or zymogen state. Conversion to the active state involves cleavage of peptide bonds. Activity measurements of proteases produce no information on the amount of zymogen present. The ability of RIA to distinguish between precursor and enzyme appears to depend to a large extent on the size of peptide removed in the activation

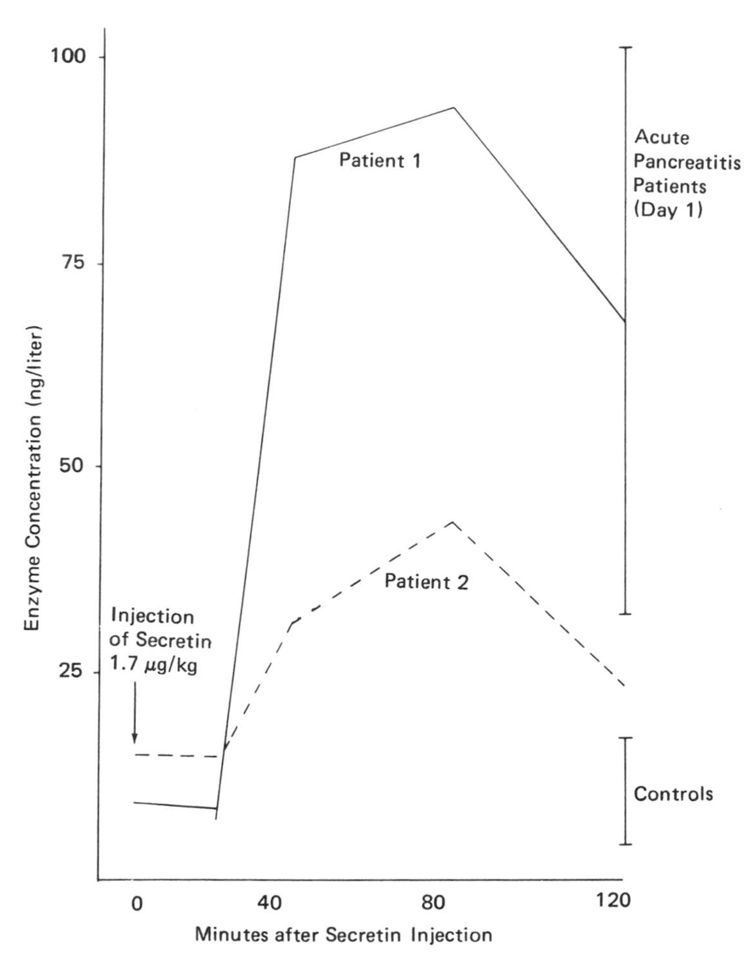

Figure 5. Serum carboxypeptidase B levels following secretin injection in two patients with acute recurring pancreatitis. (After Geokas *et al.*, reference 33.)

Table 11. Degree of Immunological Cross-reactivity between Proteases and Their Proenzymes Compared with Amount of Proenzyme Removed during Activation[a]

Enzyme	Cross-reaction of antiserum to enzyme with zymogen	Zymogen removed during activation (%)
Trypsin	+++	3
Chymotrypsin	+++	8
Plasmin	+++	16
Pepsin	−	21
Thombin	−	46
Carboxypeptidase B	−	50
Carboxypeptidase A	−	63

[a] Adapted from Aron, reference 2.

process. Thus antibodies against carboxypeptidase A do not appear to cross-react with procarboxypeptidase A of which the enzyme itself is only a small fraction. It is much more difficult to differentiate immunologically between trypsin and trypsinogen, where the activation stage involves merely the removal of a hexapeptide. Immunological relationships between some proteases and their zymogens are shown in Table 11.

6.2. Enzyme Determinations for the Diagnosis of Myocardial Infarction

The determination of serum levels of aspartate aminotransferase (AST), lactic dehydrogenase (LDH), and creatine phosphokinase (CK) has proved of value in the diagnosis of myocardial infarction. Later developments have included the assay of the LDH isoenzymes 1 and 2 employing, for example, 2-oxobutyrate as substrate, and the assay of the MB fraction of CK by procedures which usually involve initial separation of the MB and MM fractions by column chromatography. A recent report has indicated that an even more specific diagnosis of myocardial infarction may be made by measurement of serum enolase and pyruvate kinase.[34]

It can be argued that the continued use of catalytic assays in the above and in related situations illustrates the general maxim that "thought is dominated by existing methodology rather than existing methodology by thought." Since these enzymes are being employed as tissue markers, it is more appropriate to first assess the criteria for a suitable marker (Table 12). Early diagnosis is greatly facilitated if the marker is not normally detectable in the circulation and an extremely sensitive and precise analytical technique is employed.

Release of the marker should be specific to the tissue being assessed and also to the type of lesion. Thus it should not be present in tissues other

Table 12. Biochemical Investigation of Suspected Myocardial Infarction—Criteria for Suitable Marker[a]

Criteria
Normally undetectable in circulation
Specific for myocardial destruction; not being released due to increased permeability; not present in other tissues
Circulating levels related directly to size of infarction and reflecting further progression; clearance from the circulation should be unaffected by renal, hepatic, or other disorders
Availability of an extremely sensitive, precise analytical technique

[a] To establish with certainty whether the patient has had an infarction, its extent, and evidence of further infarction.

than the heart, nor should it be released as a result of increased membrane permeability, but only if cellular destruction has occurred.

Assays based on the catalytic activity of enzymes have, to date, failed to meet these criteria and also fail to determine the presence of proenzymes, enzyme subunits, or enzyme fragments, despite the fact that these would be equally effective markers.

An RIA has been developed for the MM isoenzyme of CK, but it does not differentiate between skeletal muscle damage and myocardial damage because this isoenzyme is common to both tissues.[35] The cross-reactivity between the MB isoenzyme and antisera raised against the MM and BB isoenzymes, which has been demonstrated by Jockers-Wretou and Pfleiderer,[6] would suggest that it may prove extremely difficult to produce an antiserum totally specific for MB. Hence, the RIA for MB, as produced by Roberts *et al.*,[36] involves the use of an antiserum raised against BB which cross-reacts with the cardiac hybrid isozyme but not the MM form. Measurements made in this assay assume all B subunits are derived from the MB isozyme. However, earlier beliefs that BB was virtually absent from the circulation have recently been challenged.[37] While RIA may resolve this issue, the presence of basal BB levels in samples might complicate interpretation. Another approach would be to establish an RIA specific for the AST isoenzyme of the mitochondria (Mito-GOT) as opposed to that found in cytoplasm (Cyto-GOT), to avoid the problem of increased membrane permeability.

Such approaches, however, reflect the natural tendency to continue along established pathways. Since a catalytically active enzyme is no longer required for the analysis, one can employ RIA for a chosen specific myocardial protein, not necessarily an enzyme or an enzyme subunit. This would be analogous to the RIAs being developed for specific fetal or placental proteins for the assessment of fetoplacental well-being or for tumor-associated proteins. A concerted effort to establish immunoassays for specific myocardial, renal, hepatic, and other tissue proteins (in preference to enzyme assays) could well yield dividends in clinical chemistry.

7. Conclusions

The present paper first emphasized the difference between catalytic assays for enzymes, which assess function, and immunoassays, which assess structure. More than 40 papers relating to the determination of enzyme levels by their concentration had been published by the end of 1975 and these have been reviewed. Based on this data a case has been advanced:

1. That bioassays will remain the technique of choice, in clinical chemistry, for those enzymes such as plasmin which effect their physiological role in the biological fluid in which they are being determined
2. That the combination of bioassay and immunoassay may prove of value in many basic biochemical studies
3. That immunoassays may partially and gradually replace catalytic enzyme assays for the many enzymes which do not effect their physiological role in the biological fluid in which they are being determined

Such assays will have particular relevance for isoenzymes and for those enzymes for which circulating inhibitors exist. Finally, it is possible that this approach will enable the development of assays for tissue markers other than enzymes.

8. References

1. B. Cinader, *Ann. N.Y. Acad. Sci., 103,* 495 (1963).
2. R. Arnon, Immunochemistry of enzymes, in: *The Antigens* (M. Sela, ed.), Vol. 1, pp. 87–159, Academic Press, New York (1973).
3. H. Luers and F. Albrecht, *Fermentforschung, 8,* 52 (1924).
4. R. L. McGeachin and J. M. Reynolds, *J. Biol. Chem., 234,* 1456 (1959).
5. J. A. Carney, *Clin. Chim. Acta, 67,* 153 (1976).
6. E. Jockers-Wretou and G. Pfleiderer, *Clin. Chim. Acta, 58,* 223 (1975).
7. R. Arnon and H. Neurath, *Proc. Natl. Acad. Sci. U.S.A., 64,* 1323 (1969).
8. D. J. Langley and J. A. Carney, *Comp. Biochem. Physiol., 55B,* 563 (1976).
9. M. R. Pollack, *Immunology, 7,* 707 (1964).
10. J. H. Keutel, K. Okabe, H. K. Jacobs, F. Ziter, L. Maland, and S. A. Kuby, *Arch. Biochem. Biophys., 150,* 648 (1972).
11. E. Maron, C. Shiozawa, R. Arnon, and M. Sela, *Biochemistry, 10,* 763 (1971).
12. R. K. Brown, R. Delaney, L. Levine, and H. Van Vunakis, *J. Biol. Chem., 234,* 2043 (1959).
13. J. P. Ryan, H. E. Appert, J. Carballo, and R. H. Davies, *Proc. Soc. Exp. Biol. Med., 148,* 194 (1975).
14. J. Carballo and R. Troiano, *Fed. Proc., 32,* 514 (abstract) (1973).
15. H. J. Kolb and G. M. Grodsky, *Proc. Soc. Exp. Biol. Med., 137,* 464 (1971).

16. N. Taniguchi, T. Kondo, N. Ishikawa, H. Ohno, E. Takakuwa, and I. Matsuda, *Anal. Biochem., 72,* 144 (1976).
17. J. G. Ratcliffe, *Br. Med. Bull., 30,* 32 (1974).
18. L. A. Moroz and N. J. Gilmore, *Nature, 259,* 235 (1976).
19. J. B. Wyngaarden, *Am. J. Med., 56,* 651 (1974).
20. B. Geiger, R. Navon, Y. Ben-Yoseph, and R. Arnon, *Eur. J. Biochem., 56,* 311 (1975).
21. H. J. Kolb and G. M. Grodsky, *Biochemistry, 9,* 4900 (1970).
22. R. C. Karn, B. B. Rosenbloom, J. C. Ward, A. D. Merritt, and J. D. Shalkin, *Biochem. Genet., 12,* 485 (1974).
23. S. B. Rosalki and J. H. Wilkinson, *J. Am. Med. Assoc., 189,* 61 (1964).
24. S. Bauminger and L. Levine, *Biochim. Biophys. Acta, 236,* 639 (1971).
25. D. Michaeli, J. D. Pinto, E. Benjamini, and F. P. de Buren, *Immunochemistry, 6,* 101 (1969).
26. H. J. Kolb, *Eur. J. Biochem., 43,* 145 (1974).
27. J. Mattenheimer, *The Theory of Enzyme Tests,* Boehringer, Mannheim, Germany (1972).
28. R. S. Temler and J. P. Felber, *Horm. Metab. Res. Suppl., Series 5* 17 (1974).
29. R. S. Temler and J. P. Felber, *Biochim. Biophys. Acta, 236,* 78 (1971).
30. J. Carballo, K. Kasahara, E. Appert, and J. Howard, *Proc. Soc. Exp. Biol. Med., 146,* 997 (1974).
31. I. M. Samloff and W. H. Liebman, *Gastroenterology, 66,* 145 (1974).
32. I. M. Samloff, D. M. Secrist, and E. Passaro, *Gastroenterology, 70,* 309 (1976).
33. M. C. Geokas, F. Wollesen, and H. Rinderknecht, *J. Lab. Clin. Med., 84,* 574 (1974).
34. M. Herraez-Dominguez, D. M. Goldberg, A. J. Anderson, J. S. Fleming, C. C. Rider, and C. B. Taylor, *Enzymes, 21,* 211 (1976).
35. G. A. Nicholson and W. J. O'Sullivan, *Proc. Aust. Assoc. Neurol., 10,* 105 (1973).
36. R. Roberts, B. E. Sobel, and C. W. Parker, *Science, 194,* 855 (1975).
37. A. C. Byrnes and A. Sheldon, *Clin. Chem., 21,* 1845 (letter) (1975).
38. R. S. Temler and J. P. Felber, 7th Int. Congr. Chem., Geneva, Evian (1969).
39. R. S. Temler and J. P. Felber, Hormones, Lipids, and Miscellaneous, Vol. 3, p. 267, Karger, Basel (1970).
40. R. S. Temler and J. P. Felber, *Immunochemistry, 7,* 875 (abstract) (1970).
41. R. S. Temler and J. P. Felber, IXth Int. Congr. on Clin. Chem., Toronto, Canada, pp. 13–18 (1975).
42. R. M. Stroud, M. Bracco, and K. Yonemasu, *Fed. Proc., 29,* 303 (abstract) (1970).
43. R. M. Stroud, *J. Lab. Clin. Med., 77,* 713 (abstract) (1971).
44. I. M. Samloff, *Gastroenterology, 60,* 713 (abstract) (1971).
45. R. A. Rush, S. H. Kindler, and S. Udenfriend, *Clin. Chem., 21,* 148 (1975).
46. R. A. Rush, P. E. Thomas, T. Nagatsu, and S. Udenfriend, *Proc. Natl. Acad. Sci. U.S.A., 71,* 872 (1974).
47. R. A. Rush, P. E. Thomas, and S. Udenfriend, *Proc. Natl. Acad. Sci. U.S.A., 72,* 750 (1975).
49. L. Tuderman, E. R. Kuuti, and K. I. Kivirikko, *Eur. J. Biochem., 60,* 399 (1975).
48. H. J. Wedner, L. N. Parker, and M. G. Rosenfeld, *Anal. Biochem., 65,* 175 (1975).
50. C. G. Strong, H. R. Tapia, V. R. Walker, and J. C. Hunt, *J. Lab. Clin. Med., 79,* 170 (1971).
51. S. F. Rabiner, I. D. Goldfine, A. Hart, L. Summaria, and K. C. Robbins, *J. Lab. Clin. Med., 74,* 265 (1969).
52. V. E. Headings and R. E. Tashian, *Biochem. Genet., 4,* 285 (1970).
53. E. Magid, J. de Simone, and R. E. Tashian, *Biochem. Genet., 8,* 157 (1973).
54. S. S. Rayyis and M. C. Geokas, *Gastroenterology, 58,* 986 (abstract) (1970).
55. M. C. Geokas and S. S. Rayyis, *Gastroenterology, 60,* 664 (abstract) (1971).

56. E. A. Bauer, A. Z. Eisen, and J. J. Jeffrey, *J. Biol. Chem.*, *247*, 6679 (1972).
57. A. Z. Eisen, G. P. Nepute, G. P. Striklin, E. A. Bauer, and J. J. Jeffrey, *Biochim. Biophys. Acta*, *350*, 442 (1974).
58. S. Iino, K. Abe, T. Oda, H. Suzuki, M. Sugiura, *Clin. Chim. Acta*, *42*, 161 (1972).
59. B. Jacoby and K. D. Bagshawe, *Cancer Res.*, *32*, 2413 (1972).
60. C. H. Chang, S. Raam, D. Angellis, G. Doillgart, and W. H. Fishman, *Cancer Res.*, *35*, 1706 (1975).
61. R. A. Rush and L. B. Geffen, *Circ. Res.*, *31*, 444 (1972).
62. R. P. Ebstein, D. H. Park, L. S. Freedman, S. M. Levitz, T. Chuchi, and M. Goldstein, *Life Sci.*, *13*, 769 (1974).
63. T. Kashimoto, D. Park, R. P. Ebstein, M. Goldstein, M. Levitz, and S. Yaverbaum, *Experientia*, *30*, 1363 (1974).
64. Y. Ben-Yoseph and E. Shapira, *J. Lab. Clin. Med.*, *81*, 133 (1973).
65. J. F. Cooper and A. Foti, *Invest. Urol.*, *12*, 98 (1974).
66. A. Foti, H. Herschman, and J. F. Cooper, *Cancer Res.*, *35*, 2446 (1975).

5

Clinical Liquid Chromatography

L. R. Snyder, B. L. Karger, and R. W. Giese

1. Introduction

Liquid chromatography (LC) has undergone considerable development during the past decade, both in column design and in instrumentation. Although further advances in performance are to be expected, current methodology is already far enough advanced to ensure a place for LC in the clinical laboratory. Consider, for example, just a few of its present capabilities: (1) assay rates for clinical determinations of 5–10/h (with further increases in assay rate expected); (2) good separations of compounds of interest from the complex mixtures that constitute serum, urine, and other body fluids; and (3) detection limits in favorable cases at the picogram level.

Present applications of LC in the clinical laboratory have been made possible by using (1) highly efficient columns packed with small (5 to 10-μm) particles, (2) high-pressure, precise pumps for fast and reproducible separation, (3) sample valves for easy and precise injection of sample, and (4) a variety of detection modules and techniques that offer high sensitivity and minimal extra-column band broadening. In the present chapter we will review not only the applications of LC in the clinical laboratory, but the technique itself from the standpoint of the special needs of clinical analysis. The more general aspects of LC have been discussed in a number of basic texts (e.g., references 1a–d). Detectors for LC and especially clinical LC are discussed in detail in Chapter 3 of this volume.

L. R. Snyder • Technicon Instruments Corporation, Tarrytown, New York 10591
B. L. Karger and R. W. Giese • Northeastern University, Boston, Massachusetts 02115

1.1. Present Importance of Clinical LC

We anticipate that modern LC will have an important over-all impact on the practice of clinical chemistry. This is largely because LC offers a reasonably simple, inexpensive, and rugged capability to satisfy many of the current needs of clinical analysis, including (1) reagent characterization, (2) research, (3) reference samples, (4) reference methodology, (5) genetic screening, (6) pediatric chemistry, (7) specialty testing, and (8) routine analysis. Some indication of this has begun to appear in the large and exponentially growing literature in this field. Nearly twice as many articles on LC were published in the journal *Clinical Chemistry* in 1977 as in the previous year. Review chapters[2,3] and a book[4] on clinical LC have also been published.

In this chapter we will examine the potential impact of LC on clinical chemistry in some detail. This review will include general considerations, such as which equipment and techniques will become most useful for clinical analysis, and a discussion of applications from the recent literature (mainly since January 1976, in order to minimize overlap with previous reviews). We will be selective, rather than comprehensive, and attempt to indicate areas of greatest current interest and future potential. Of the general areas of clinical chemistry, specialty testing and routine analysis will receive major emphasis.

1.2. Comparison and Competition of LC with Alternate Techniques

Special challenges exist in clinical chemistry for the introduction of any new technique. Many previous procedures have failed, either because they provided too much information, or because there was a significant difference in their performance in the hands of experts vs. their use in actual practice. Isoelectric focusing is a good example of a methodology failing to gain general acceptance in clinical chemistry laboratories for both of these reasons. These considerations should not limit modern LC, however, because it is easy to use and its conditions are readily altered over wide limits to give the specific information desired.

An obvious barrier to the future acceptance of clinical LC is the tendency to use and retain methodology that "works," even when more reliable techniques become available. There are good reasons for this. A less-reliable procedure may be preferred because it is simpler, faster, and/or less expensive. Also, considerable effort is often involved in any changeover to a new procedure, including troubleshooting, correlation of old vs. new methods, and rechecking of normal ranges.

Under these circumstances, one must at present look for applications where current methodology such as colorimetry, electrophoresis, immunoassay, and other forms of chromatography have obvious shortcomings which LC might overcome. As it turns out, there are many such potential applications of LC. The section on applications will provide greater detail; however, a few general comments are in order. Consider that: (1) colorimetric procedures generally are widely susceptible to interferences; (2) electrophoresis has remained a demanding and semiquantitative technique; (3) immunoassays can be expensive and are limited in their ability to distinguish within certain groups of drugs, drug metabolites, isoenzymes, and alternate or closely related forms of many of the hormones; (4) gas chromatography (including GC–mass spectrometry) is limited to the analysis of thermally nonlabile, nonionic, relatively low-molecular-weight materials (with derivatization offering only a partial relaxation of these requirements); and (5) tedious, classical chromatographic procedures, such as thin-layer chromatography and classical LC, are widely employed in clinical chemistry, but are slower, less efficient, and less quantitative than modern LC. Thus, major opportunities for clinical LC do exist.

Besides its potential as an analytical tool per se, LC also can be used at present as a cleanup procedure prior to application of some other analytical technique. For example, cross-reactivity and/or inhibition in competitive-binding assays can be decreased by a preliminary LC separation of the sample, e.g., 17α-hydroxyprogesterone in plasma by competitive protein binding[5] and melatonin in urine by RIA,[6] each after reverse-phase LC.

1.3. Advantages and Limitations of LC

The following list abbreviates (and therefore oversimplifies) the advantages of LC, but gives a general idea of the reasons for its broadly established and rapidly increasing popularity. LC

1. offers rapid, high-sensitivity, quantitative analysis
2. can analyze for several substances in the same run
3. can utilize the advantages of internal standards
4. usually separates interferences, but at a minimum can reveal their presence in almost all cases
5. overcomes the fundamental sample limitations (molecular size, charge, thermal stability) of GC
6. often avoids the derivatization requirements of GC
7. utilizes a mobile phase (liquid) which is active in the separation and whose properties can be changed, thus allowing much more separation capability than in GC where the mobile phase (gas) is inert

8. interfaces smoothly with on-line chemical steps before and after the column to facilitate automation of pretreatment and detection steps
9. readily incorporates valving arrangements such as parallel columns to multiply sample throughput and chromatographic capability

On the other hand, no technique is without limitations. LC

1. can only do one sample at a time
2. often requires the pre-column cleanup of clinical specimens
3. has less detection capability overall than GC
4. requires further improvement in the reproducibility and stability of commercial columns

2. General Considerations

In this section we will provide a brief discussion of the general technique of liquid chromatography, particularly with reference to factors that play an important role in clinical applications. This includes a discussion of LC columns and equipment and various peripheral techniques or concerns of special interest in clinical analysis. Further discussion of some of these areas is interwoven into the following section on clinical applications of LC.

2.1. Columns

The principles governing separation by LC are now well established and have been described in detail elsewhere (e.g., reference 1a). However, a brief summary of some important points is appropriate, especially as these relate to clinical analysis. In particular, clinical analysis often requires rapid sample turnaround and the ability to process large numbers of assays in a cost-effective manner. Therefore, the minimization of separation time in clinical LC is of prime importance.

The relative separation or resolution of two adjacent chromatographic bands can be defined by the function R_s, which is in turn related to the conditions of separation:

$$R_s = \frac{1}{4}\underbrace{[k'/(1 + k')]}_{\text{(i)}}\underbrace{N^{1/2}}_{\text{(ii)}}\underbrace{(\alpha - 1)}_{\text{(iii)}} \qquad (1)$$

R_s should equal at least 1.0 for quantitation by peak-height measurements, and in general it is best if R_s is at least 1.25. R_s can be adjusted for adequate separation by varying the three factors (i–iii) in turn. Term (i) is usually controlled by varying the strength of the mobile phase through changes in its composition. For every separation there is an optimum value of the capacity factor k', which is usually between 2 and 5. Term (ii) can be maximized by attention to the column plate number N, a measure of how efficiently the column separates samples in general. For very fast analysis rates, as in clinical LC, it is essential to generate N values of 500–3000 while holding separation time to a few minutes. Term (iii) is determined by the separation factor α or the selectivity of the LC system. Values of α are varied by changing the composition of the mobile phase or (less frequently) the stationary phase or temperature. Here we will focus on the role of the column in providing adequate resolution for clinical separations.

2.2. Phase Systems

At present there are a variety of phase systems available for high-performance LC. At first glance, the selection of mobile and stationary phase conditions for a given separation may appear complex. Fortunately, there is a present trend toward the use of a single stationary-phase type—chemically bonded *n*-alkyl reverse-phase packings[7]— for the majority of clinical chemistry applications. Variation of the mobile phase in turn permits a very broad range of substances to be chromatographed. The use of a single stationary phase for most applications offers simplicity and convenience. In this section we will discuss the most commonly used phase systems and explore the well-earned populatiry of the reverse-phase packings.

2.3. Adsorbents

The polar adsorbents silica and alumina were among the earliest stationary phases employed in LC.[1a] Both of these adsorbents are available from several commercial suppliers, in either packed columns or as the free packing. The small-particle supports (5–10 μm in diameter) are available either as irregular or spherical particles. Both particle types appear to give similar column performance; however, spherical particles generally result in somewhat higher permeability (i.e., lower pressure drop) and may provide more stable columns.

The retention mechanism and the structural selectivities possible with polar adsorbents are now well understood.[8] These adsorbents work best for nonpolar or moderately polar substances. Ionic compounds are generally difficult to chromatograph (however, see reference 9). Polar adsorbents can resolve substances based on differences in the positions of atoms or groups within the molecule. Thus, these packings are particularly useful in the separation of isomeric compounds (e.g., ortho, meta, para isomers).[8] In addition, functional-group selectivity can also be achieved.

For several reasons, polar adsorbents are becoming less popular relative to the bonded packings for reverse-phase LC. First, control of the mobile phase for separations on silica and alumina is more critical; traces of polar solvents (e.g., water) can greatly affect retention.[8] This is less true of the bonded reverse-phase packings. Second, the time necessary for a solvent change can be excessive, due to slow equilibration of mobile phase and adsorbent. This can increase the time required to carry out gradient elution, or to change from one mobile phase to another. Finally, and most importantly, polar and ionic impurities elute very slowly from adsorbent columns, in contrast to their rapid elution from reverse-phase columns. As a result these impurities tend to accumulate on adsorbent columns, gradually degrading their performance.

2.4. Bonded Reverse-Phase Packings

A variety of chemically-bonded reverse-phase packings are commercially available. Their synthesis involves the reaction of activated silica with reagents such as alkyldimethylchlorosilanes:

$$(O)(O)\!\!>\!\!Si{-}OH + R{-}Si(CH_3)_2{-}Cl \rightarrow (O)(O)\!\!>\!\!Si{-}O{-}Si(CH_3)_2{-}R$$

R is most often C-18 (octadecyl) but also can be C-2, C-4, or C-8. A hydrophilic surface (silica) is thereby converted to a hydrophobic one. Substances elute in a general order of decreasing polarity, and mobile-phase strength increases with decreasing polarity (e.g., tetrahydrofuran is a stronger mobile-phase solvent than water). This is the "reverse" of the situation in adsorption chromatography. Mobile phases typically used in reverse-phase chromatography consist of miscible water/organic solvent mixtures.[7]

We should distinguish between monomeric (bristle) phases, as illustrated by the above alkyldimethyl surfaces, and polymeric phases formed when di- or trichlorosilanes (or alkoxysilanes) react in the presence of water.[7] Monomeric phases are preferred for their faster mass-transfer kinetics (i.e., higher performance columns).[10]

An important aspect of bonded-phase packings is the accessibility of remaining silanol groups on the silica surface. For steric reasons, it is impossible for all of these silanols to react initially, and new silanol groups can be formed by hydrolysis of chloro groups when di- or trichlorosilanes are used for bonding. These silanols, if accessible to molecules of mobile phase or sample, can adversely affect both the performance and stability of the packing. Thus they may cause nonreproducible retention (from packing to packing), and they may also accelerate the removal of the bonded-phase layer by mobile phase especially in aqueous salt solutions of higher pH.[11] Therefore, maximum column lifetime (an important consideration in clinical LC) is favored by a minimization of accessible silanol groups—or by maximization of the coverage of the silica surface by bonded phase.[11-13] Depending on the chain length of the alkyl group, surface coverages of 3–4 μmol/m^2 represent maximum values.[11] Following the initial synthesis of a bonded-phase packing, a second silanization is often performed, in order to reduce further the number of exposed silanol groups.

Retention and selectivity in reverse-phase LC are now under intense study. Much empirical data is available from early work in paper chromatography (e.g., reference 14), and especially from more recent work on columns (e.g., reference 15). Since water is almost always a component of the mobile phase, and since its presence dominates the solvent properties of the mobile phase, hydrophobic interactions (e.g., references 16, 17) are quite important. Compounds of increasing polarity or increasing molecular weight generally elute later.

The variation of mobile-phase selectivity (for change in α values) in reverse-phase LC was for a long time believed to be rather limited. However, recent studies (e.g., references 18, 19) have shown that binary and especially ternary solvents as mobile phase can provide a wide variation in sample α values. Figure 1 illustrates the shift in relative retention that can be achieved by change of mobile phase in reverse-phase LC.

Reverse-phase chromatography can also be a powerful tool for the separation of ionic or ionizable substances, and it competes strongly with ion-exchange chromatography. Generally, in the reverse-phase mode, ionic substances elute more rapidly than corresponding neutral species. For weak acids and bases, this means that pH can be a useful parameter in the separation of these substances. As an example, Figure 2 shows the change in capacity factor (k') as a function of pH for biogenic amines and their acid metabolites (reverse-phase LC). The acids are seen to elute more rapidly in basic media, whereas the bases elute more rapidly in acidic media. Ionization control is limited, however, as the bonded phases must be used between pH 2 and 8. Higher pH's attack the silica matrix, whereas lower pH's cause hydrolysis of the bonded phase.

For those substances that are strong acids or bases (i.e., maintain their ionic character throughout the pH range), as well as those substances which can be made ionic through pH control, an alternative means of separation is possible based on reverse-phase ion-pair chromatography.[20,21] A counterion of charge opposite to the ionic sample molecule is added to the mobile phase in order to retard the ionic species. The ion-pair process can

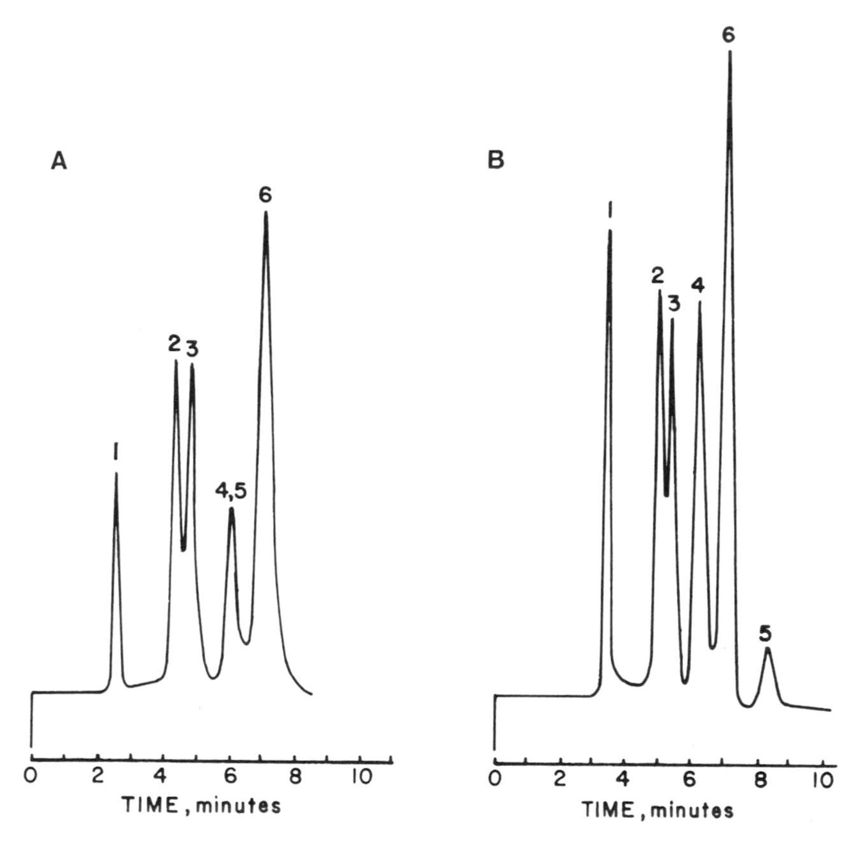

Figure 1. Reversed-phase separation of selected C_5 compounds. Column, 30 cm × 4.5 mm ID μBondapak C_{18}; Temperature, 25°; velocity, 0.33 cm/s; 1 = cyclopentanone, 2 = 3-pentanol, 3 = 2-pentanol, 4 = 3-methyl-1-butanol, 5 = methyl butyrate, 6 = *n*-pentanol. Detection, differential refractometer at 8×. (A) Mobile phase: *n*-propanol–water (15:85, v/v). (B) Mobile phase: acetonitrile–water (16:84, v/v). (Reprinted from reference 17 with permission.)

be written as shown in the equation below:

$$A^+_{aq} + B^-_{aq} \rightleftharpoons (A^+, B^-)_{org} \qquad (2)$$

where A^+_{aq} represents the ionic sample species to be retarded in the mixed aqueous mobile phase, and B^-_{aq} is the counterion added to this phase. For a given bonded phase, the extent of retention will be a function of the counterion structure and concentration, the composition and ionic strength of the mobile phase, and the pH of the mobile phase. *N*-Hexyl sulfonate and tetrabutylammonium have been typical anionic and cationic counterions, respectively. Ion-pair reverse-phase chromatography has become popular for the separation of ionic substances, since good efficiency is possible with reasonable control over retention.

As an extension of ion-pair chromatography, Knox has recently introduced soap chromatography, in which the counterion is a detergent molecule. Examples are cetyltrimethylammonium[9] and 1-dodecylsulfonate counterions.[22] In the reverse-phase mode, Knox has shown that there is a significant hydrophobic adsorption of the counterion onto the bonded phase, giving essentially a noncovalent ion exchanger. As seen in Figure 3 in the separation of catecholamine standards and derivatives, remarkably

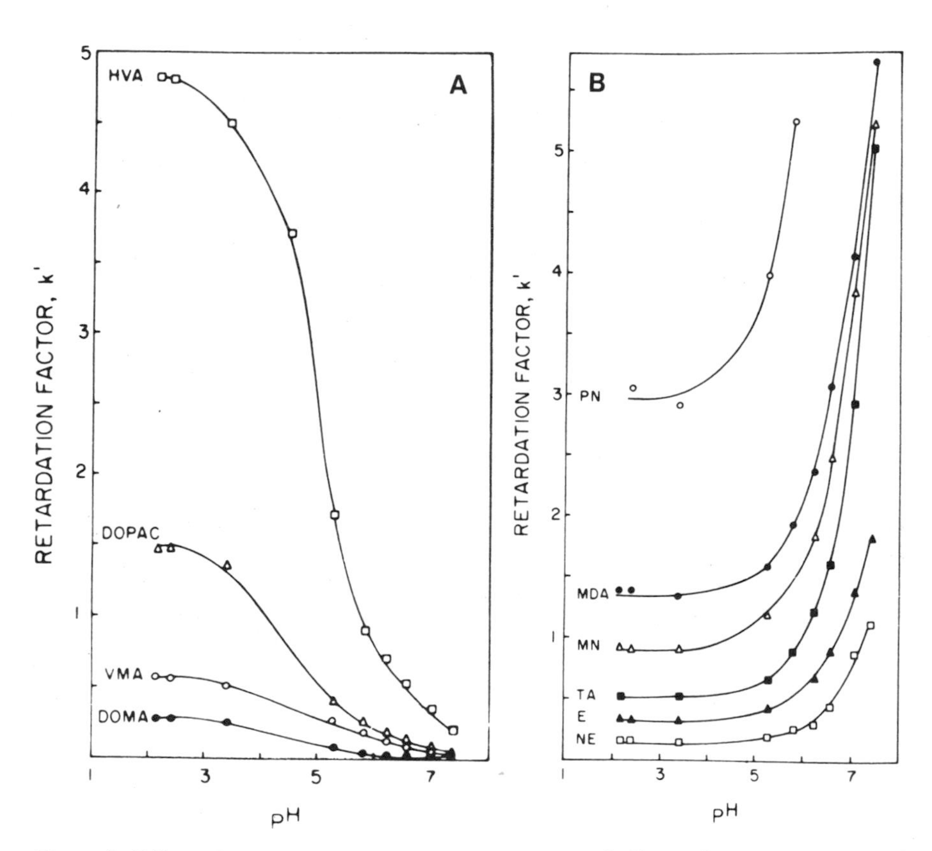

Figure 2. Effect of pH on (A) the retention of acidic metabolites and (B) the retention of biogenic amines. Column: Partsil 1025 ODS; eluent, 0.1 mol/liter phosphate buffers; flow rate, 1 ml/min; inlet pressure, 1500 psi (100 atm); temperature, 25°C. (A) DOPAC, 3,4-dihydroxyphenylacetic acid; HVA, homovanillic acid; VMA, vanilmandelic acid. (B) E, epinephrine; MDA, 3-*O*-methyl-dopamine; MN, metanephrine; NE, norepinephrine; PN, paranephrine; TA, tyramine. (Reprinted from reference 81 with permission.)

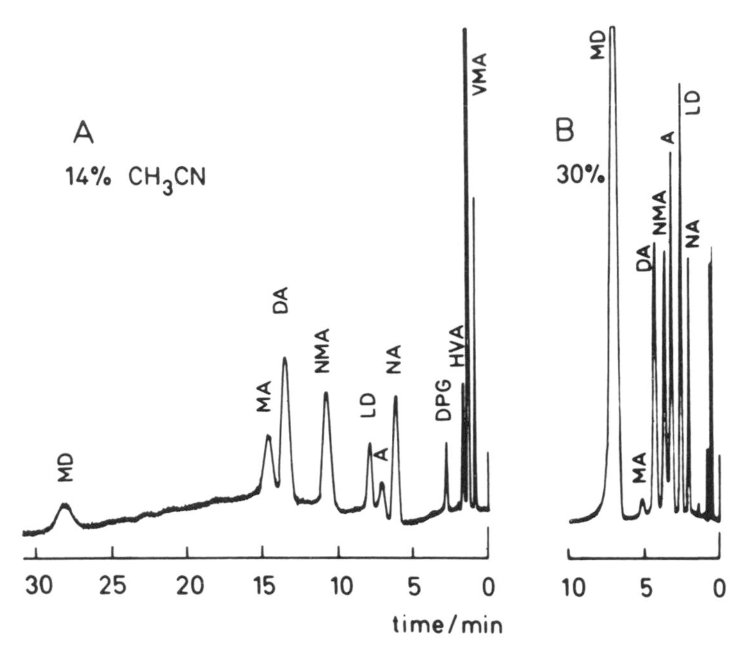

Figure 3. Soap chromatography of catecholamines and derivatives. Packing, ODS/TMS silica; detection, 280 nm, 0.01 absorption units full scale; eluent, water–acetonitrile–sodium lauryl sulfate–sulfuric acid, 70:30:0.02:0.04 (v/v/w/v). The peaks are MD (α-methyl dopa), MA (metadrenaline), DA (dopamine), NMA (normetadrenaline), A (adrenaline), LD (L-dopa), and NA (noradrenaline). The quantity of NA injected was around 20 ng. (Reprinted from reference 22 with permission.)

good separations can be achieved. Soap chromatography expands the scope of the reverse-phase packing to highly polar and ionic species that are ordinarily difficult to chromatograph with high efficiency. The detergent can be readily removed from the bonded phase, thereby returning the column to its original state for use in conventional reverse-phase LC.

Present reverse-phase LC applications are largely confined to the use of the bonded-phase packings. However, these packings are somewhat restricted by the narrow pH range (2–8) over which they can be used. Furthermore, there are other disadvantages of these materials. As an alternative, Guiochon and co-workers[23–25] are currently exploring the possibility of using graphitized carbon and silica particles as reverse-phase packings. While these materials have not yet been developed to the point where they offer comparable performance to the bonded-phase packings, eventually they might provide a much more stable substitute that can be used at both low and high pH's.

2.5. Polar Bonded Phases

Some other groups that have been bonded to the silica surface include:

$\equiv Si(CH_2)_3NH_2$ $\quad$ $\equiv Si(CH_2)_3OCH_2CH(OH)CH_2OH$

$\equiv Si(CH_2)_3CN$ $\quad$ $\equiv Si(CH_2)_3OCH_2CH(OH)CH_2NH_2$

These bonded phases provide a polarity which is intermediate between the alkyl-bonded phases for reverse-phase LC, and the polar silica surface for adsorption LC. The polar-bonded phases can be used in either a normal-phase (adsorption) mode or in a reverse-phase mode with water–organic solvent mixtures as mobile phase. In the normal-phase mode these packings resemble silica, but give weaker retention. The reverse-phase mode is of special interest, since polar, water-soluble compounds such as sugars and antibiotics can be separated.[26]

2.6. Ion-Exchange Packings

Ion-exchange resins based on porous polystyrene beads have found significant application in the clinical laboratory. These packings are still used in amino acid analyzers for the measurement of peptides and amino acids in physiologic fluids, such as urine, and for the high-resolution separation of other constituents of urine. However, these separations generally require several hours and are therefore not widely used. The older ion-exchange resins give relatively poor column efficiencies, and they are being superceded by bonded-phase ion exchangers of various types:

$$\equiv SiCH_2-C_6H_4-SO_3^-Na^+$$

$$\equiv SiCH_2-C_6H_4-{}^+NR_3Cl^-$$

$$\equiv Si(CH_2)_3NH_3^+Cl^-$$

Unfortunately, these materials have low exchange capacity (e.g., less than 1 meq/g vs. 5 or more meq/g on the resins) and poor stability, and so far they have not been very popular. Workers in clinical analysis will generally prefer one of the bonded-phase packings discussed above (non-ion-exchange alkyl or polar) for the separation of ionic species.

2.7. Liquid–Liquid Chromatography

In the late 60s and early 70s, many workers used pellicular packings coated with immiscible liquid phases for modern LC. These applications of liquid–liquid chromatography have now largely been superceded by the small-particle bonded-phase packings described above. Nevertheless, the liquid-coated packings still retain several advantages: (1) easy preparation of reproducible stationary phases, (2) easy replacement of the liquid phase for column regeneration or change in stationary phase, and (3) higher sample loadings in some cases.

Columns for liquid–liquid chromatography can be used in either normal-phase or reverse-phase modes. For normal-phase separations, silica packings of the same type used as adsorbents are employed, and the stationary phase is a polar liquid. For reverse-phase LC, silanized silica packings are employed, and the stationary phase consists of a relatively nonpolar liquid. The liquid stationary phase can be coated in one of two ways: by addition to the packing before filling the column, or by *in situ* coating of an already packed column (e.g., reference 1a). In addition, either heavily-loaded[28] or lightly-loaded[29] packings can be used.

Separations by liquid–liquid chromatography are based on solubility differences in the two-phase system. In addition, ion-pair chromatography is readily achieved both in normal-phase and reverse-phase modes. Both approaches are being used at the present time for trace analysis of drugs in physiological fluids (e.g., reference 30).

2.8. How Column N Values are Varied for a Given Application

The column plays a central role in any LC separation, determining both the separations that are possible and the speed at which a given assay can be carried out. As discussed above, separation and resolution R_s is determined by k', α, and N. The mobile phase is used mainly to control k' and α, while N is largely determined by the column and the available separation time t. The theory of column efficiency as a function of different experimental variables is now well understood (e.g., references 31–36 and review of reverence 1a), and experimental results for "good" columns are generally in close agreement with this theory. Therefore, a good understanding now exists on how N varies with experimental conditions. Here we will largely omit the various details of this relationship and instead concentrate on certain practical conclusions of interest to clinical analysis by LC.

Two general aspects of column efficiency will be considered. First, we will examine how the efficiency of a given column or column packing changes as we vary different experimental conditions. This will allow us to optimize the performance of a given column or column packing for a particular application. Second, we will look briefly at how N changes with the type of packing, and what limiting values of N are possible for a given separation time t and column pressure P.

Given a particular mobile phase and column type, we are usually faced with the problem of achieving some adequate final resolution of critical compounds in the sample. This means that we must be able to vary N in a predictable manner. There are basically three ways in which N can be increased:

1. Decrease the pressure drop P across the column (or decrease the mobile phase flow rate F, which is equivalent), while holding column length L constant.
2. Maintain P constant, while increasing L (e.g., add additional column lengths in series).
3. Increase L and P together, while holding separation time t constant.

The relative increase in N for a given change in L, P, or t as above is predictable, if we specify our initial experimental conditions (see discussion of reference 37). For purposes of illustration, we will assume a representative set of conditions (reverse-phase separation at 25°, 25-cm column of 10-μm particles, t_0 = 100 s). If our starting value of N is N_1, let the value be N_2 after some change in conditions. Table 1 illustrates the dependence of N_2/N_1 on P (L constant), L (P constant), or L and P (t constant).

For the case of varying P with L constant, which involves only a change in pump flow rate, an increase in N by a factor of 3 is seen to

Table 1. Dependence of N_2/N_1 on P, L, or L and P

N_2/N_1	P_2/P_1 (L const.)	L_2/L_1 (P const.)	L_2/L_1 (t const.)
0.2	7	0.4	0.1
0.5	2.5	0.7	0.3
1.0	1.0	1.0	1.0
1.5	0.5	1.3	4
2	0.3	1.5	~50
3	0.1	2.0	—[a]
5	—[a]	2.7	—[a]
10	—[a]	5	—[a]

[a] No conditions yield this increase in N_2.

require a decrease in P by a factor of about 10. This means that the separation time t will be increased tenfold. Increases in N by more than threefold cannot be achieved by simple decrease in P (for this particular set of initial separation conditons).

For the case of varying L with P constant, column length L is increased by adding further lengths of column while flow rate F is reduced in proportion to the increase in L (for constant P). The separation time t will increase as $(L_2/L_1)^2$. Thus, for a threefold increase in N, L must be increased by a factor of 2.0 (see table), and the time t increases by 2^2 = fourfold. Note that the increase in separation time is much less for an increase in L (fourfold) than a decrease in P (tenfold), for the same increase in N. Normally, any required increase in N is possible by an increase in L in this fashion, although very large changes in N may involve a prohibitive increase in separation time (e.g., 5^2 = 25-fold for increase in N by tenfold).

Our third option for increased N is to increase L and P together, while holding t constant. In this case, P increases as L^2. Thus, for a 1.5-fold increase in N, L must be increased fourfold (see table), which requires an increase in P by 16-fold. Not only are large increases in N in this way prohibitive in terms of the pressure required, but as seen above, a more than two-fold increase in N is not even possible. Therefore this option for increased N is not often used.

Column efficiency can also be varied by changing the viscosity of the mobile phase. Generally, a decrease in mobile-phase viscosity results in higher values of N; thus, an increase in separation temperature (which results in lower mobile-phase viscosity) generally increases N. As a corollary to the above discussion, for a given packing (of given type and particular size) column efficiency N will be largest for a given separation time t when a maximum pressure P is used, and column length L is then adjusted to give the required value of t. Alternatively, a required value of N can be achieved in the shortest time t by again operating at maximum column pressure, and adjusting L to give the required value of N.

2.9. Maximum Attainable *N* Values in LC

The N value of a column packing varies with particle size d_p, with porous vs. pellicular packings, and with the nature of the stationary phase.[1a] Small-particle porous packings are now used almost exclusively for clinical LC separations, as these provide the fastest and most efficient separations. For well-prepared packing materials, there is usually little difference in column efficiency between packings that use different stationary phases. This leaves particle size as the major variable that determines N values as a function of P and t. A treatment summarized in

Table 2. Column N Values for Representative Conditions and Different Values of Particle, Size and Separation Time

d_p (μm)	N for different values of t			
	10 s	1.5 min	15 min	2.5 h
1	4700	12,000	17,000	18,000
2	3300	15,000	41,000	64,000
5	1200	7,500	39,000	150,000
10	400	3,200	21,000	110,000
20	100	1,100	8,700	58,000

reference 1a yields the column N values shown in Table 2 for representative conditions (reverse-phase separation at 70°) and different values of particle size d_p and separation time t, assuming a maximum column pressure of 5000 psi. In each case L is varied to give the required value of t (assumed equal to 10 t_0 or $k' = 9$).

The above data—which are based on theoretical calculations—are interesting in several respects. First, we see that very rapid LC separations are possible (e.g., 10 s total), yet with fairly high N values (e.g., about 5000 plates). Second, we see that for any given separation time and pressure, there exists an optimum particle size for which N is maximum. For separation times of about 1.5 min, this optimum value of d_p is about 2 μm. Generally, the optimum value of d_p increases with separation time, varying from 1 μm for very fast separations to about 5 μm for relatively slow separations. Third, very large values of N are favored by larger values of d_p, but in this case long separations are involved. Thus, the maximum N possible with 1 μm particles and 5000 psi pressure is about 18,000 plates, increasing to about 8 million plates for $d_p = 20$ μm (not shown, $t \gg 2.5$ h).

So far it has not been found possible to routinely pack columns with $d_p < 5$ μm so as to yield the above performance values. Thus, the values in the above table that fall above the dashed line ($d_p < 5$ μm) are at present somewhat speculative. The data above assume sample molecules with molecular weights in the 200–500 range. Generally lower N values will be found for larger molecules. In principal, however, similar N values can be obtained for large molecules as for small, if the particle size used for large sample molecules is reduced two- to threefold for each tenfold increase in molecular weight.[1a]

If particles smaller than 5 μm ultimately prove popular in clinical LC, the very narrow bands leaving the column will require specially designed equipment to prevent serious extra-column band broadening and loss in over-all N. Detector flowcells will require a decrease in size to perhaps 0.5–

1 μl total volume, fast-response data-handling devices will be necessary (e.g., 0.1 s full-scale), and sample injection techniques will require careful optimization.

While the above discussion has emphasized the possibilities of small-particle separations ($d_p < 5$ μm), it should be stressed that present columns of 5-μm particles appear quite adequate for most clinical LC applications. Thus from the preceding table, we see that 5-μm particles allow 1200–7500 plates for separation times of 10 s to 1.5 min. Most present applications of clinical LC require no more than 1500–3000 plates, which means that such separations can be carried out at rates of 120–300/h if all conditions are optimized. So far actual separations have generally fallen short of this potential, so that typical separation times are 5–10 min, i.e., about a factor of 10 slower than expected. However, a narrowing of the gap between theory and practice during the next few years seems inevitable.

2.10. Column Handling and Lifetime

Long-term column stability is obviously desired for high-volume clinical analysis. Column replacement at frequent intervals is expensive ($200–300/column) and may require reoptimization of separation conditions for new columns. Experience has shown that column lifetime can be significantly increased by attention to some simple preventive measures.

First, columns should be subjected to as little unnecessary stress as possible, since this tends to cause settling of the packing and creation of a void space at the top of the column—with a resulting disastrous drop in column efficiency. While this phenomenon is less serious with well-packed columns, the following guidelines are applicable to all columns:

1. Avoid dropping, banging, or otherwise "shocking" the column.
2. Avoid rapid pressure changes in use, e.g., do not change the mobile phase flow rate by more than 1 ml/min.
3. Consult the supplier's recommendations for other precautions concerning the handling of the column.

Another area of possible degradation is the accumulation at the top of the column of particles from the sample or mobile phase. Such accumulation will result in a significant increase in column-pressure drop and a decrease in performance. A thin screen or mesh should be placed at the top of the column to prevent particles from entering the column. If the pressure rises, the screen should be replaced. It is also good practice to prevent particles from reaching the top of the column. Samples, especially biological specimens, should be filtered prior to injection into the liquid chromatograph. Moreover, as bacterial growth can occur in an aqueous mobile phase, filtering of aqueous mobile phases is also recommended. A

0.5-μm filter can also be placed downstream from the pump, just prior to the injector.

Aside from particulate matter (including bacterial growth), organic impurities in the water used for reverse-phase separations can be retained at the column inlet. This leads to a gradual decrease in both retention times and column efficiency. Alternatively, if gradient elution is employed, these same impurities may elute during the separation and cause instability of the detector baseline. The water used in reverse-phase LC separations, particularly for typical clinical assays at the trace level, must be exceptionally pure. The deionized water commonly used in clinical laboratories may contain excessive organic impurities and require further treatment before use. Filtration of the water through either carbon or XAD-2 resin is one approach, depending upon the LC application and the quality of the water prior to pretreatment. The Milli-Q system sold by the Millipore Corp. (for about $1000) is also finding increased use for preparing LC-quality water.

Solvents other than water for use as LC mobile phases should also be of high purity. Several firms (e.g., Burdick & Jackson) now sell special solvents for use in LC. Just as solvent impurities can accumulate on an LC column, so can impurities from injected samples. In some cases these solvent and/or sample impurities can be removed from the column periodically, by "column regeneration": the washing of the column by a very strong mobile phase (e.g., isopropanol or THF for reverse-phase columns).

Finally, clinical samples such as serum, urine, or saliva almost always require some pretreatment or cleanup prior to their injection. The various techniques used have already been discussed in Chapter 3 (see also reference 1a). With attention to the above recommendations, thousands of clinical samples can be analyzed on a given column before it requires replacement.

2.11. Equipment

The basic components of a high-performance liquid chromatograph consist of (1) solvent reservoir, (2) pump, (3) injection device, (4) column, (5) detector, and (6) data-handling device. For each component there are a wide variety of choices, depending on the uses to which the equipment will be put. In this section we will briefly examine the various LC components currently available, paying particular attention to the special requirements of the clinical laboratory (e.g., simplicity, reliability, precision). More detailed information is given elsewhere (e.g., references 1 and 38).

2.11.1. Solvent Reservoir

The function of this module is to (1) degas the mobile phase and (2) hold its composition constant during the emptying of the reservoir. Degassing of the mobile phase is often unnecessary, and if degassing is employed, then care should be taken to avoid any evaporation of the mobile phase (which would change composition, in the case of mobile-phase mixtures). A better approach[39] is the continuous sparging of the mobile phase during its use by means of a small flow of helium. This results in the removal of gases such as nitrogen and oxygen, and their replacement by smaller concentrations of less-soluble helium. For various reasons, this can significantly stabilize the detector baseline, allowing lower detection levels. Removal of oxygen from the mobile phase also increases column life for polar bonded-phase packings. Note also that removal of oxygen is important when using fluorescence detection.

Since a 1% change in mobile phase composition in reverse-phase separations (e.g., from 40 to 41% volume) yields a 5–10% change in retention times,[40,41] it is obviously important to keep the composition of the mobile phase unchanged. The solvent reservoir achieves this by suitable design which prevents any evaporation.

2.11.2. Pumps

First, consider the maximum operating pressure of the pump. As previously discussed, column performance (i.e., separation speed) is maximized by operating at the maximum pressure drop and adjusting column length for the required number of theoretical plates N. From this point of view, the higher the maximum operating pressure of the pump, the better. On the other hand, the advantage of, for example, a 10,000-psi pump over a 5000-psi pump is somewhat marginal, while the practical problems of working at higher pressures mount rapidly as pressure increases above 10,000 psi. For these reasons, a reasonable maximum pump pressure for clinical LC appears to be about 5000 psi, and many commercial pumps with ratings of 5000+ psi are available.

Second, pulsations within the pumping system affect the noise level of the detector, thereby raising detection limits in trace assays. While thermostatting the column effluent before it enters the detector,[42] or using tapered flowcells,[43] can reduce the impact of pump pulsations, pulse-free pumps are generally desirable.

Finally, it is essential in high-throughput analysis that the pump maintain constant operating conditions. Run-to-run reproducibility, as reflected in the retention time, can often be better than 0.5% with commercial pumps.[40,41,44] Day-to-day reproducibility (which may involve pump resetting) can be of the order of ~1%, depending on the pump selected. These values are adequate for clinical assay, if they can be maintained over long periods of time. However, it is important in the selection of a pump to pay attention to flow constancy and resettability.

We can classify pumping systems into two types: constant pressure and constant volume (e.g., reference 1a). In the constant-pressure mode, a change in the condition of the column, frit blockage, or mobile-phase viscosity will change the flow rate. On the other hand, constant-volume pumps maintain a constant flow rate regardless of such conditions. For clinical applications, constant-flow-rate pumps are clearly better for reproducible retention and quantitation.

There are two main types of constant-pressure pumps[1a]: gas displacement and pressure amplifier. Because these pumps are of quite limited interest for clinical application, we will not discuss them further. There are likewise two main kinds of constant-volume pumps: syringe-type and reciprocating-type. In the syringe-type pump a large piston is mechanically or electronically driven through a large-volume chamber (250–500 ml), with the flow rate being controlled by the rate of movement of the piston. When the piston chamber has emptied, the piston is recycled and the volume is filled with mobile phase. These pumps can be reasonably precise devices, are pulseless (especially under isocratic operation), and can operate at high pressures. However, a change in mobile phase (for another assay) is relatively inconvenient, and these pumps suffer from other disadvantages unless special precautions are taken (e.g., references 45 and 46).

At present it appears that the reciprocating-type pumps are best suited for clinical application, particularly those giving relatively pulse-free flow. The mobile phase is pumped by one or more diaphragms or pistons. In the early days of LC,[1a] a single reciprocating piston emptied and filled a small-volume piston chamber on 180° cycles. This approach obviously led to very large pump pulsations which could significantly affect detector noise unless significant pulse dampening was applied. In recent times, dual reciprocating piston pumps have largely replaced the single piston arrangement. Here, two pistons are operated in an asymmetric manner in order to compensate for the different speeds of emptying the piston chambers. By this approach, and the use of small-volume pistons and/or flow-feedback control, pulsation can be reduced to quite low levels.

Instrument suppliers are moving increasingly toward the use of microprocessor-controlled pumps, which include many of the above features.[47] Such systems seem particularly applicable to the needs of clinical LC systems, which will also benefit from full automation (see below).

2.11.3. Injectors

These are discussed in some detail in Chapter 3 (as well as in reference 1a), and we will avoid repetition as much as possible. The sample-injection device of choice for clinical LC is the conventional six-port sample valve, used either manually or with automatic injection. These sample valves allow precisely controlled injection volumes and are therefore favored for quantitative clinical analysis.

While the injection of sample dissolved in the mobile phase is favored for many reasons (see discussion of Chapter 3), there are exceptions to this. If the solvent containing the sample is weaker than the mobile phase used in the separation (e.g., an aqueous sample solution in reverse-phase LC), larger injection volumes (with increased detection sensitivity) are possible.[48,49] The reason is that the sample components of interest will be strongly adsorbed to the packing at the column inlet and thereby concentrated into a smaller volume. In favorable cases, several milliliters of sample solution can be injected in this fashion, without excessively broadening the final chromatographic bands.

On the other hand, small volumes (up to 25 μl) of stronger solvents can be injected without adverse effect. For example, 25-μl samples in *n*-hexane as solvent have been successfully injected in reverse-phase separations.[50] This approach can be useful in clinical LC separations, for at least two reasons: (1) where the pretreatment scheme yields the final sample in an organic solvent–extract solution (e.g., chloroform solvent, to be injected on a reverse-phase column); and (2) in order to achieve better solubility of the sample in the injection solvent.

2.11.4. Detectors

A discussion of LC detectors, with emphasis on applications of clinical interest, has been treated exhaustively in Chapter 3.

2.12. Procedures and Techniques

2.12.1. General Considerations

Clinical samples for analysis by LC normally contain a large number of components, with those compounds present at lower concentrations being of primary interest. This places a considerable burden on the over-all separation/detection scheme, in order to avoid interference by one compound in the measurement of another. Previous workers who have reported various clinical assays by LC have been well aware of this problem, and potential interferences should be carefully checked out for each new procedure. Therefore, an important aspect of any clinical LC scheme is its over-all specificity or ability to discriminate among the myriad compounds present in the sample.

An over-all analytical scheme for clinical LC may involve a series of separation steps followed by detection of the components at the levels desired. Specificity can play an important role in each of these steps. For purposes of discussion, we can define five types of discrimination possible in an LC clinical analysis: physical, chemical, chromatographic, mechanical, and detector.

In *physical discrimination* we take advantage of physical separation processes during sample pretreatment. Thus, we can effect an initial separation or elimination of potential interferences by such procedures as liquid–liquid extraction, with or without pH adjustment to eliminate acids, bases, or neutral compounds. For a general discussion of this area, see Chapter 3 as well as reference 1a.

In *chemical discrimination* we convert either interferences or the compounds to be assayed into derivatives which alter the separation or detection properties of the compounds. This kind of specificity for change in chromogenic properties of compounds of interest in fact forms the basis of classical clinical analysis (e.g., determination of glucose by neocuproine reaction). Numerous examples are provided in Chapter 3.

In *chromatographic discrimination* we try to make use of differences in chromatographic distribution to achieve separation. Here we frequently take advantage of the chemical properties of the substances, e.g., differences in ionization equilibria, differences in charge-transfer complexation ability, etc. The selection of phase systems (see previous section) is often based on these chemical differences.

Mechanical discrimination can be a powerful tool when used in chromatography. Here, particular interferences are simply switched out of the system prior to their reaching the detector. As an example of how this might be used, consider the chromatogram in Figure 4A. We wish to determine the minor or trace components in the presence of the major component. Unfortunately the minor component elutes on the tail of the major component. A solution to this problem is illustrated in Figure 4B,

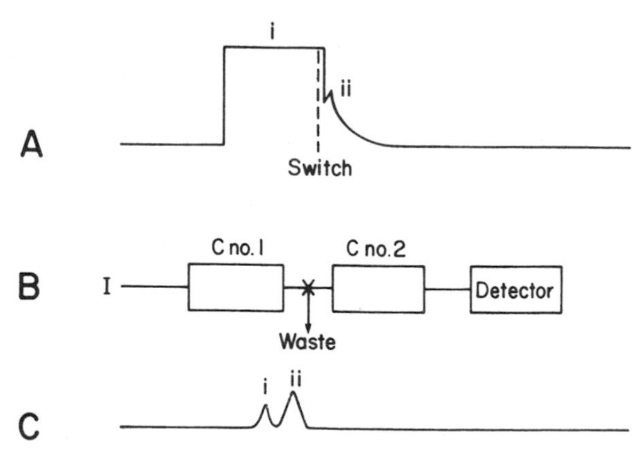

Figure 4. Removal of frontal interference by column switching.

in which two columns are coupled in series with a switching valve between them. In operation, the valve would be open to waste while the major component is eluting from the first column. At the time indicated in Figure 4A, the valve is switched for flow through the second column. In this way we drastically change the concentration ratio of interference to component, so that analysis as shown in Figure 4C can be achieved. The use of switching valves in this manner to enhance separation can be useful.[51–53] Moreover, it is easily automated and could therefore be used in a clinical LC system. See also the discussion of Figure 1 of Chapter 3.

Finally, *detector discrimination* can be a powerful means of solving particular clinical problems. The more selective the detector, the less demand is placed on the chromatography[54] (a sufficiently specific detector would not need chromatography at all).

The selectivity of the detector can also be used in a second manner for determining whether interferences are present under particular chromatographic bands. A simple, yet powerful, technique is peak ratioing, in which the absorbance of a band is read at two wavelengths. If a single compound is under the band, then the absorbance ratio should equal that of the standard substance itself.[55] By scanning the entire chromatographic band, and confirming that the absorbance ratio is identical across the band, this technique can be extended to rule out even minor concentrations of interferences with fairly similar spectral properties.

2.12.2. Derivatization

In the early days of modern liquid chromatography it was often stated that substances requiring derivatization for GC analysis (e.g., highly polar or ionic species) could be chromatographed directly in a convenient manner by LC. While this characteristic of LC still remains true, it is now recognized that derivatization can be an important tool, particularly for more difficult separation or detection problems.[55] Derivatization already plays a significant role in clinical LC.

Derivatization (i.e., chemical modification of either the sample components or interferences) can enhance the power of LC in a variety of ways: (1) convert a component to a UV or fluorescent-active product; (2) simplify the cleanup steps; and (3) less frequently, change the chemical and therefore chromatographic properties of the substance (e.g., polarity decrease of highly polar substance).

Depending on the particular problem, either pre- or post-column derivatization (with reaction detectors) may be used. Pre-column derivatization allows considerable freedom in the selection of conditions and derivatizing agent, as reaction time need not be a determining factor. The disadvantages of the pre-column approach are (1) the possible formation of side products which can interfere with the chromatographic analysis and (2) the conversion of solutes into more similar forms which therefore are more difficult to separate. Post-column derivatization overcomes most of the disadvantages of the pre-column approach. The solvent and background remain constant from run to run, side products are less likely to interfere or even to occur, and the separation is first achieved on the original components. However, the use of reaction detectors places limits on the time available for the reaction before band dispersion becomes significant. In addition, the reagent selected must obviously not be detected under the conditions used for the analysis of the derivative. Thus, both pre- and post-column derivatization complement each other, and the particular approach selected will obviously be a function of the problem.

Pre-column derivitization is normally carried out off-line, using conventional reactions that are reviewed in Chapter 3 and reference 55. In post-column derivitization, a reaction detector provides for the controlled addition of reagents to a column effluent in a continuous manner, with provision for on-line mixing for some time *t* at temperature *T*. An example of reaction detection already familiar to many clinical laboratories is the use of ninhydrin reaction in conjunction with automated amino acid analyzers for physiologic fluids. Here, the colorless amino acids leaving the column are converted into strongly absorbing products that can be photometrically measured at either 440 or 570 nm. For a discussion of the kinds of reactions previously used in reaction detectors, see reference 55.

A major objective in the design of all reaction detectors for high-performance LC is to maintain minimum spreading after the bands leave the column. For fast reactions that require incubation times of 30 s or less (e.g., Fluram reaction with amines to form fluorescent product), this can be achieved by simply combining column effluent and reagent after the column, and then passing the mixture sequentially through a multiturn coil of small inside-diameter tubing and into the detector. In this manner the reaction mixture is properly mixed prior to measurement in the detector, which is usually a photometer or fluorometer, but can in principle be any kind of detection system. For examples and further discussion of this approach, see references 56 and 57.

A second approach, for reaction times of up to 3 min or so, is to use a small column packed with glass beads following the effluent–reagent tee. The hold-up time of this secondary column is equal to the desired incubation time *t*. The particle size of the beads in the secondary column must be small—again for minimal extra-column band broadening, which creates a need for higher pressures in pumping reagents into the system. An example of an LC separation based on this approach is shown in Figure 5A, for the separation of the creatine kinase (CK) isoenzymes. In this case the reagent mixture consists of the substrates and indicating enzymes used

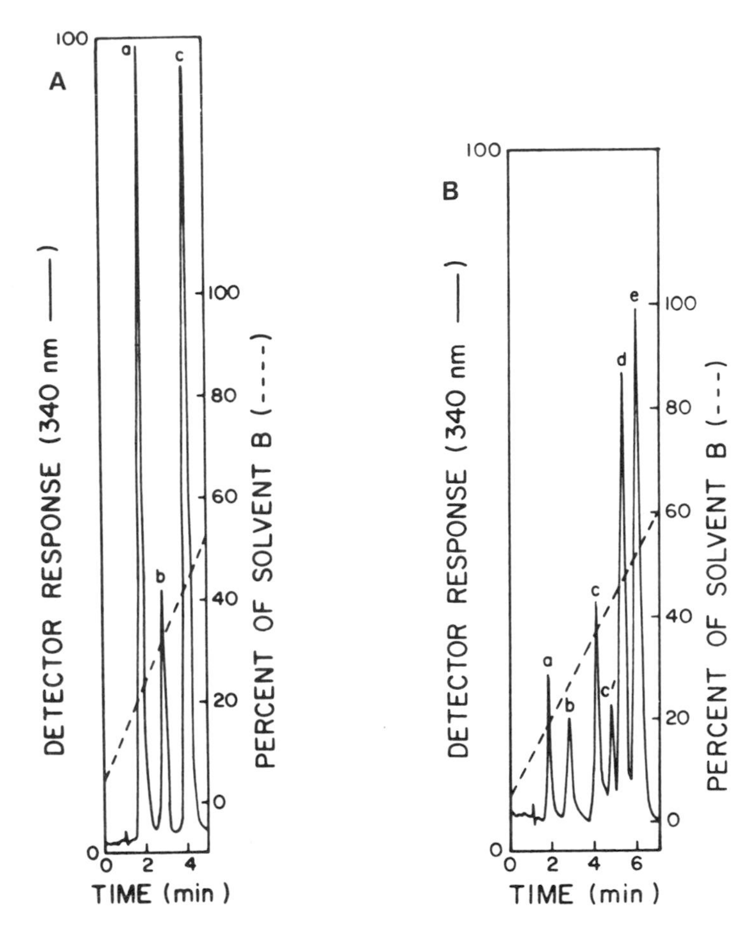

Figure 5. (A) Separation of CPK isoenzymes using a post-column enzyme detector. Column, 250 × 4 mm ID stainless steel; packing, DEAE-Glycophase/CPG (250-Å pore diameter, 5- to 10-μm particle size); temperature, 25°. Solvents: (A) 0.05 M Tris, 0.05 M NaCl, 10^{-3} M mercaptoethanol, pH 7.5; (B) 0.05 M Tris, 0.03 M NaCl, 10^{-3} M mercaptoethanol, pH 7.5; flow rate, 4 mm/s (3 ml/min); pressure, 2500 psi; a = CPK_3; b = CPK_2; c = CPK_1. (B) Separation of LDH isoenzymes using a post-column enzyme detector. Column, 250 × 4 mm ID stainless steel; packing, DEAE-Glycophase/CPG (250-Å pore diameter, 5- to 10-μm particle size), temperature, 25°. Solvents: (A) 0.025 M Tris, pH 8.0, (B) 0.025 M Tris, 0.2 M NaCl, pH 8.0; flow rate, 4 mm/s (3 ml/min); pressure, 2500 psi; a = LDH_5; b = LDH_4; c = LDH_3; d = LDH_2; e = LDH_1. (Reprinted from reference 206 with permission.)

in the Rosalki CK determination (creatine phosphate, ADP, glucose, hexokinase, etc.). CK isoenzyme activity is then indicated by the formation of NADH which absorbs at 340 nm. For further examples and discussion of this approach, see references 58 and 59.

A final approach to reaction detectors makes use of air segmentation, e.g., autoanalyzer systems. Column effluent is segmented by air bubbles as it leaves the column and is then combined with reagents in a tee. The segmented mixture is then passed through a length of incubation tubing, and finally to a debubbler (if necessary) and detector. Reaction times of as much as 20 min can be accommodated with this approach, if the entire LC/reaction-detector system is carefully designed. An example of a continuous-flow reaction detector is provided in Figure 6, for the detection of separated polyamines via the ninhydrin reaction. For additional examples and discussion of continuous flow reaction detectors, see references 60 and 61.

For some LC applications, reaction detectors appear to be virtually indispensable. The above example of the detection of separated enzymes or isoenzymes in the presence of other sample constituents is a case in point. Here it would be almost impossible to separate the enzymes of interest from all other sample components, and the low concentrations of these enzymes would preclude their detection by direct photometric means. Reaction detectors in this case take advantage of the specific catalytic properties of the enzyme.

2.12.3. *Analytical Aspects*

HPLC has been shown to be a precise tool for quantitative analysis. With proper precautions, peak height or peak area can be measured with relative standard deviations of 0.5 to 1.0%.[41,44,62] These values indicate that the precision of an LC clinical assay will usually be controlled by the sample cleanup steps (e.g., extraction). Thus, Rocco *et al.*[63] found relative deviations of peak height of 2–3% for the analysis of procainamide and *N*-acetylprocainamide from serum using an internal standard. In a separate study of the same substances without internal standards, Shukur *et al.*[64] found errors of 8–10%. However, even deviations of 10% are acceptable for many clinical assays by LC.

A frequent question is whether to use peak-height or peak-area measurements in the quantitative analysis of clinical samples. Each technique has its separate advantages (see references 1 and 65). In most cases it appears that greater precision is possible with peak-area measurements. However, peak-height determinations are less subject to possible interferences. On balance, typical clinical assays by LC probably place more emphasis on accuracy than precision.

Retention reproducibility has already been mentioned in our discussion of solvent reservoirs and pumps. To reiterate, retention can often be as precise as 0.5% relative standard deviation from run to run. However, the resetting of separation conditions from day to day can lead to 1% or greater variation in retention from such factors as pump resettability, temperature control of the column, and mobile-phase composition. Special

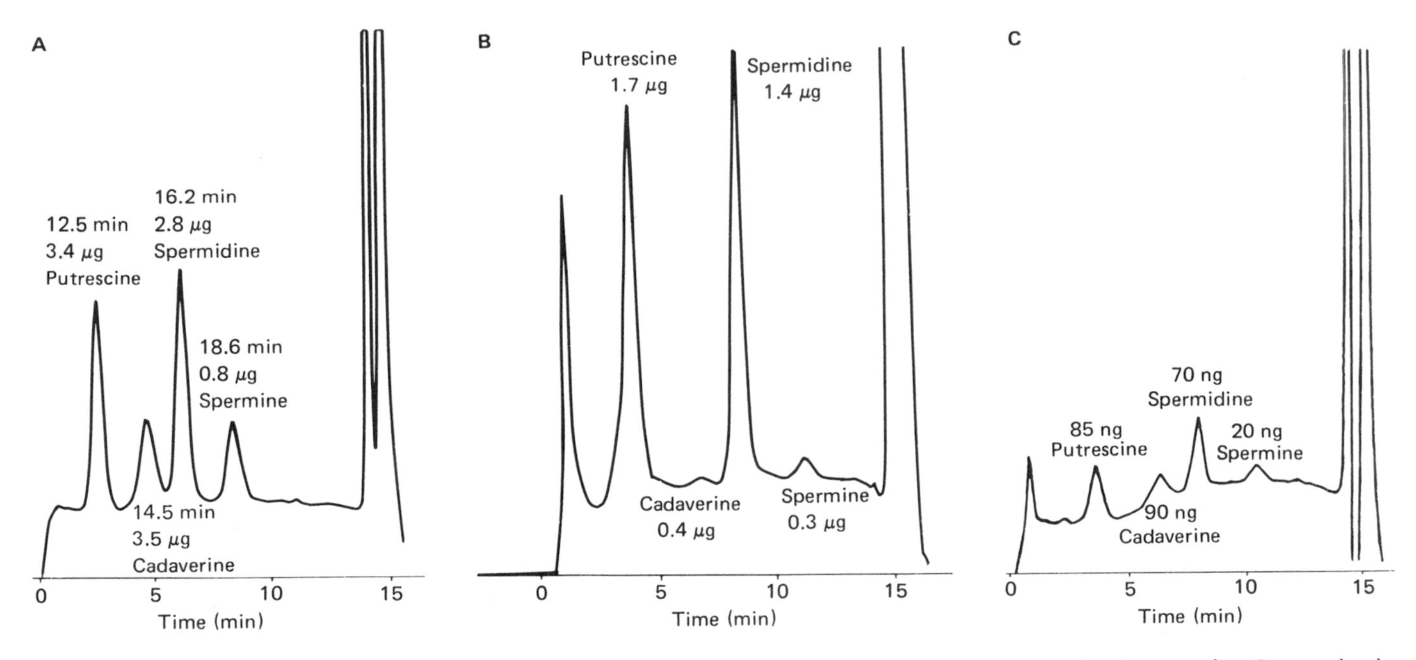

Figure 6. Assay for polyamines in body fluids; (A) polyamine standards, (B) polyamines in hydrolyzed urine sample, (C) trace-level polyamine standard. Resin: Technicon C-2, 8% cross-linked, sulfonic acid type, 8- to 12-μm diameter. Column: 45 × 4 mm. Temperature: 80°C. Elution: buffered, ionic strength gradient. Detection: colorimetric ninhydrin reaction. (Reprinted from reference 110 with permission.)

care should be taken to reproduce the makeup of the mobile phase, as retention in reverse-phase separation is particularly sensitive to this variable.

A more serious concern is the reproducibility of commercial reverse-phase columns. In our experience,[50] retention reproducibility of columns from the same manufacturer is usually no better than 10–15%; at times more serious variations occur. Thus, each new column must be checked carefully and calibrated, where appropriate. Variations from manufacturer to manufacturer with supposedly the same packing material (e.g., bonded reverse-phase C_{18}) can be very large (up to 30%), not only in absolute retention but also in relative retention.[50] While variations in retention may partly be attributed to differing surface areas of the silica particles, the differences in relative retention also represent variable bonded-phase characteristics. One must expect different separation conditions from column to column, when different commercial sources are selected. It is hoped that in the future manufacturers will define the specifications of their columns more carefully.

As we have already noted, Chapter 3 of this volume covers the topic of trace LC analysis. This is obviously an important subject in clinical analysis, and the reader is advised to see that chapter for details.

2.12.4. Gradient Elution and Related Techniques

Most clinical assays by LC have involved the determination of a single compound or group of similar compounds whose retention times fall within a narrow range of values. In these cases, the compound or compounds of interest generally elute at the end of the chromatogram with values that fall in the range $1 < k' < 10$. Such assays are best carried out by isocratic elution, where the conditions of separation are maintained the same throughout the analysis. In other cases, we may want to determine several compounds of widely different elution properties, or the compound of interest may be followed by late-eluting bands. Here isocratic elution is not advantageous, for reasons that are well known to chromatographers: excessive separation time, broadening of late-eluting bands to the point of nondetectability, etc.

Gradient elution or solvent programing has been been used occasionally in clinical LC to solve problems of the above type. An example is seen in Figure 5 for the separation of the creatine kinase and lactate dehydrogenase isoenzymes. Here the k' values of the various isoenzymes differ greatly, so that isocratic separation is not feasible. Although gradient elution is a powerful technique for problems of this type, or for rapid purging of late-eluting sample constituents when the compound of interest has an intermediate k' value, the technique has its disadvantages for clinical application. First, gradient elution and its necessary equipment is generally complex and/or expensive, adversely affecting system reliability and cost. Second, quantitation via peak-height measurements is less precise. Finally, column regeneration is generally required after each gradient elution separation—and this adds to total analysis time.

One alternative to gradient elution that seems particularly suited to clinical LC is "coupled-column" operation[51,52,66] which is another example of the previously discussed column-switching technique. The use of coupled columns essentially substitutes stationary-phase programing for the mobile-phase programing used in gradient elution. In the simplest arrangement of valve and columns, shown in Figure 7, column 1 might consist of a short *n*-alkyl chain length (e.g., C_2) bonded-phase column and column 2 would then be a long chain-length (e.g., C_{18}) column. Sample retention is considerably reduced on column 1 vs. column 2.

In normal operation of the system of Figure 7, the valve is first set to connect column 1 to column 2. Sample is then injected, and compounds with low k' values are allowed to elute from column 1 and enter column 2. The valve is then switched to allow direct elution of remaining sample components from column 1 into the detector. In this manner, strongly retained compounds are quickly eluted from column 1 into the detector. Weakly retained sample components that were allowed to pass into column 2 are more strongly retained in that column, for good resolution with the single mobile phase used in coupled-column LC. Therefore, following elution of all compounds from column 1 into the detector, the valve is again switched for the subsequent elution of remaining compounds from column 2. In this way all compounds are separated in an optimum k' range (e.g., 2–5) from their respective columns.

The advantages of coupled-column LC over gradient elution include the following: (1) very simple and inexpensive equipment that is easily automated; (2) a relatively stable system, inasmuch as solvent composition is not changed during the assay; (3) better quantitation by peak height; and (4) no need for column regeneration.

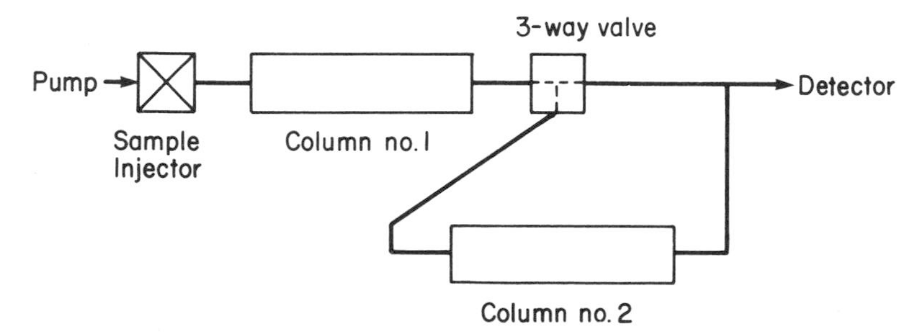

Figure 7. Coupled-column arrangement.

While the technique of coupled columns has so far seen limited application, the necessary column packings and switching valves have only now become readily available. It is anticipated that some clinical assays will benefit from the application of this technique.

2.12.5. Automation

Maximum precision and the avoidance of sample misidentification are greatly favored by automation in clinical analysis. Also, the automation of high-volume tests generally leads to a reduction in cost-per-sample and more efficient utilization of laboratory personnel. For a number of clinical applications of LC, it is clear that the necessary sample volume to justify full automation already exists in the larger laboratories.

The necessary components for a totally automated system for clinical LC are shown in modular form in Figure 8. This scheme provides for six separate steps between introduction of samples and reporting out of final results. While a given LC assay might dispense with one or more of these individual steps, many present methods involve each of the functions shown. Figure 8 assumes that the six modules are directly interfaced for completely automated operation, but this need not always be true. For example, sample pretreatment (step 2) might be done off-line, with the treated samples then introduced into an automated system which does not include this step. The following discussion considers each of the six steps of Figure 8 in turn, with attention to the interfacing of adjacent modules into a fully automated system.

Sampling. This step needs no further discussion, since it is basic to other automated clinical assays. Commercial sampler/injection systems (steps 1 and 3) for use in LC are available from several instrument companies. Samplers designed for interfacing to sample-pretreatment modules

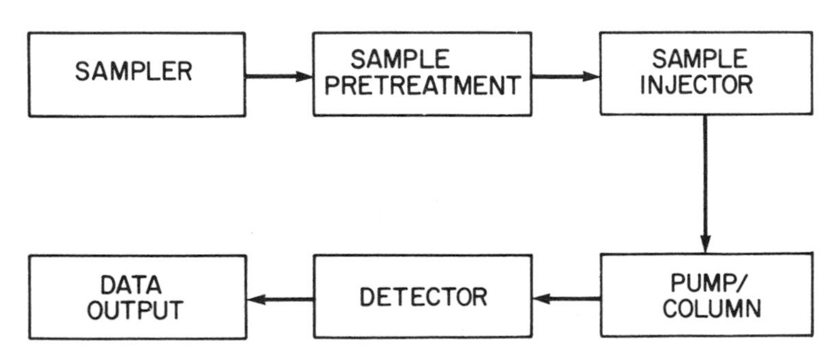

Figure 8. Totally automated LC system.

based on continuous flow analysis (see below) are already used in the clinical laboratory.

Sample Pretreatment. Present LC methods for clinical analysis normally require some form of pretreatment prior to injection. This pretreatment can include one or more of the following operations: dilution, extraction, evaporation to dryness, redissolution of evaporated samples, deproteinization, pH adjustment, derivative formation, filtration, preseparation by gel filtration, etc. The partial automation of many of these operations is well known in the literature, and various such devices are commercially available. However, complete automation of any one or more of these steps, with the capability of interfacing the over-all pretreatment module to both the sampler and to the injection module, can at present be achieved only with continuous-flow (e.g., autoanalyzer) systems. Fortunately such systems are already well known to the clinical chemist. Further discussion here of sample pretreatment will therefore focus on the continuous-flow approach. For a general discussion of autoanalyzers and continuous-flow analysis, see references 67 and 68. For a specific discussion of the use of autoanalyzer components with LC, see references 50, 69, and 69a.

The dilution and/or extraction of serum or urine can be achieved by continuous flow analysis, as discussed in reference 70. The major problem encountered is the final separation of the organic extract phase from the original sample phase, since emulsions often result particularly in the case of serum or urine. However, this problem can be minimized by attention to the design and construction material of the phase separator (see reference 70). Other operations, such as filtration, derivative formation, etc., can also be carried out with autoanalyzer components.

Injection. Pretreated sample leaves the autoanalyzer pretreatment module as an air-segmented stream. This stream can be debubbled in the conventional manner and fed to an automatic six-port sample injection valve, operated by a timing mechanism that is coordinated with the sampler. Several companies now supply such automated sampling valves for high-pressure LC.

LC System and Detector. The equipment for the LC system (pump, column, etc.) has already been discussed. A general discussion of LC detectors is provided in Chapter 3.

Data Reporting. In a clinical LC system, this can be as simple as the recorder chart of the analog chromatogram. Alternatively, the latter analog signal can be further processed by a microprocessor in the LC system, or interfaced into a laboratory computer system. The extent of data processing can vary from simple calculations of concentration (based on peak height or area) to confirmation of peak identity and absence of interferences via an analysis of the peak shape and its exact retention time.

3. Applications

The possible applications of LC which impact on the clinical laboratory fall either directly or indirectly into several areas:

1. reagent characterization
2. research
3. reference samples
4. reference methodology
5. genetic screening
6. pediatric chemistry
7. specialty testing
8. routine analysis

We have chosen in this chapter to emphasize only the last two topics: specialty testing and routine analysis. Further, primary attention generally will be given to those LC applications which may be important in the immediate future, i.e., the next 2–5 years. Before examining these applications, however, some general comments on areas 1–6 are appropriate.

Many *reagents* used in routine clinical analysis are in an inadequately characterized form. LC offers a unique tool for the analysis of both content and purity of such compounds as coenzymes, dyes, and surfactants, both as nominally pure materials and in mixture with other reagents. One example, the characterization of the reduced form of nicotinamide adenine dinucleotide (NADH), an important coenzyme, suffices to establish the utility of LC in this area (e.g., references 71 and 72).

There is a great need for more *research* involving clinical chemistry. Basic studies are needed to define further the clinical biochemistry of various disease processes, in order to improve diagnosis, prognosis, and treatment. The clinical chemistry laboratory plays an important role in these efforts, and LC is a useful tool for some of these studies.

LC can provide *reference methodology*. For example, ion-exchange LC involving detection at 254 nm and confirmation by GC/MS has been shown to provide a specific assay for creatinine body fluids.[73] This method is discussed further below. Similarly, LC separation followed by specific electrochemical detection constitutes a potential reference method for uric acid in serum.[74] Preparative LC should help to provide reference samples for the clinical laboratory.

Genetic screening involves testing for inherited diseases and includes prenatal diagnosis. The most widely employed tests emphasize detection of biochemical abnormalties prior to the onset of symptoms. For example, statewide testing programs exist for the detection of phenylketonuria in newborns based on elevated urinary levels of phenylalanine. Similarly, decreased levels of thyroxine in cord blood signal the likelihood of cretinism. The complications from each of these conditions are preventable, if therapy is begun early. This is the major justification for early diagnosis. Unfortunately, many more genetic diseases are diagnosed only after symptoms arise. Therefore, it is desirable to improve diagnostic and therapeutic procedures to the point where presymptom testing for more of these conditions becomes practical.

Genetic screening is a very large field, particularly when it is realized that abnormalities of the metabolism of amino acids, carbohydrates, lipids, nucleic acids, porphyrins, hormones, and vitamins are included. Generally, these abnormalities arise as a consequence of mutant enzymes. However, mutations of nonenzymatic proteins are also important, leading to additional genetic diseases, e.g., sickle-cell anemia, which involves an abnormal hemoglobin. As pointed out in a recent review, almost 2000 genetic disorders are known, many of which are harmful.[75] There are many opportunities for LC in this field, and some important analyses already have been achieved, e.g., those for porphyrins, amino and urinary acids, and lipids. Porphyrin analysis is discussed later in this chapter. Amino acid analyzers are widely used for confirmation and monitoring of aminoacidurias. The new generation of these analyzers involving small-particle columns and elevated pressure is another example of modern LC. The use of reverse-phase LC has allowed much more rapid separation of urinary acids[76] than was previously achieved by ion-exchange LC. For lipid analysis, developments are currently confined to research laboratories. However, the ability of LC to detect Farber's disease and homozygotes and heterozygotes of type A Niemann-Pick disease has been demonstrated.[77,78]

For screening of large populations, high throughput with low cost is essential. For evaluation of high-risk groups or individuals already beset with symptoms, and for monitoring of treatment, the need for high throughput is much reduced, since far fewer analyses are required.

Pediatric chemistry places special emphasis on the small sample size that is available from infants. Thus microanalytical procedures are favored, especially those featuring reliability, accuracy, and precision (because of the reduced opportunity to repeat procedures by the same or confirmatory methods). Since microscale verisons of many tests are not adequately sensitive or reliable for pediatric work, alternate methodologies often are desirable. Because high sensitivity, reliability, microscale sample capability, and broad scope are its basic strengths, LC is well suited to tackle some of the analysis problems in this area. A book[79] and a review[80] covering the special needs of the pediatric laboratory are available.

Specialty testing and *routine analysis* by LC are the main topics of the present chapter. While major emphasis is placed on system requirements for large-volume automated LC applications, the flexibility and versatility of LC should not be overlooked for application to specialty tests that involve only a few samples per week. The cost of developing a new assay by LC is usually small (in contrast to methods based on immunoassay), and many different assays can be carried out on the same equipment and columns. The following sections provide detailed reviews of (1) biogenic amines, (2) drugs, (3) hemoglobin AIc, (4) hormones, (5) isoenzymes, (6) lipids, (7) porphyrins, (8) vitamins, (9) organic acids, (10) ubiquinone and (11) nucleotides, nucleosides, and bases. A note added in proof covering these topics is presented at the end of the chapter.

3.1. Biogenic Amines

3.1.1. Catecholamines

In clinical practice, "catecholamines" includes mainly three substances: dopamine, norepinephrine, and epinephrine. Each has a catechol nucleus (a benzene ring with two adjacent hydroxyl groups) and an amine function (see below). Important sites of production or action include the brain, the postganglionic sympathetic nerves, and the adrenal medulla. The major metabolite of dopamine is homovanillic acid (HVA), and the major metabolite of epinephrine/norepinephrine is vanillylmandelic acid (VMA), each of which occur in the urine (both conjugated and free). Various intermediate metabolites such as normetanephrine and metanephrine also are encountered.

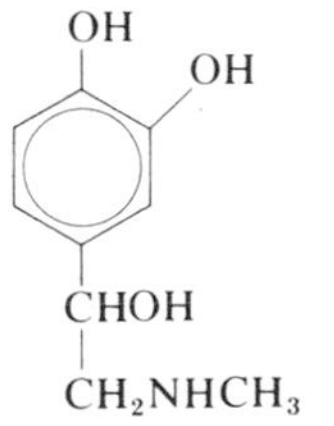

OH, OH, CHOH, CH_2NHCH_3

Epinephrine

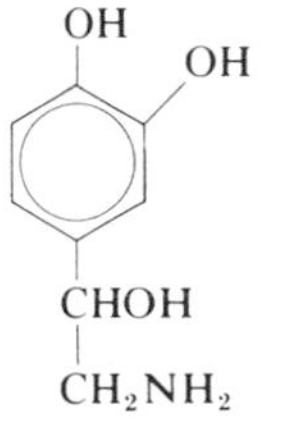

OH, OH, CHOH, CH_2NH_2

Norepinephrine

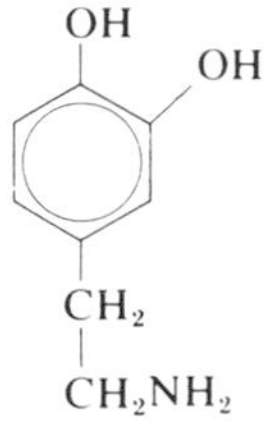

OH, OH, CH_2, CH_2NH_2

Dopamine

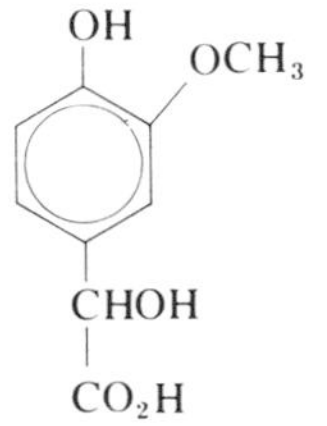

OH, OCH_3, CHOH, CO_2H

Vanillylmandelic acid (VMA)

OH, OCH_3, CH_2, CO_2H

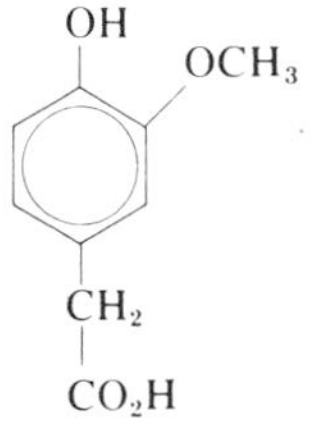

Homovanillic acid (HVA)

Measurement of the catecholamines and their metabolites provides an indication of tumors of the tissues involved in catecholamine production, for example, the tumors pheochromocytoma, neuroblastoma, and ganglioma. In these conditions, the catecholamines and their metabolites tend to be greatly elevated in both the serum and urine. Although assay for total levels of these compounds is diagnostically useful, additional clinical information can be obtained from measurements of the specific amines and metabolites.

Of all the biogenic amines, the catecholamines and their major metabolites vanillic acid and homovanillic acid have been most investigated by LC. The ability of LC to rapidly separate synthetic mixtures of the catecholamines has been amply demonstrated (e.g., references 81–83). This discussion will be mostly confined to analyses of actual samples. As a generalization, the early popularity of ion-exchange chromatography for these analyses has given way to improved procedures utilizing reverse-phase LC.

In one of the earliest papers describing the LC analysis of catecholamines in urine, Kissinger reported quantitation of norepinephrine, L-dopa, epinephrine, and dopamine within 10 min on a bonded-phase cation-exchange resin, using electrochemical detection.[84] More recently, Kissinger has reported an improved methodology based on the use of a microparticulate reverse-phase column modified by an anionic detergent.[85] A chromatogram is shown in Figure 9. The cleanup procedure includes an adsorptive extraction of the catecholamines with alumina.

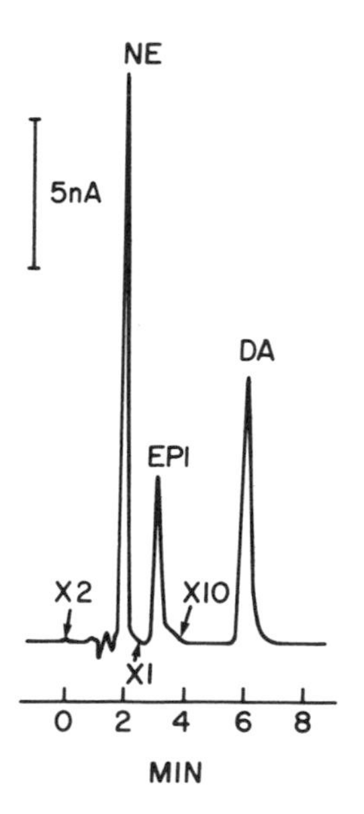

Figure 9. Chromatogram of norepinephrine (NE), epinephrine (E), and dopamine (DA) isolated from human urine by ion exchange followed by alumina. Column: 30 cm C_{18} Permaphase; mobile phase: sodium octylsulfate in citrate/phosphate buffer; detection: thin-layer amperometry operated at +0.720 V vs. an Ag/AgCl reference electrode. Concentrations in the urine were NE: 60 μg/liter, E: 22 μg/liter, DA: 510 μg/liter. (Reprinted from reference 85 with permission.)

Vanillylmandelic acid (VMA) has been quantitated in urine using an initial sodium periodate oxidation to vanillin.[86] Extracted vanillin is then quantitated by electrochemical detection after elution from a C_{18} reverse-phase column.

An LC assay for urinary homovanillic acid, the principle metabolite of L-dopa and dopamine, has been reported.[87] A sequence involving solvent extraction, TLC, and then LC with electrochemical detection was used. The related metabolite, 3,4-dihydroxylphenylacetic acid, has been quantitated similarly.[88]

An LC procedure employing a packing of porous styrene–divinylbenzene for both vanillylmandelic acid and homovanillic acid in urine has also been reported, using detection at 280 nm.[89,90] The disadvantages of the procedure were that gradient elution was required, the separation time was long (50 min), and an interference with HVA was encountered after coffee ingestion. However, only a simple organic extraction of acidified urine was used to prepare the sample for LC analysis.

Mell and Gustafson[91] extracted catecholamines from 150–250 ml of urine with alumina and then analyzed a dilute acetic acid eluate by reverse-phase LC using 280 nm detection. Significant peaks from a normal serum were seen for norepinephrine, dopamine, normetanephrine, and metanephrine. Less efficient chromatography was used here than in the work cited earlier for some of these analytes,[85] and the metabolite dihydroxyphenethyl glycol may be present under the peak for norepinephrine.[92]

Urinary normetanephrine, metanephrine, and 3-methoxytryamine have been quantitated by reverse-phase LC with amperometric detection.[93] Potentially interfering catecholamines were eliminated by complexation with borate in the cleanup procedure. A low-pH mobile phase rather than an ion-pair mode was used to avoid peak tailing.

LC procedures for catecholamines involving derivatization before the column (for fluorescence detection) also have been reported; either a fluorescamine[94] or dansyl[95,96] group was introduced. In the latter case, determination down to 50 pg was cited. For the homovanillic acid analysis cited previously, the electrochemical detection limit was 100 pg.[87] It therefore would appear that both fluorescence and electrochemical detection have the potential to allow the LC analysis of catecholamines in plasma samples.

Because of the complexity of most urine samples, including the large number of possible catecholamines and/or related compounds present in urine, it is especially important to look for possible interferences in these assays. Thus, one study[83] showed a well-separated peak in urine (soap chromatography) at the position where epinephrine was expected. Further analysis of this peak, however, established that it was in fact some other compound.

Thus, there has been much work in this area. Urine has been the only physiological sample which has been analyzed in practical terms by LC, but reasonable detection capability for plasma samples is in hand. (An analysis of L-dopa and dopamine in serum by LC is not an exception because therapeutic rather than endogenous concentrations were involved.[162]) In regard to the analysis of plasma samples, it is interesting that no practical immunoassays for the catecholamines such as epinephrine in this fluid have appeared. The probable reason is that these substances are too small as haptenic groups. This void has been filled, however, by a radioenzymatic procedure (e.g., reference 97), in which individual catecholamines are quantitated with a sensitivity of 5–10 pg by a procedure in which [^{3}H]methyl groups are enzymatically transferred to the catecholamines followed by a TLC separation step and scintillation counting. A review emphasizing derivatization of biogenic amines (including catecholamines) to facilitate chromatographic analysis has recently appeared.[98]

3.1.2. 5-Hydroxyindoles

Serotonin (5-hydroxytryptamine) is a vasoconstrictor and smooth-muscle stimulant. It is the active component of a series of 5-hydroxyindoles, all of which are either precursors or metabolites of serotonin. The diagnostic significance of the 5-hydroxyindoles is that their excessive production and excretion is a biochemical indication of malignant carcinoid tumors, a type of neoplasm particularly associated with the gastrointestinal tract. In these cases, more than 99% of the total 5-hydroxyindoles excreted into the urine are normally in the form of 5-hydroxyindole acetic aicd (5-HIAA), the final metabolic product of serotonin. The major part of 5-HIAA is excreted free, with some conjugated to sulfate. In other cases, the total urinary 5-hydroxyindoles are comprised of significant fractions of compounds other than 5-HIAA.

The quantitation of 5-HIAA in urine using low-pressure, cation-exchange separation (approx. 18 min) followed by an on-line reaction with *o*-phthalaldehyde for fluorescence detection was reported several years ago.[99] Six indoles (including three 5-hydroxyindoles) have been isocratically separated by reverse-phase LC and detected by fluorescence without derivatization in less than 20 min.[100] LC analysis with fluorescent detection of 5-HIAA in actual samples of urine[100a] and both urine and cerebrospinal fluid also has been reported,[100b] as has the analysis of four indoles in cerebrospinal fluid.[100c] In the latter work, detection limits of 12–70 pg were reported.

3.1.3. *Polyamines*

There is increasing interest in the clinical assay of polyamines in urine and serum, in connection with the management of cancer patients. The chemical structures of the polyamines and their normal ranges in serum and urine are shown in Table 3. It was initially reported[102] that the total concentrations of polyamines in urine are elevated for most cancer patients, which led to the hope that the assay for polyamines in urine or serum might have screening value for the detection of occult cancer. Apparently, cancer cells are rich in the polyamines, and the death of these cells as a result of normal body processes leads to elevated concentrations of these substances in serum and urine. Although subsequent studies (e.g., reference 103) did not support screening analyses for cancer detection, more recent work (e.g., references 104 and 105) suggests that polyamine measurement can be used to monitor the effectiveness of chemotherapy for cancer, much as carcinoembryonic antigen (CEA) assays are used in conjunction with various types of cancer treatment. Following initial therapy, polyamine levels in serum and urine normally show a distinct elevation as a result of cancer-cell death. The polyamines then tend to return to normal levels when treatment is successful.

Polyamines have been measured by electrophoresis, gas chromatography, and LC. Electrophoresis is less sensitive than LC or GC, and the sample derivatization required in GC is time-consuming and inconvenient. Thus, LC is now the procedure of choice (e.g., references 103, 104, and 106–109). These assays to date all involve ion-exchange chromatography, following sample pretreatment. Usually, a hydrolysis step is included to deconjugate the polyamines. In these studies, several factors have made the assays fairly slow (e.g., 40 min to 2 h per sample): (1) relatively poor column efficiencies, (2) the need to separate small amounts of polyamines from large amounts of amino acids, and (3) the use of reaction detection (ninhydrin or fluorescamine).

More recently, this general procedure has been improved.[110] First, a pre-separation of the sample amino acids and polyamines is effected on a

Table 3. Polyamines in Body Fluids

Polyamine	Structure	Normal range[101] Urine (mg/24 h)	Serum (nmol/ml)
Putrescine	NH_2—$(CH_2)_4$—NH_2	1–4	0
Spermidine	NH_2—$(CH_2)_3$—NH—$(CH_2)_4$—NH_2	2–3	0.2–0.5
Spermine	NH_2—$(CH_2)_3$—NH—$(CH_2)_4$—NH—$(CH_2)_3$—NH_2	0–1	0

pre-column or loadable cartridge. This is done off-line, prior to leading the cartridges-plus-samples onto an automatic amino acid analyzer. Second, a salt-gradient solvent program is utilized for optimum resolution and spacing of the polyamines. Third, a two-column system is used, with regeneration of one column occurring during separation of sample on the alternate column. This effectively reduces assay time by half. Finally, the reaction detector uses air segmentation for minimal extra-column band broadening; this allows an analysis rate of approximately 6 samples/h. Chromatograms from various polyamine samples by this techinque are shown in Figure 6. Further increase in assay rate should be possible, as by using bonded-phase small-particle packings in place of the present porous resins. A review on assay procedures, including LC, for polyamines in urine, serum, and cerebrospinal fluid has appeared.[110a]

3.1.4. Creatinine

This compound is a common metabolic waste product produced at a relatively constant rate in the body. It is measured in urine to assess the completeness of 24-h collections and to provide a better guide to the excretion rate of other substances than is provided by urine volume. Measurement in serum, when accompanied by urine values, allows calculation of the creatinine clearance rate, which is a sensitive and highly regarded indicator of glomerular filtration rate. Serum creatinine levels alone also are useful, since they are elevated in moderate to severe renal disease (because of reduced excretion).

The Jaffe reaction, in which alkaline picrate reacts with creatinine to form a yellow color, is widely used to measure creatinine in both serum and urine. However, the procedure is relatively nonspecific. Although less convenient, two assays for creatinine in urine and serum by LC have been reported to date, both of which offer greater (if not total) specificity. The first involves an ion-exchange LC separation of creatinine prior to an online Jaffe reaction[111] In the second assay, the sample is first cleaned up by adsorption and cation exchange. The creatinine fraction is then injected onto a small-particle C_{18} column. Detection at 254 nm shows a well-resolved peak for creatinine from both serum and urine samples.[73] The isolation procedure is quantitative and the detection limit for creatinine is 10 μg. The peak for this substance can also be collected, trifluoroacetylated, and further confirmed by gas chromatography–mass spectrometry, which constitutes a potential reference method.

3.2. Drugs

Clinical drug analysis can be organized into four major and several minor categories, as shown in Figure 10. The major categories are thera-

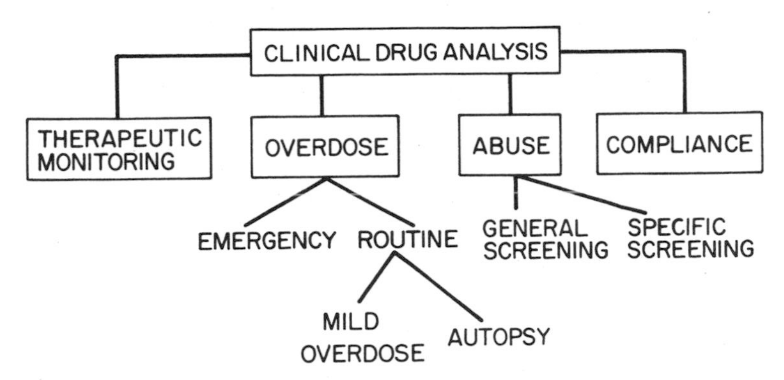

Figure 10. Categories of clinical drug analysis.

peutic monitoring, overdose, abuse, and compliance testing. This particular scheme is based on the reasons for doing the tests. It is useful to be aware of this point of view, because the purpose of a determination is related to the preferred assay methodology.

Therapeutic monitoring is becoming the most common reason for clinical drug analysis. Although the number of different drugs involved is much smaller than in the other categories of Figure 10, the frequency of requests is significantly greater and is growing steadily in regard to both current and new analyses.

The purpose of this testing is to permit therapy to be based on concentrations of the drug or its metabolites in a physiological sample from the patient, usually serum, rather than on less dependable or less desirable parameters such as dosage or toxicity symptoms. For most drugs, the relationship between the amount given and the concentration in blood and tissues varies greatly from one patient to another. This relationship is influenced by completeness of absorption, distribution and binding processes, body size and composition, and rates of metabolism and excretion. All these factors show much individual and temporal variation due to genetic and environmental factors, consequences of disease, and coadministration of other drugs. However, the therapeutic and toxic effects of many drugs correlate with their levels in serum.

Overdose testing may be requested for either emergency or routine cases. For emergency analysis the emphasis is on rapid identification or confirmation (qualitative analysis) of the drug(s), rather than on quantitation. For more routine analysis, however, such as a patient with nonthreatening symptoms, or a patient under therapeutic control, the emphasis shifts to confirmation analysis and quantitation. Various factors including clinical symptoms often suggest at least the type of offending agent, and, in the more severe cases, general treatment cannot wait for toxicological analysis. The diagnostic role of the laboratory therefore is not as great as might be anticipated. Nevertheless, specific treatment depends critically on the type of drug in many cases,[112] and clinical symptoms have been claimed to be unreliable for diagnosis in a large percentage of overdose cases.[113] Thus, toxicological data are important, but largely for reasons such as confirmation, identification of the specific agent, assessing severity, and following the progress of the patient.[114]

Abuse analysis emphasizes addiction but also includes isolated instances of drug ingestion. If the physician suspects or knows a specific drug or drug class is being abused, then the analysis request to the laboratory can reflect this by an indication for quantitative analysis. When less is known about the drug, then general qualitative testing is more appropriate. A book on this subject is available,[115] as well as a related review on urine analysis.[116]

Compliance testing seeks to determine whether the patient is taking a drug as prescribed, if at all. In other words, the question is whether the patient is following the directions of the physician. Drug testing, particularly in more accessible fluids, such as urine or saliva, obviously is one approach. The extent of this problem and various ways to deal with it have been reviewed.[117] It has been estimated that about 30% of low results in monitoring of anticonvulsant therapy were due to noncompliance in one study.[117a]

3.2.1. Methodology

Several techniques other than LC are widely used for clinical drug analysis: spectrophotometry, fluorometry, TLC, GC (including GC/MS), and immunoassays (for example, EMIT® and RIA). We will next examine the strengths and weaknesses of these methods vs. LC.

Spectrophotometry (including calorimetry) and fluorescence may be considered together because of various similarities. These methods have certain specificity limitations. However, their lack of sophistication and complexity is often an advantage for commonly encountered drugs, because it means that the assays are rugged and satisfactorily performed by routine personnel. Also, these analyses often are used to confirm results from other, more qualitative techniques such as TLC. Nevertheless, many drug analyses are beyond the capability of these methods because of limitations of specificity or sensitivity.

TLC for drug analysis offers important advantages, particularly: (1) simultaneous analysis of many samples on the same plate, (2) rugged conditions for separating most of the drugs in the classes encountered commonly, and (3) low cost. For general or specific screening purposes, this technique probably is the method of choice whenever sensitivity is adequate and only semiquantitative information is sought. TLC therefore is well-suited for drug abuse and overdose screening. However, for therapeutic monitoring, where quantitative information (often at low levels) is required, TLC is not useful. Also, interfering spots can be encountered on TLC plates, because of the limited resolution and specificity available, so that a confirmatory test such as spectrophotometry or GC is often used (this also quantitates the answer). Although some workers and companies have claimed that TLC with extra steps such as alternate mobile phases can confirm results on its own, in practice this procedure is not often used.

Gas chromatography and GC/MS are widely used in more sophisticated laboratories for drug analysis. Neither method is inherently a rapid technique. This is because of time for instrument warm-up, limitation to one analysis at a time, and return of the oven to starting conditions whenever temperature programing is employed. Frequent claims in the literature that these are rapid techniques refer to an instrument that is warmed up and ready for the next injection. The main advantages of GC are its high sensitivity and its ability to quantitate several drugs in a single run. When the detector is a mass spectrometer, the analysis now usually becomes qualitative rather than quantitative for clinical drug analyses, but the ability to identify instantly a large number of possible drugs is achieved because the mass spectrum usually is fed automatically into a computer library containing a large number of spectra (e.g., 1000 or more) for identification by matching. Thus, GC/MS is an important tool for emergency testing in which uncommon agents are present or whenever other techniques, for various reasons, are unsuccessful. Some laboratories have used GC/MS for primary screening of all emergency samples and have found that certain drugs are involved most of the time. It has been profitable to carry out simpler and less costly procedures such as TLC or GC with simple detection (e.g., flame ionization) whenever possible; thus, GC/MS is being used more for severe emergency or problem samples. Its utility for overdose drug analysis has been reviewed.[118]

A general disadvantage that is often cited for GC is its inability to analyze ionic, thermally labile, or high-molecular-weight substances. However, the vast majority of drugs are volatile, stable, low-molecular-weight materials, so that GC and GC/MS have broad applicability. Even the fairly polar drugs can be analyzed directly. However, peak tailing and adsorption losses of polar drugs tend to make such analyses less reliable than analysis after derivatization (e.g., acetylation, methylation) to form less polar products. On-column derivatization suffers from column deterioration and therefore is not a general answer to some of these problems, as has been pointed out.[119] Also, conjugated drug metabolites (e.g., drug–glucuronate, drug sulfate) require acidic or enzymatic hydrolysis prior to GC analysis. These problems, along with (1) the need for skilled personnel, (2) a significant initial investment, (3) sample cleanup and frequent derivatization requirements, (4) excessive analysis times in some cases, (5) occasional problems with column life and reproducibility, and (6) requirement for column change-over for some analyses, have all limited the use of GC and GC/MS for drug analysis. However, GC and GC/MS play and will continue to play a major role in this field.

Immunoassays such as EMIT and RIA enjoy wide and increasing use in all types of drug analysis. Their advantages and disadvantages relative to each other and versus other techniques such as GC and LC for drug testing have been discussed previously.[120,121] Both techniques involve analysis by competitive binding and thus are indirect. The readout is enzymatic activity for EMIT, whereas it is radioactivity (either β or γ counting) for RIA. Both procedures offer elimination of extraction and cleanup steps (although predilution of samples may be required for some RIA analyses) and high sensitivity (especially for RIA). EMIT is exceedingly rapid for drugs in the milligram per liter range, such as the commonly tested anticonvulsants and theophylline in serum, offering sample turnaround times of about 3 min.

A general disadvantage of both EMIT and RIA is their susceptibility to specific and nonspecific interferences. As an example of a specific interference, *p*-hydroxyphenobarbital, a major metabolite of phenobarbital, interfered in a radioimmunoassay of the latter in both serum and urine, unless the phenobarbital was selectively extracted into chloroform.[122]

3.2.2. *Therapeutic Monitoring: General*

Most applications of drug analysis by LC have involved therapeutic monitoring, particularly for anticonvulsants, antiarrhythmics, and theophylline. Aside from the extensive need for monitoring of these particular drugs, LC analyses for them have been particularly effective because these drugs are good UV absorbers and have therapeutic levels in the microgram per milliliter range. Thus, these drugs represent a relatively easy analysis for LC. Development of LC assays for other drugs which are therapeutically effective at lower levels (e.g., nanogram per milliliter level), or which are less UV-active, obviously will be more challenging. Specific drugs and references are included in Table 4. (For background discussions, see references 123, 124, and 124a.) Apart from the assay of these drugs in serum, other assays of drugs in physiologic fluids at the present time are for the most part on a research rather than clinical basis. Assays of drug metabolites in urine are often reported, but these are mainly of interest in connection with pharmacokinetic or bioavailability studies. Although extensive work already has been carried out to establish therapeutic ranges for total drugs, some interest has been expressed in determining the concentration of the

Table 4. Current and Potential Therapeutic Drug Monitoring by LC[a]

Drug function	Drug	Therapeutic range (μg/ml)	References
Anticonvulsants	Phenytoin	10–20	117a, 124, 128, 130, 134, 135, 138, 141, 141a
	Phenobarbital	20–40	124, 130, 135, 139–141, 262
	Ethosuximide	40–100	128, 130, 135
	Carbamazepine	6–10	128, 130, 132, 135, 137, 140, 142, 143, 262
	Primidone	8–12	128, 130, 135, 141
	Valproic acid	50–100	176, 262
	Mephenytoin	5–16	176, 181–183, 262
	Benzodiazepines		144, 262
	Trimethadione	20	262, 266a
	Paramethadione		266a
	Methsuximide	0.1–1.4	262, 266a
	Ethotoin		266a
	Phenacemide		262
Antiarrhythmics	Procainamide	4–8	145–149
	Lidocaine	1–6	145
	Quinidine	4–8	150
	Propranolol		150a,b
	Diisopyramide		151
	Acebutolol		151a
Bronchodilator	Theophylline	10–20	129, 131, 152–158
	Dyphylline		179, 180
Antisclerotic	*p*-Aminobenzoic acid and metabolites		166, 167
	Clofibrate		167a
Antiarthritic	Phenylbutazone and metabolites		168
Antiuricemic	Allopurinol and metabolites		169
Antiintraocular tension	Acetazolamide		170
Antithrombotic	Sulphinpyrazone		171
Anticarcinogenic	5-Fluorouracil, methotrexate		172, 288

(Continued)

Table 4. ***(Continued)***

Drug function	Drug	Therapeutic range (μg/ml)	References
Antiinflammatory	Diftalone and metabolites		173
Analgesic	Carprofen		173a
	Phenacetin		174, 174a
	Acetaminophen		174a, 175, 175a
Antibiotics	Gentamicin		176, 176a,b
	Tobramycin		
	Mefruside		176c
	Cephalosporins		176d
Chemotherapeutics	Adriamycin		177, 177a,b
	Daunorubicin		
	Amikacin		
Antifibrinolytic	ε-Amino caproic acid		178
Diuretics/antihypertensives	Hydrochlorothiazide		159–161
Antiparkinsonism	L-Dopa (and dopamine)		162
Antipsychotics	Perphenazine		163, 176
	Fluphenazine		
	Chlorpromazine		
Antidepressants	Amitriptyline		164, 164a,b,c
	Nortriptyline		
	Chlorimipramine		165
	Imipramine		
Antimalarials	Mefloquine		165a
Anesthetics	Flunitrazepam		165b

[a] This list mostly presents drugs for which LC assays have been established in serum or urine. A few of the references simply give basic information about the drugs. (See Note Added in Proof.)

drug which is free in serum, as opposed to the sum of free plus protein-bound drug. Probably, as has been suggested (e.g., references 125 and 125a), free drug concentrations are a more reliable index of patient status. Whether or not assays for free drugs will be generally adopted remains to be seen; they should probably be determined whenever the therapeutic status of the patient is markedly inconsistent with the total serum level of the drug.

Measurement of drugs in saliva rather than serum does offer a non-invasive approach. A drawback is that the pH of the saliva varies considerably (5.8–7.8), and the salivary epithelium is impermeable to the ionized forms of drugs.[126] Thus, salivary concentrations of drugs with pK_a values in this range will depend on the pH of the saliva at the time of sampling. Nevertheless, the salivary levels of the anticonvulsants recently have been shown to offer an approach to their monitoring.[127] Included in this report was a method to correct for the effect of salivary pH on the concentration of phenobarbital (pK_a of 7.2), since, as expected, salivary phenobarbital concentrations varied widely as a function of salivary pH.

Because most drugs are nonpolar, or can be rendered so by pH adjustment, they generally can be extracted from serum with an organic solvent such as chloroform or ethyl acetate. This step separates the drug from some of the interfering substances (including proteins) in the sample. The organic extract usually is evaporated to dryness and then redissolved in a smaller volume of some solvent that is compatible with the subsequent LC separation, e.g., methanol/water for a reverse-phase column. Other approaches to LC sample pretreatment have included charcoal extraction of the drug (e.g., reference 128), precipitation of serum proteins by acetonitrile or methanol, followed by direct injection (which is particularly attractive wherever possible, e.g., references 129 and 130), and ultrafiltration (e.g., reference 131). A few workers have chosen to inject serum samples directly onto the column; however, frequent injections of large-volume samples (e.g., 10 μl or greater) tend to shorten the life of most columns now used in clinical LC, apparently because of protein (and lipid?) accumulation on the column. Analyses placing relatively little protein on the column (per injection) allow more assays per column than analyses involving injections containing greater amounts of proteins.

Columns used in therapeutic drug assays have included reverse-phase (usually C_{18}–silica), silica, and ion-exchange packings, with either pellicular or porous particles of 5–40 μm diameter. However, most recent assays, and particularly the faster ones, have used 5- to 10-μm reverse-phase packings. Several authors (e.g., references 128, 129, 131, and 132) have commented that with suitable precautions it is possible to use the same column for 300–600 assays *without* noticeable column degradation. Some workers have found that a simple deproteinization of the sample is adequate (e.g., references 129 and 130) although most use a more involved extraction scheme. A good pre-LC cleanup of serum samples is, in any case, recommended.

Alternatively (or in addition to sample cleanup), guard columns can be used to protect the main analytical column from sample contamination and to extend column life. Guard columns are typically 5-cm lengths of the same inside diameter as the main column, packed with a 30- to 40-μm pellicular packing of the same type used in the main column (e.g., pellicular C_{18}). The guard column is connected between the sample injection unit and the main column and is replaced periodically as needed. Its function is to protect the main column from buildup of particulate or strongly retained components in the sample. For a detailed description of the use of guard columns and their effect on over-all column efficiency, see reference 130a.

In most reported assays of serum for therapeutic drugs (cf. Table 4), analysis times are short and apparent plate numbers for individual peaks are not greater than 500–1000 plates. As a result, possible interferences or overlapping peaks can easily go unnoticed. Several studies have shown that in most of these LC methods the possibility of significant interference is unlikely. However, this possibility must always be kept in mind, and the shapes and exact (or relative) retention times of the peaks should be checked. As discussed earlier, wavelength ratioing can be used to ascertain if more than one peak is under a chromatographic band.

3.2.3. Therapeutic Monitoring: Anticonvulsants (Antiepileptic Drugs)

At the present time, the analysis of this class of drugs represents a major potential for LC in the clinical laboratory. It is estimated that over two million patients in the United States presently take one or more of the five major anticonvulsants: phenytoin, phenobarbital, ethosuximide, carbamazepine, or primidone. For a general review of the subject of serum assays for the anticonvulsants, see reference 133 for colorimetric and GC assays and reference 134 for LC assays. Table 4 lists most of the anticonvulsants in clinical use today, as well as references to reported LC assays, plus the commonly accepted therapeutic ranges. The latter "normal ranges" are of interest here mainly to show the required sensitivity for each assay, which is usually of the order of 1 μg/ml. These relatively large concentrations of the therapeutic drugs in serum, in conjunction with their generally high UV absorptivity, mean that detection is normally not a problem; sample derivatization for increased detectability therefore is not needed.

The general approach to the pretreatment and LC separation of these compounds is described above. The simultaneous separation of all five major anticonvulsants in 8 min has been reported.[117a,130] This procedure employs a simple pretreatment of sample (solvent precipitation of protein, followed by centrifugation), and requires only 25 μl of serum.

This assay has been found to work well in practice. A shortcoming is that it does not resolve phenylethylmalonamide, the major metabolite of primidone, from primidone, as reported by the original authors.[130] The optimum column temperature has been found to decrease with use. The eluted drugs are routinely detected by their absorbance at both 200 and 254 nm, as shown in Figure 11. A similar, more recent assay for the same drugs has been reported which claims to offer some improvements,[135] although requiring 15 min/assay. Results from the analysis of routine samples have yet to be reported.

The separation and analysis of the common barbiturates has been

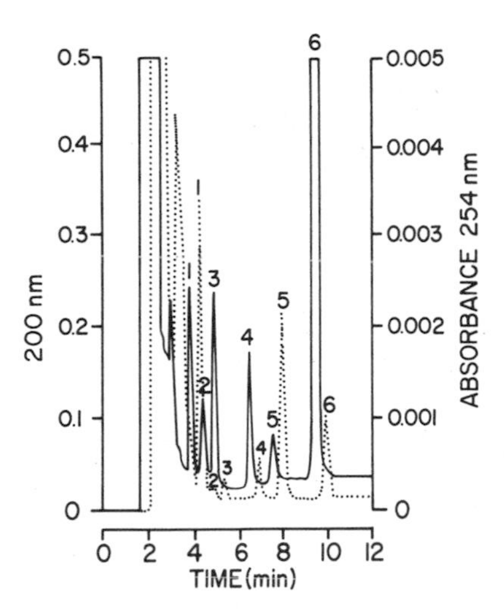

Figure 11. Analysis of anticonvulsants in serum. Column: 300 × 4 mm μBondapak C_{18}; mobile phase: 50/50 mixture of acetonitrile/0.01 M potassium phosphate buffer, pH 8.0, at a flow rate of 0.6 ml/min; detection: 200 nm (solid line) and 254 nm (dotted line); temperature: optimum value decreases with use (range 30–14°C) and is checked daily; peaks: (1) phenobarbital; (2) primidone; (3) ethosuximide; (4) phenytoin; (5) carbamazepine, and (6) cyheptamide internal standard. (Reprinted from reference 117a with permission.)

described.[136] Measurement at 220 nm necessitated an involved sample pretreatment (detection specificity generally decreases at lower wavelengths). The acidic barbiturates were first extracted from acidified serum with an organic solvent and then back-extracted twice with intermittent pH adjustment.

Assays for other anticonvulsants (e.g., benzodiazepines) are noted in Table 4, and these determinations generally parallel the procedures used to assay the major anticonvulsants.

Metabolites of the anticonvulsants are of occasional interest. Thus carbamazepine-10,11-epoxide is an active form of the parent drug carbamazepine and should be assayed along with the latter (e.g., references 132 and 137). Hydroxylated metabolites and conjugates of the anticonvulsants are found in urine (e.g., *p*-hydroxyphenytoin, reference 138), but are of research interest only.

3.2.4. Therapeutic Monitoring: Antiarrhythmics

These drugs, also listed in Table 4, previously have been assayed mainly by GC or by nonspecific methods. More recently several papers have described the LC assay of the major antiarrhythmics, procainamide, lidocaine, and quinidine, as well as the experimental drug diisopyramide. Pretreatment and LC separation procedures are similar to those used for the anticonvulsants, with detection variously at 205 and 254 nm. The *N*-acetyl metabolite of procainamide also possesses antiarrhythmic activity and should be assayed along with procainamide.[146–149] Serum levels of this active metabolite as much as nine times greater than that of the parent procainamide have been seen.[147] Figure 12 illustrates one LC assay for procainamide plus *N*-acetyl procainamide.

3.2.5. Therapeutic Monitoring: Theophylline (Bronchodilator)

This compound is widely used in the treatment of asthma and emphysema. Several reported LC methods for its assay in serum (Table 4) attest to the importance of this determination. Pretreatment and LC separation procedures resemble those for the anticonvulsants. Detection at 280 nm provides adequate sensitivity and eliminates many potential interferences. An assay can be carried out with as little as 5 μl of serum. For a relatively simple procedure, see the method described in reference 158 (illustrated in Figure 13). Because of interferences (e.g., from caffeine in coffee) and other problems, GC and spectrophotometric procedures for theophylline have been unsatisfactory.

3.2.6. Therapeutic Monitoring: Other Drugs

Many other drugs of therapeutic interest have been assayed by LC in serum or urine, but few of these determinations have yet entered the routine clinical laboratory. Presumably this reflects lack of agreement on the need to monitor these drugs during therapy. In other cases, the primary goal of such assays is clearly in connection with pharmacokinetic or

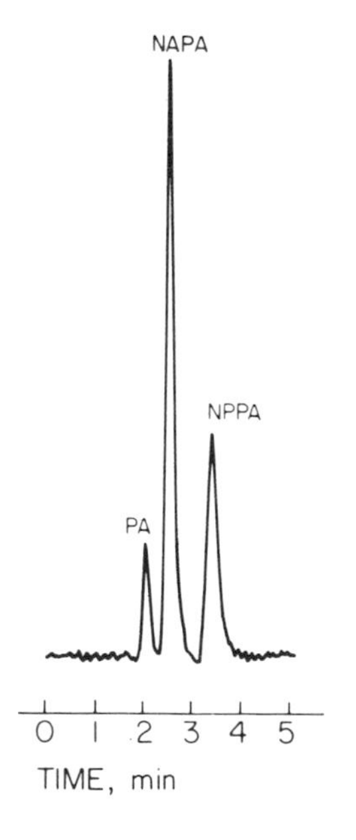

Figure 12. Assay for procainamide (PA) and *N*-acetyl procainamide (NAPA). Serum from a patient with a very high weight ratio (nearly 9/1) of NAPA to PA. Column: 300 × 4 mm μBondapak C_{18} (reverse phase); mobile phase: 40/59/1 by vol methanol–water–acetic acid, pH adjusted to 5.5 with NaOH; flow rate: 2 ml/min. (Reprinted from reference 147 with permission.)

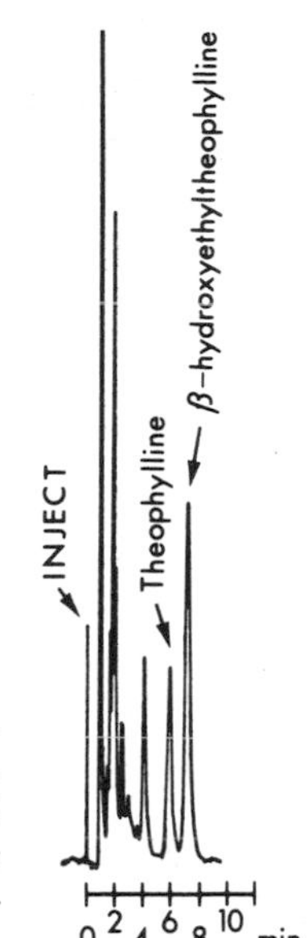

Figure 13. Routine determination of theophylline in serum. Concentrations: 10.7 mg/liter theophylline; 36.7 mg/liter internal standard. Conditions: column, 300 × 4 mm μBondapak C_{18}; mobile phase, 7% (volume) acetonitrile/aqueous acetate (pH = 4.0); 2.0 ml/min flow rate; detection at 254 nm. (Reprinted from reference 158 with permission.)

bioavailability studies. Nevertheless, many of these LC assays should eventually develop into clinically useful tests. A few such applications of possible future clinical interest are described below.

The thiazides, especially hydrochlorothiazide (HC) are widely used as diuretics and antihypertensive agents. Significant side effects are often noted in treatment with HC, suggesting a need for monitoring. Several assays for HC in serum by LC have been reported,[159–161] covering the therapeutic range of roughly 100–700 ng/ml. Each of these assays involves solvent extraction on serum samples as small as a few hundred microliters, followed by reverse-phase LC on small-particle columns. Detection is at 270–280 nm.

L-Dopa is commonly used for Parkinson's disease. In some cases inhibitors, e.g., carbidopa, having extracerebral dopadecarboxylase activity are coadministered. In these cases it appears desirable to monitor the serum concentrations of both L-dopa and dopamine, the latter serving as an index of the effectiveness of the inhibitor. Recently an LC assay for drug levels of these two catecholamines in serum was described.[162] The demanding sample pretreatment included hydrolysis of conjugates to the free drugs, deproteinization, and alumina extraction. LC analysis then involved pellicular cation-exchange separation followed by electrochemical detection.

The phenothiazines are neuroleptics used in treatment of psychiatric out-patients. It has been suggested[163] that certain of these drugs should be monitored in order to avoid relapse of the patient. An LC assay for the compounds perphenazine and fluphenazine in serum has been reported.[163]

An appreciable relapse rate also exists among psychiatric patients under treatment with the tricyclic antidepressants such as amitriptyline and nortriptyline.[164] It appears that blood levels of these drugs vary widely among different individuals with the same dosage rate. Most assays are carried out at present by GC (e.g., reference 164), but recently an LC assay was reported for chlorimipramine and its active metabolite desmethyl chlorimipramine in plasma.[165] The latter assay featured ion-pair separation of the two compounds, followed by detection at 254 nm. The other tricyclic antidepressants can also be quantitated by this system. These drugs are normally not administered in combination, which simplifies the LC analysis. Further references to LC analyses of the tricyclic antidepressants are cited in Table 4.

Many other drugs have been assayed in serum or urine by LC, but the clinical significance of these assays currently is unclear. Some of the drugs obviously are experimental, others are used clinically, but any need for therapeutic monitoring remains to be established. Table 4 includes several such drugs that have been assayed in serum or urine by LC, drugs for which LC assays should be forthcoming, and also the drugs which already have been discussed.

3.3. Hemoglobin AIc

Hemoglobin AIc is a minor component of hemoglobin found in normal individuals but elevated two- to threefold in patients with diabetes mellitus. Limited studies have suggested that the level of hemoglobin AIc reflects the mean blood sugar concentration over a previous period of time, e.g., possibly a few months. Thus, it has potential value in regard to the diagnosis, prognosis, and monitoring of diabetes mellitus.[165c,d] An LC assay for hemoglobin AIc has been reported which employs a cation-exchange column and takes 27 min.[165] A fraction of hemoglobin which contains hemoglobin AIc and is easier to isolate is called "total fast hemoglobin" based on its electrophoretic behavior. It was measured in 11 min by these workers and was also found to be elevated in diabetes mellitus.

3.4. Hormones

Except for the catecholamines, already discussed in the category of biogenic amines, little has been achieved so far in regard to the practical analysis by LC of hormones in physiological fluids. The few procedures that have been published have mostly fallen short of being immediately applicable to clinical chemistry, either because they are limited to analyses of standards only or of pathological samples containing concentrations far above physiological values. Some workers have used modern rather than classical LC to separate hormones prior to analysis by radioimmunoassay or a related technique, but this largely represents research, rather than practical clinical chemistry (e.g., reference 184). Nevertheless, much effort is likely to be given to practical hormone analysis by LC in the next few years, now that ultrasensitive detectors for LC (e.g., fluorescence and electrochemistry) have become commercially available. Many of the hormones will probably require derivatization in order to be measurable at physiological levels, even by these detectors. Exceptions include a few hormones which are present in somewhat higher concentration (e.g., cortisol in serum), or which can be conveniently and rapidly concentrated from a large volume of urine so that direct detection is possible. Also, the inherent electroactivity of certain hormones (e.g., the catecholamines) allows direct detection at physiological levels without derivatization.

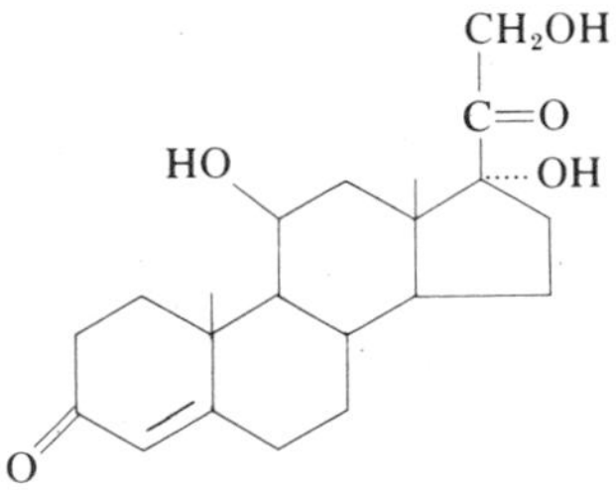

CH_2OH
C=O
HO
OH
O

Cortisol

3.4.1. Steroids

Steroid hormones in plasma and their conjugated metabolites in urine are mainly measured to assess the status of the adrenal cortex, the pituitary (which regulates the adrenal cortex), the gonads (ovaries and testes), and the fetal–placental unit. In other words, all the tissues which play a major role in steroid hormone biosynthesis are considered.

The analysis of steroid groups (e.g., 17-ketosteroids) in urine has been carried out by classical colorimetric procedures such as the Zimmerman reaction. These older, less specific assays have gradually given way to protein-binding assays (PBA) for plasma steroids and to gas chromatography (after hydrolysis and derivatization) for specific urinary steroids. In some cases the PBA procedures for plasma steroids are adequately specific and free from interference. However, there is a definite need for LC procedures for the various urinary steroids because of the tediousness of corresponding GC methods.

Cortisol is the one steroid hormone in plasma which is present in sufficiently high concentration (~20–250 μg/liter in normal plasma) to allow quantitation with a UV detector. As a result, the only LC procedures reported thus far for endogenous steroid hormones in human plasma have been for cortisol.

In one of the first such procedures to be reported, plasma was treated with ethanol, internal standard (prednisolone), and sodium sulfate, and then extracted with methylene chloride. The separated organic layer was taken through washing and concentration steps and then injected onto a silica column. Detection at both 239 and 254 nm gave peaks for cortisol and prednisolone after approximately 18 and 28 min, respectively, as shown in Figure 14. Comparison with PBA analysis for a series of patient samples, normal and diseased, revealed a higher average value for cortisol by LC (11.1 μg/dl) than by PBA (7.1 μg/dl), with values from several samples being at least twofold higher by LC. A patient on decortin had an apparent false-positive result for cortisol (6.8 μg/dl) by PBA, but no cortisol was detected by LC.[185] A procedure very similar to this was published more recently.[185a]

In a slightly different procedure for cortisol,[186] 0.5 ml of plasma was extracted with methylene chloride after addition of [^{3}H]cortisol (5000 counts, as an internal standard), methanol, and aqueous sodium hydroxide. The aspirated organic layer was blown down, reconstituted with methylene chloride, and a portion was injected onto a silica column and eluted using dichloromethane/methanol. Detection at 254 nm showed a peak for cortisol at about 6 min. This peak was collected into a scintillation vial for counting. Quantitation then was based on the recovery of radioactivity (60–90%) and the peak area. The advantages of this analysis are speed (30 min total analysis time) and minimization of interferences. A disadvantage is the large sample volume (RIA for cortisol requires only 10–25 μl of serum.)

LC analyses also have been reported for steroids in urine. The choice here is either to assay the unhydrolyzed steroid metabolites directly or after a hydrolysis step (the hydrolysis step removes the conjugated glucuronate and sulfate groups). Either approach in principle offers an easier analysis than GC, because the latter requires a tedious derivatization step after hydrolysis in order to render the steroids adequately volatile and thermally stable for the GC separation.

Direct analysis of the conjugated estrogenic steroids in urine by LC has been investigated, but several difficulties have been encountered. These steroid conjugates can be extracted from urine with XAD-2, eluted with methanol, and separated to some degree by various LC phases such as ion exchange on ECTEOLA–cellulose. UV detection was found to be subject to many interfering peaks, as shown in Figure 15, which compares the elution pattern by UV versus the actual elution of steroids via off-line

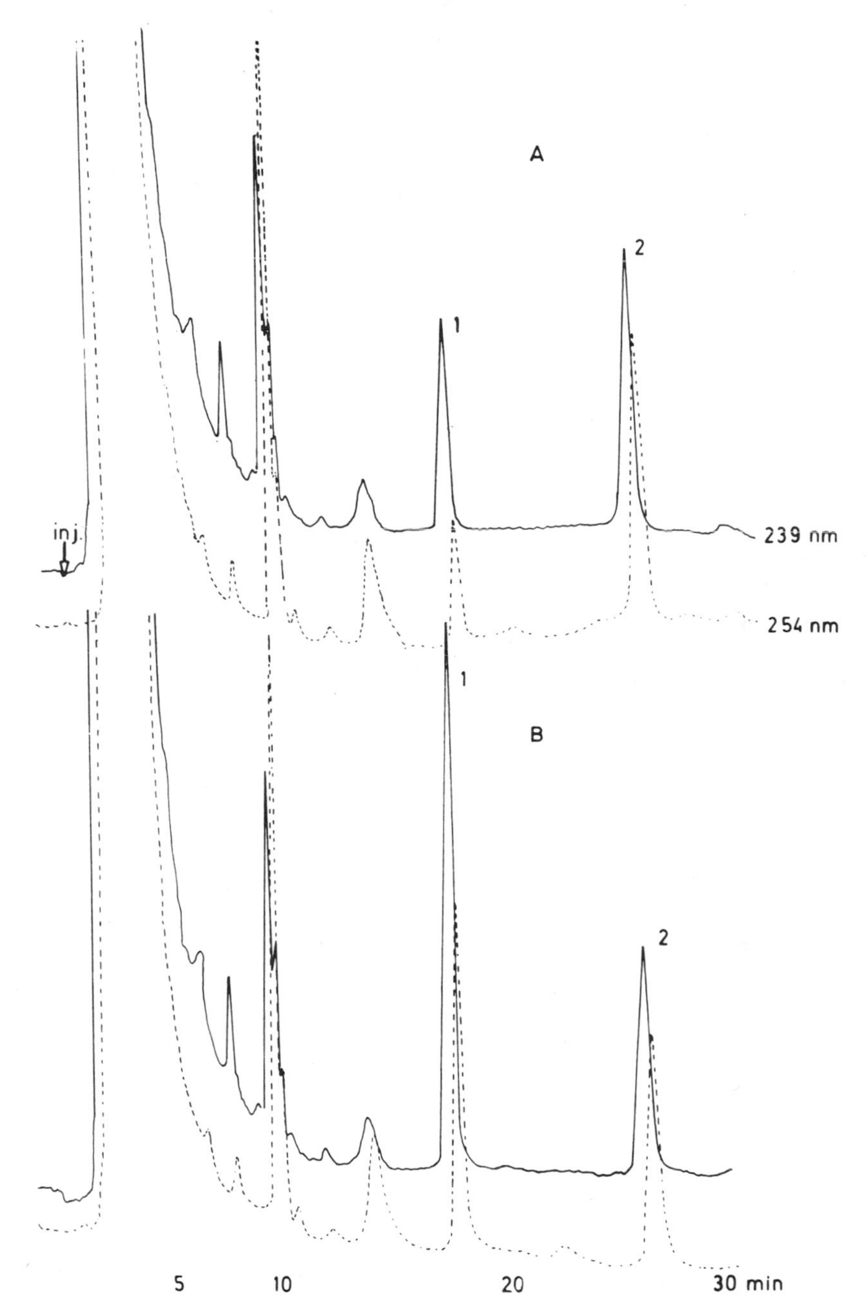

Figure 14. (A) Assay for cortisol in serum. Plasma extract of a healthy child. (B) Chromatogram of the sample sample as in A after addition of 113 ng of cortisol prior to extraction. 1 = cortisol; 2 = internal standard (prednisolone). Column: Zorbax-Sil, 250 × 2.1 mm; eluent: dichloromethane–ethanol–water; flow rate: 0.45 ml/min; temperature: 20°; recording at 239 nm, Schoeffel SF 770 detector, 0.02 AU; recording at 254 nm, Chromatronix 200 detector, 0.01 AU. (Reprinted from reference 185 with permission.)

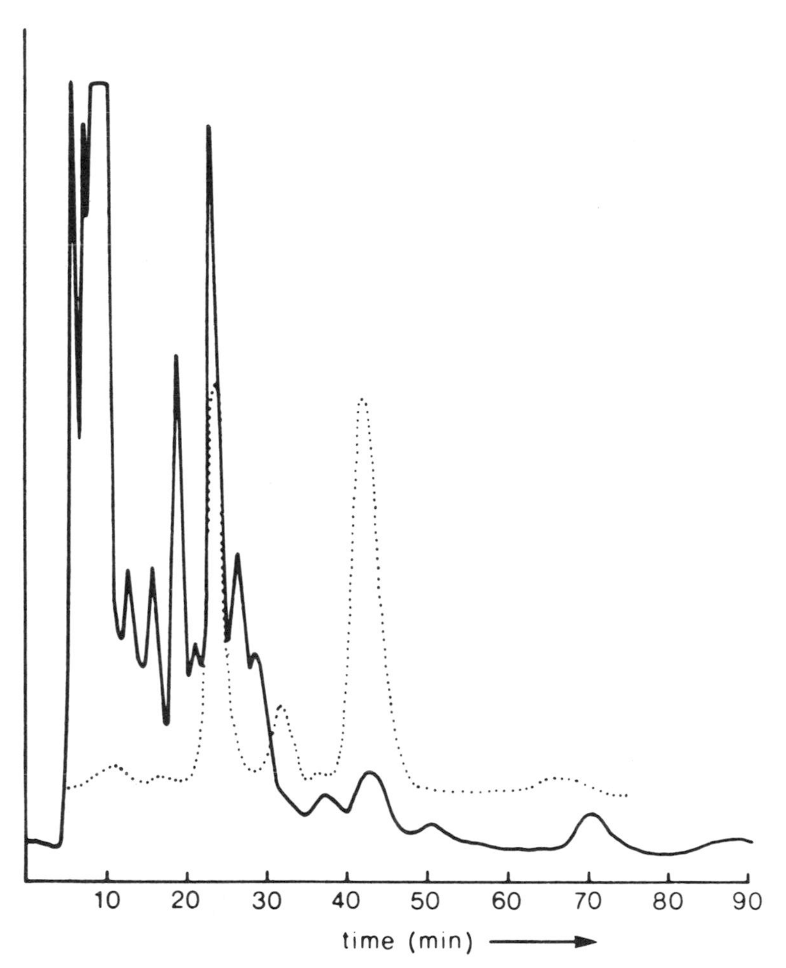

Figure 15. Assay for conjugated estrogens in urine. Methanolic XAD-2 extract of human pregnancy urine. Solid line, absorbance at 220 nm; dotted line, chromatogram obtained by off-line batch determination of estrogenic material. Phase system: ECTEOLA–cellulose (anion exchange) B 300, 11 μm; 0.025 M perchlorate–0.01 M phosphate, pH 6.8. Temperature, 70°; pressure, 20 bar. (Reprinted from reference 187 with permission.)

Kober reaction.[187] Another difficulty is that each type of steroid metabolite (e.g., estriol) can be encountered conjugated to either glucuronate or sulfate, and alternate sites of attachment tend to occur when more than one hydroxyl group is present on the steroid. This adds considerable complexity to the chromatograms. Nevertheless, various partial separations of conjugated estrogen standards have been achieved, utilizing different phase systems. An on-line, segmented-flow Kober reaction has been introduced to enhance detection specificity.[188] Chromatograms from urine samples were presented, although the largely unresolved peaks were not identified. Others have used reverse-phase LC to separate standards of free 17-ketosteroids and their sulfates and glucuronides.[189]

Analysis of urinary estrogens by LC after a hydrolysis step also has been investigated.[190] A 40-ml sample of pregnancy urine (in which the urinary estrogens are considerably elevated) was hydrolyzed with concentrated hydrochloric acid, and the estrogens were extracted into diethyl ether. Concentrated extracts were then analyzed both on silica and on octadecylsilica, with detection at 280 nm. The chromatograms are shown in Figures 16 and 17B, with analysis much more successful on the reverse-

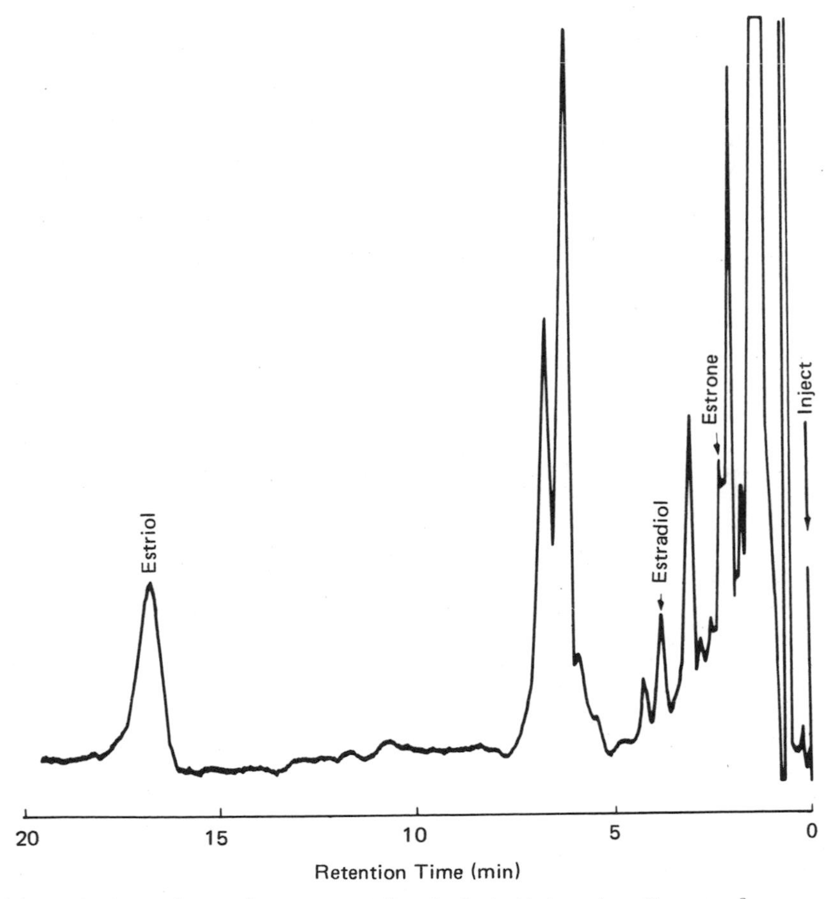

Figure 16. Assay for total estrogens (after hydrolysis) in urine. Extract of pregnancy urine separated on Partisil-5. Mobile phase, 5% (v/v) ethanol in *n*-hexane; flow-rate, 5 ml/min; detector, 280 nm range; range, 0.1 absorbance unit. (Reprinted from reference 190 with permission.)

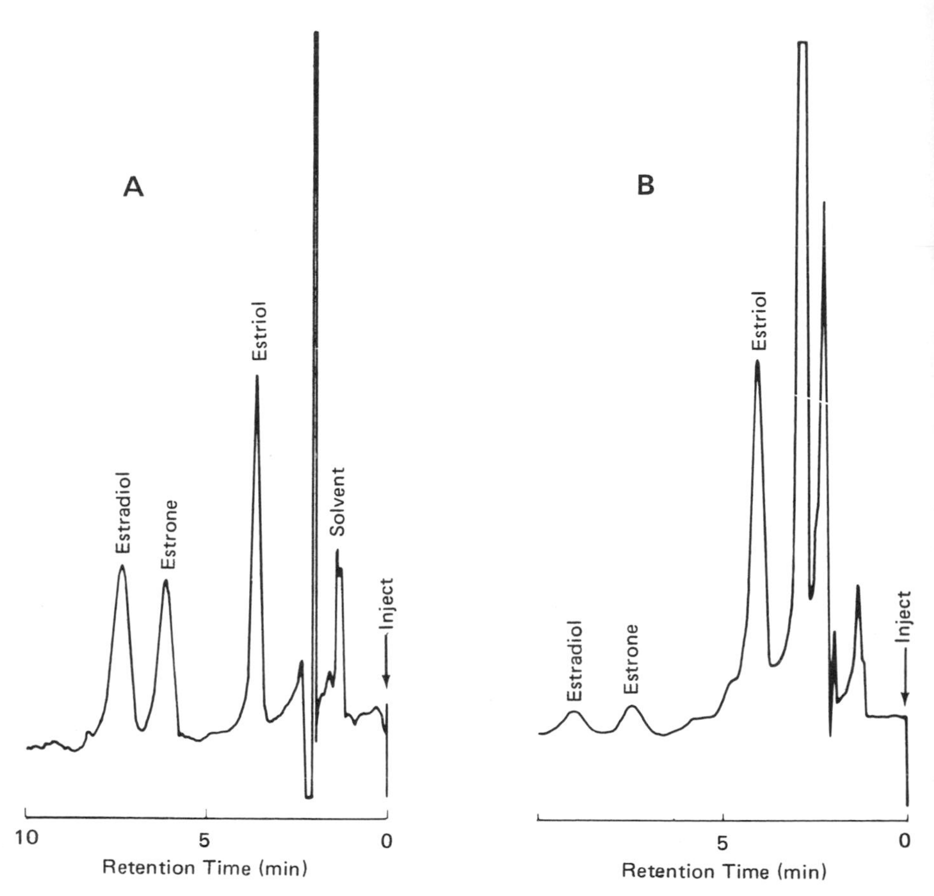

Figure 17. (A) Chromatogram of estrogen standards. (B) Separation of estrogen mixture of Figure 16 on Partisil-10 ODS. Mobile phase, methanol–0.1% aqueous ammonium carbonate (55:45, flow rate 2 ml/min; detector range, 0.25 absorbance units. (Reprinted from reference 190 with permission.)

phase system. In an analogous procedure, a peak for estriol was seen when an ether extract of hydrolyzed pregnancy urine was analyzed by amino bonded-phase LC using detection at 217 nm.[191] A procedure for determining free cortisol in urine also has been reported.[192] A silica gel column was used with detection at 254 nm.

3.4.2. *Thyroid Hormones*

In one of the few efforts to assess accuracy in routine testing of hormones, it was found that most laboratories overestimated T_3 and underestimated T_4.[193] This suggests an opportunity for reference analysis and perhaps even routine analysis of thyroid hormones by LC. The ability of LC to separate standards of the thyroid hormones in less than 2 min has been demonstrated,[194] but considerable work remains to set up a reliable LC procedure for quantitation of thyroid hormones in physiological samples. Major problems to be solved are sample cleanup and detection of the low concentrations.

3.4.3. Polypeptide and Protein Hormones

Now that attention is being directed toward protein separations by LC (see following section), it is probable that similar analyses for polypeptide and peptide hormones will be attempted. For example, the tendency of these compounds to circulate in the body in several molecular forms, as well as the presence of similar (immunologically cross-reactive) compounds in serum, suggests the use of LC separation in conjunction with some form of immunoassay. However, the next several years are likely to see only research efforts in this area, due to the special problems involved in the detection of the minute amounts of these substances present in serum.

3.5. Isoenzymes

The determination of various serum isoenzymes is one of the faster-growing areas in routine clinical testing (e.g., reference 195). Serum enzymes have found heavy use over the past two decades as markers of tissue damage. For example, death of heart cells releases their characteristic enzymes into the bloodstream. Because a given enzyme is often found in more than one organ, there is ambiguity in assigning elevated serum concentrations of that enzyme to a particular organ. Fortunately, many enzymes exist in multiple molecular forms called isoenzymes, which show greater organ specificity. For example, while the enzyme creatine kinase (CK) is found in heart, muscle, and brain tissue, there is a specific CK isoenzyme (CK-MB) that is mainly present in heart tissue. Elevated concentrations of CK-MB in serum are therefore highly indicative of a heart attack (and its extent), and the assay for CK-MB is commonly used to confirm this diagnosis. Other examples of the organ specificity of various serum isoenzymes are listed in Table 5.

Despite the importance of serum isoenzymes in the differential diagnosis of a number of disorders, present methods for their assay are unsatisfactory. These methods include electrophoresis, kinetic assay employing differential inhibition, and immunoassay. Electrophoresis has been frequently criticized for its insensitivity and inconvenience. Inhibition and immunoassay methods generally lack specificity; for example, several such methods for the assay of CK-MB report the MB (heart) and BB (brain)

Table 5. Isoenzymes of Clinical Significance

Enzyme	Tissue specifity of a given isoenzyme[a]	References
Acid phosphatase	Prostate	195, 196
	Spleen	
Alkaline phosphatase	Liver	195
	Bone	
	Intestine	
	Kidney	
	Placenta	
Amylase	Pancreas	
	Salivary glands	
Creatine kinase	Muscle (MM)	195
	Heart (MB)	
	Brain (BB)	
γ-Glutamyl transferase	Liver	198
	Bile duct	
	Pancreas	
Lactate dehydrogenase	Heart (LD_1)	195
	Liver (LD_5)	

[a] Allows major organ or tissue differentiation.

isoenzymes as a sum, with the assumption that BB concentrations are negligible. However, several other studies (e.g., references 199–203) have shown that this assumption can lead to diagnostic error. Specific assays for CK-MB are preferred.

More recently, low-pressure LC assays for the various isoenzymes have become increasingly popular. This is particularly true of the determination of CK-MB (e.g., reference 204). While these methods, in principle, overcome some of the objections raised against earlier isoenzyme assays, they are relatively slow. In view of the increasing volume of such assays required of the clinical laboratory, this is a major drawback. More seriously, unless special precautions are taken, these LC methods often give unacceptable errors (e.g., reference 205). A frequent problem associated with the use of stepwise gradient elution in these assays is the poor separation of adjacent isoenzyme bands leaving the column.

The use of high-performance LC with continuous gradient elution plus on-line reaction detection permits the reliable and rapid (4 to 6 min) separation of the isoenzymes of CK and lactate dehydrogenase.[206] However, these applications have not been reduced to actual clinical procedures. Two major problems have existed in the development of high-pressure LC assays for the isoenzymes in serum: the column and the detector. Column packings suitable for the high-pressure LC separation of proteins have only recently been reported (see reference 207 for a review), and it appears that these materials are not yet adequately stable for routine clinical use. Apart from stability, what are required for separations of the isoenzymes are wide-pore, small-particle ion exchangers that do not denature or irreversibly adsorb protein molecules. One approach to this problem has been to covalently coat the surface of porous silica particles with a layer of glycerol-like molecules, to which diethylaminoethyl (DEAE) groups are attached for anion exchange. The resulting product[207] behaves much like the commonly used DEAE-agarose gels, except that its rigidity allows its use in high-pressure LC. Hopefully, packings of this type can be made reliable and stable for the routine clinical separation of isoenzymes in serum.

The second problem in LC isoenzyme assays is the detection of separated isoenzymes as they elute from the column. The isoenzymes are present in very small concentrations, and coelute with much larger amounts of other serum proteins. Complete separation of the isoenzymes of interest from these serum proteins is not practical. Selective detection of separated isoenzyme bands can be achieved, however, based on their activity,[208,209] as discussed in an earlier section dealing with reaction detectors.

Figure 5 shows two examples of what has so far been accomplished in the LC assay of isoenzymes in serum. In Figure 5A, the three CK isoenzymes in a synthetic mixture are separated in about 4 min. In Figure 5B, the five LDH isoenzymes are separated in about the same time. Reaction-detection of NADH product was used in each case, with separation on the column described above (DEAE-silica).

3.6. Lipids

Lipid testing in routine clinical chemistry is largely confined to serum determinations of triglycerides and cholesterol. Adequate colorimetric or fluorometric assays exist for each of these substances, with little need for LC methods. While LC has been used for the separate determination of free and total cholesterol,[210] the same assay is more conveniently carried out enzymatically, with cholesterol esterase alternately included or omitted in the reaction mixture.

Certain less-common procedures in lipid analysis provide special opportunities for LC. One of these is the measurement of lecithin in amniotic fluid. Lecithin is a principal component of pulmonary surfactant, a material that coats the air sacs of the lungs and lowers surface tension, thereby allowing the lungs to breathe without alveolar collapse. Synthesis of lecithin by the fetus first becomes significant during the 32nd to 36th weeks of gestation. It appears in amniotic fluid at the same time, and its level serves to assess fetal lung maturity, that is, to predict the likelihood of respiratory distress syndrome (RDS).

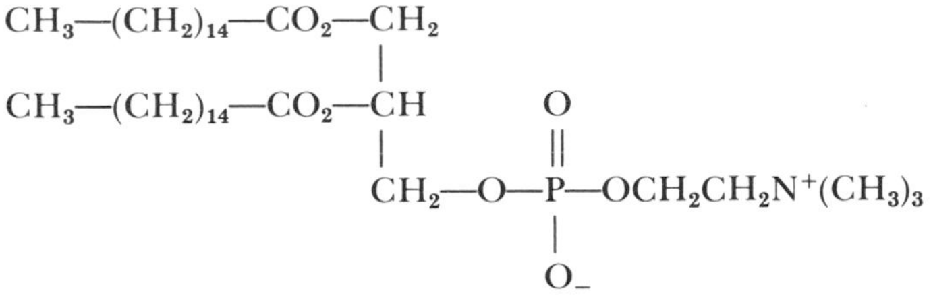

$$
\begin{array}{l}
CH_3{-}(CH_2)_{14}{-}CO_2{-}CH_2 \\
\qquad\qquad\qquad\qquad\quad | \\
CH_3{-}(CH_2)_{14}{-}CO_2{-}CH \qquad O \\
\qquad\qquad\qquad\qquad\quad | \qquad\quad \| \\
\qquad\qquad\qquad\qquad CH_2{-}O{-}P{-}OCH_2CH_2N^+(CH_3)_3 \\
\qquad\qquad\qquad\qquad\qquad\qquad | \\
\qquad\qquad\qquad\qquad\qquad\quad O_-
\end{array}
$$

Dipalmitoyl lecithin

Amniotic fluid lecithin usually is measured by TLC.[211] Because the related substance, sphingomyelin, has a relatively constant concentration in amniotic fluid during the 32nd to 36th weeks of gestation, and is also resolved by TLC, it serves as an internal standard. Hence, lecithin/sphingomyelin (L/S) ratios are reported. Values less than 2 indicate a high likelihood of RDS, while ratios greater than 3–5 (depending on the method) indicate fetal lung maturity.

Lecithin and sphingomyelin have been measured in amniotic fluid by LC.[212] Centrifuged amniotic fluid was extracted with methanol/chloroform, and the organic layer washed and evaporated. The residue was then dissolved in ethanol and chromatographed on silica using a mobile phase of acetonitrile/methanol/water, with detection at 203 nm. The ability of the procedure to assess fetal lung maturity was not determined. It is probable that detection at 203 nm is really unsuitable for this assay, as surface-active lecithin contains mainly palmitic acid,[213,214] and the unsaturated compounds absorb much more strongly at this wavelength. As a result, the authors[212] suggest the need for an alternate detection scheme (e.g., total phosphorus in the column effluent, as by a reaction detector). Further development of this analysis, therefore, is necessary before it can provide clinically useful L/S ratios. However, LC should ultimately be the method of choice for this analysis.

The analysis of bile acids offers another opportunity for LC. These cholesterol derivatives are excreted in the bile to facilitate digestion and absorption of lipids in the intestine. They also are found in small concentration in serum. Their measurement in both fluids is of clinical interest, largely to assess heptobiliary status. Their low level in serum (less than 1 μmol/liter) has made it necessary to use high-sensitivity techniques, such as radioimmunoassay (e.g., reference 215) or gas chromatography (e.g., reference 216). Although detection at 210 nm allows a detection limit of approximately 50 ng for bile acid standards by reverse-phase LC,[217] even greater sensitivity will be necessary for their direct analysis of serum. Analysis of bile acids in duodenal aspirates and bile poses less of a detection problem for LC. Both silica and reverse-phase LC systems were evaluated, and several chromatograms of human bile samples were presented in the first such report on this subject.[218]

More recently, an analysis of the six major conjugated bile acids in both duodenal aspirates and bile has been achieved by reverse-phase LC with refractive-index detection.[219] Sample pretreatment involved only the addition of methanol.

LC should be advantageous for lipid analysis wherever TLC or GC methods are applicable, because of the advantage of LC in the separation of oxidizable, thermally unstable, and/or nonvolatile compounds—all of which properties generally apply to lipids. However, the detection of this class of compounds remains an unsolved problem for their LC analysis. It is probable that derivatization before or after separation will be required for practical lipid assays by LC. For a recent review covering some advances in derivatization of lipids for LC analysis, see reference 220. The liquid chromatography of lipids, including applications of modern LC to biological samples, has been reviewed.[221,221a] Further discussion of lipid analysis in clinical chemistry by LC can be found in the section on genetic screening.

3.7. Porphyrins

Porphyrins and their precursors (especially δ-aminolevulinic acid and porphobilinogen) are of current interest in clinical chemistry for two main reasons: porphyrias and lead poisoning. Chromatographic analysis is not required for assessment of the latter condition, since a number of simple tests (e.g., erythrocyte protoporphyrin or erythrocyte aminolevulinic acid dehydratase) are available which are adequate for screening purposes. The porphyrias, on the other hand, while rather uncommon, comprise a number of different disease states where diagnosis is aided by separation of the elevated metabolites involved. This has been achieved to some extent by organic extraction procedures and classical ion-exchange chromatography, but much more rapidly and completely by modern LC. Detection is not a problem, because the porphyrins are highly colored and fluorescent. Several articles on the direct application of LC to porphyrin analysis in physiological samples were recently reviewed by Dixon.[4] In almost every case, methyl or ethyl esters were formed (porphyrins possess two to eight carboxyl groups) prior to analysis, because of a preference by most workers for the solubility and chromatographic characteristics of these derivatives. All of the separations were carried out on silica, except for the use of reverse-phase in one case.[222]

A more recent paper by Evans *et al.* describes several improvements.[223] Better isomer separation was achieved, along with expanded reliance on isocratic rather than gradient elution, largely through use of higher-efficiency microparticulate silica packings. Normal and porphyric urines were analyzed in this study, and detection was at 400 nm with mesoporphyrin-IX as internal standard. As before, the authors chose to convert the porphyrin free acids into their corresponding methyl esters.

In another recent study, small-particle silica was also employed, but the naturally occurring free acids (not esters) were chromatographed.[224,225] Rapid analysis times were obtained (less than 8 min for a group of standards) and fluorescence detection allowed as little as 30 pg of porphyrins to be detected in urine. Both normal and porphyric urines were examined, as shown in Figure 18. Sample preparation involved evaporation of urine (0.1–10 ml), extraction of the residue into mobile phase, and injection onto the LC column.

Recent studies suggest that the porphyrias can be differentiated based solely on LC analysis of porphyrins in both urine and feces.[226,226a] The prophyrins in these samples were converted to their methyl esters prior to chromatography on silica. LC elution profiles of feces and urine samples from a patient with porphyria cutanea tarda symptomatica are shown in Figure 19.

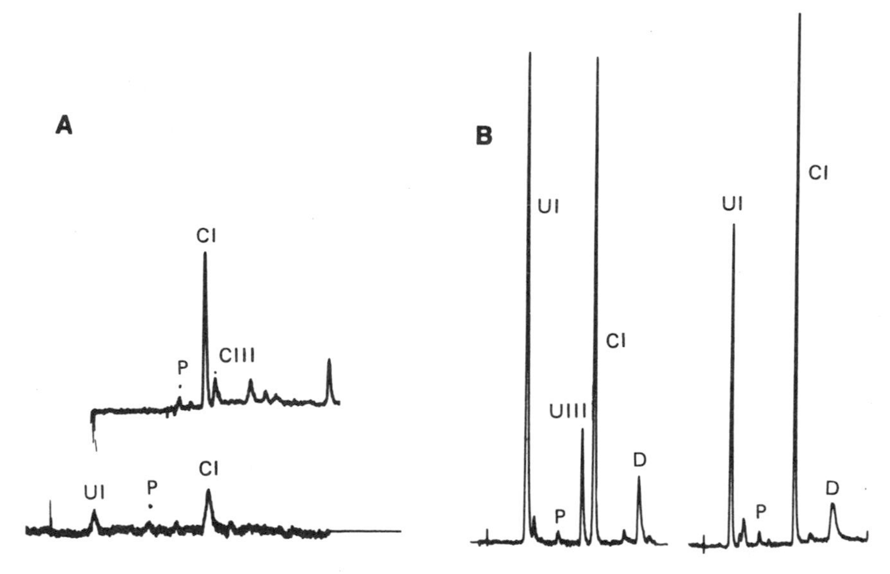

Figure 18. Separation of porphyrins in urine. Column: 13 μm silica A, 250 × 4.6 mm. Solvent: isocratic 0.3% H_2O in acetone, pH 7.60, using tributylamine, 2 ml/min; temperature: 40°C; detector: LC-1000; excitation: 403 nm; emission: 627 nm. (A) Normal urine, 100-μl specimen; (B) two erythropoietic urines, 10 μl each. Samples were blown to dryness, dissolved in mobile phase, and injected. (Reprinted from reference 225 with permission.)

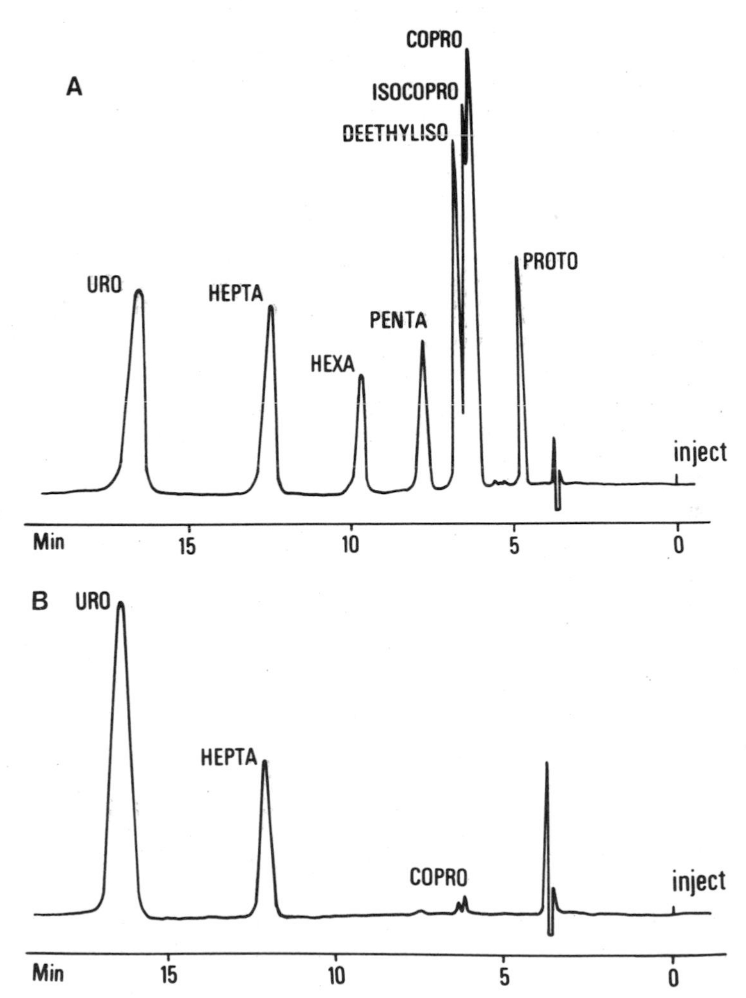

Figure 19. HPLC ester profiles for porphyria cutanea tarda symptomatica: (A) fecal porphyrins, (B) urinary porphyrins. Conditions: Each chromatogram was developed on a μ-Porasil column (300 × 4 mm) at a flow rate of 1.5 ml/min and a pressure of 500 psi. Detection of fractions was by absorption at 404 nm. Meso: mesoporphyrin dimethylester; Proto: protoporphyrin dimethylester; Deethyliso: deethylisocoproporphyrin tetramethyl ester; Copro: coproporphyrin tetramethylester; Hepta: heptacarboxylic porphyrin heptamethylester; Isocopro: isocoproporphyrin tetramethylester; Penta: pentacarboxylic porphyrin pentamethylester; Hexa: hexacarboxylic porphyrin hexamethylester; and Uro: uroporphyrin octamethylester. (Reprinted from reference 226 with permission.)

3.8. Vitamins

The vitamins are divided into two major groups: the fat-soluble vitamins (A, D, E, and K) and the water-soluble vitamins (B series and C). For most vitamins the levels in physiological samples are quite low (e.g., nanogram/milliliter level in serum), necessitating the protein-binding assays (PBA) or microbiological assay. However, for several vitamins much higher concentrations occur, e.g., vitamins A and C in serum are about 0.5 mg/liter and 10 mg/liter, respectively.

In the practical clinical laboratory, the most frequently requested vitamin tests are folic acid and vitamin B_{12}. Both analyses are carried out by PBA. Requests for analyses of vitamin A and C generally are the next most common. The higher levels of these vitamins permit non-PBA procedures such as colorimetry to be carried out. Analyses for most of the remaining vitamins are requested much less frequently, for reasons related to clinical usefulness, analytical difficulty, and the general availability of indirect procedures.

Efforts to analyze vitamins in physiological samples by LC currently are underway, although little progress has been reported thus far. The inherent fluorescence of several of the vitamins will no doubt be taken advantage of whenever possible. These LC efforts, if successful, are likely to expand the usefulness of clinical vitamin analysis. This is because the general biochemical or nutritional roles of the vitamins are reasonably well understood in most cases, so that a knowledge of the vitamin status of an individual could be used for both diagnostic and therapeutic purposes.

The extended unsaturation and relatively high serum levels of vitamin A has allowed UV detection at 328 nm (λ_{max}) to be used to quantitate the retinol form of this vitamin after separation by LC. (Retinol is the primary form of vitamin A in normal fasting blood.) A chromatogram is shown in Figure 20. The internal standard is all *trans*-9-(4-methoxy-2,3,6-trimethylphenyl)-3,7-dimethyl-2,4,6,8-tetraenol. The vitamin was separated from the serum matrix by extraction with organic solvents and then chromatographed on silica. The procedure would appear to be highly advantageous relative to other techniques, since it showed good sensitivity, linearity, and precision.[227] Another procedure on silica also has appeared for vitamin A serum.[227a]

The ability to separate rapidly the standards of some of the alternate forms of folic acid[228] and of vitamin B_6[229] by LC has been demonstrated. As an example of fluorescence detection for direct vitamin analysis by LC, a well-resolved chromatographic peak for riboflavin in freshly voided urine was reported recently.[230]

Since vitamin D_3 is a prohormone, measurement of this vitamin and its conversion products in physiological samples should be clinically useful.

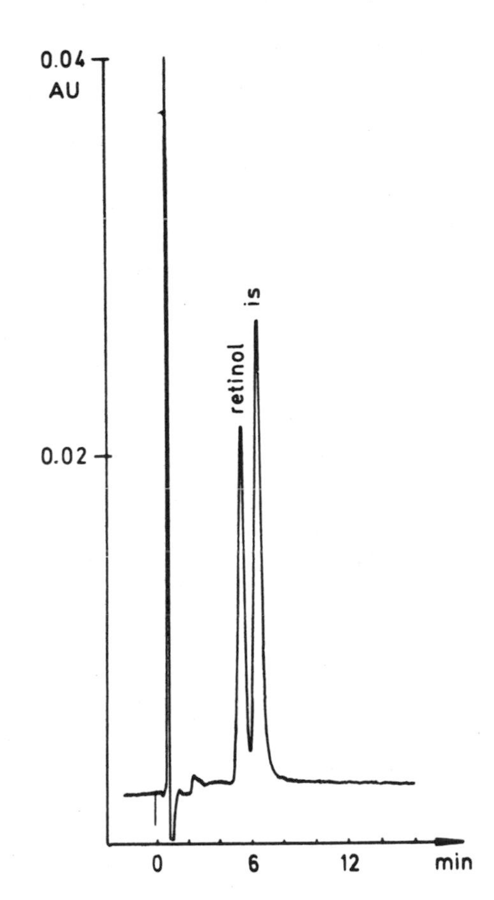

Figure 20. Assay for retinol in serum. Serum extract to which internal standard has been added. Column: 150 × 2 mm MicroPak Si-10; mobile phase: petroleum ether–dichloromethane–isopropanol (80:19.3:0.7 by vol), 0.5 ml/min; detector: 328 nm, 0.04 absorption units full scale; temperature: ambient. (Reprinted from reference 227 with permission.)

The actual hormone is 1,25-dihydroxy-D_3 which arises after D_3 is first converted to 25-hydroxy-D_3. Sufficient concentrations of 25-hydroxy-D_3 normally are present to allow measurement by means of LC, utilizing detection at 254 nm for the *cis*-triene chromophore of this substance. In this procedure, radioactive 25-OH-D_3 is added to a plasma sample, followed by extraction, chromatography on two different column systems, and injection onto an LC column. The effluent of the 25-OH-D_3 region is collected and counted to assess recovery.[231,231a] Peaks for both 25-OH-D_2 and 25-OH-D_3 were seen as shown in Figure 21. Similar work has been reported elsewhere.[232,232a] Other workers have achieved the LC analysis of not only 25-OH-D in plasma, but also D in 24,25-dihydroxy-D.[232b] A pre-column cleanup procedure was employed and different LC conditions were used for each analyte. Detection was by UV.

Tocopherols (vitamin E) have been determined in serum by LC.[233] The sample was treated with ethanol to precipitate the proteins and the

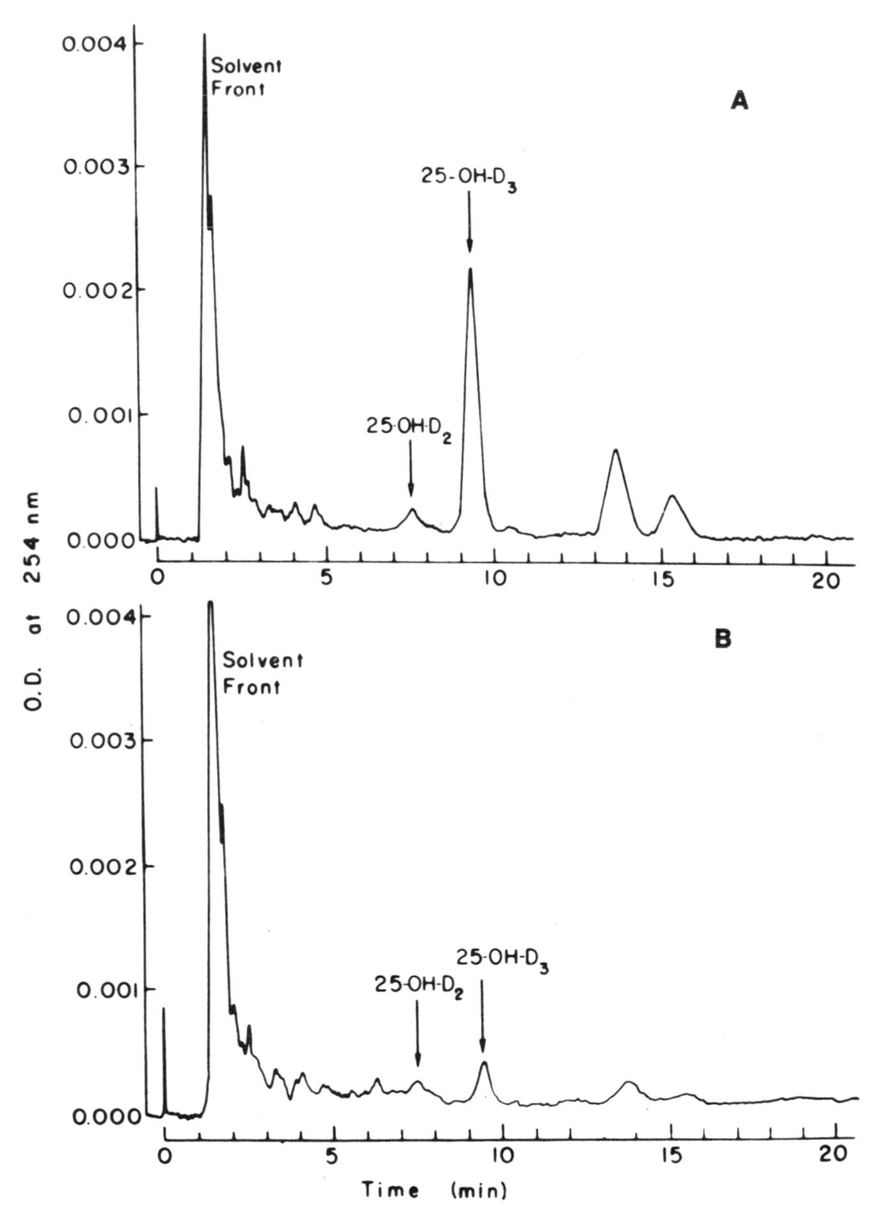

Figure 21. Typical high-pressure liquid chromatography profiles obtained with human blood plasma samples with (A) high and (B) low 25-OH-D levels; 25-OH-D_2 and 25-OH-D_3 elution positions are indicated. Late-eluting compounds absorbing at 254 nm were unidentified. Plasma extracts from 8-week vitamin D-deficient chicks showed no 254-nm absorbing peaks in the 25-OH-D regions. The chromatography was carried out on a 300 × 4 mm silicic acid column with 2.5% isopropanol in hexane. (Reprinted from reference 231a with permission.)

filtrate was extracted with hexane. An aliquot of the evaporated extract was analyzed by LC with fluorescence detection at 325 nm after excitation at 298 nm. Lipids were reported not to interfere.

3.9. Organic Acids

Lactate, β-hydroxybutyrate, and acetoacetate have been determined by ion-exchange LC after acidification and organic extraction of whole blood.[234] Peak quantitation was based on simultaneous UV absorption and refractive-index measurements. The organic acids pyruvate and α-ketoglutarate were determined in urine by gradient elution reverse-phase LC.[235] Urine samples were derivatized with *o*-phenylenediamine to convert these keto acids into quinoxalones, which were then quantitated by LC after organic extraction. Pyruvic acid/creatinine ratios were higher in diabetes and renal disease. Over a hundred acidic urinary constituents were separated within 30 min by using 5-μm octadecylsilica columns and gradient elution with increasing acetonitrile concentration in dilute phosphoric acid solution at 70°.[76] Hippuric acid in urine has been determined by ion-exchange LC.[235a]

3.10. Ubiquinone

Ubiquinone (coenzyme Q) has been determined in serum by LC on a reverse-phase column utilizing either direct absorption detection at 275 nm or pre-column reaction with ethylcyanoacetate to obtain a fluorescent product which was quantitated on this basis.[236]

3.11. Nucleosides, Nucleotides, and Bases

Analysis of nucleotides, nucleosides, and their bases in serum and urine by LC is of current interest, particularly in regard to the possibilities of detecting cancer and monitoring subsequent chemotherapy. Although preliminary data is available which encourages such studies (e.g., references 237 and 238), much work remains to be done. Recent references to LC studies pursuing this objective are 239, 240, 241, and 242.

4. Note Added in Proof

This section covers the application literature mostly from the beginning of November, 1977, to the end of March, 1978. The large number of citations for this five-month period provides further evidence for the current intense interest in clinical LC. The order of topics is the same as that in Section 3.

A short review on clinical and biochemical LC recently appeared.[243] In the category of *reagents*, paired-ion LC was used to determine the 4-nitrophenol content of 4-nitrophenyl phosphate, a substrate for alkaline phosphatase analysis.[244] *Genetic screening* was extended by an ion-exchange separation of thirty to forty UV-active metabolites in urine.[245] A low-pressure LC system involving on-line reaction detection with diphenylpicrylhydrazyl was reported for screening melanoma mines.[246]

In regard to *catecholamines,* LC analysis of dansyl derivatives has continued,[247] an *o*-phthalaldehyde reaction detector was reported,[248] and two radioenzymatic procedures which each included an HPLC separation step appeared.[249,250] Similarly to the analysis of 5-HIAA (an *indole*) in urine cited earlier,[99] a cation exchange separation (ca. 18 min) followed by on-line reaction detection with *o*-phthalaldehyde was reported for 5-hydroxytryptophan in plasma.[251] A similar analysis for 5-HIAA in urine also appeared.[252] Three articles have presented LC procedures for *tryptophan* in physiological samples,[252a,b,c] and a short review on the LC analysis of tryptophan metabolites in urine was published.[252d] All of the more recent analyses of *polyamines* by LC have involved pre-column derivatization, either with tosyl chloride,[253] *o*-phthalaldehyde,[254] quinoline-8-sulfonic acid,[255] or dansyl chloride.[256,257] LC (and GC) analysis of polyamines has been reviewed.[258,258a]

More recent LC work on *creatinine* has involved detection at 215 nm.[259,260]

A review on *drug analysis* by LC that covers the literature from 1970 to 1976 has appeared,[261] as has a book on the quantitative analysis and interpretation of antiepileptic drugs by various methods including LC.[262] Further literature on the anticonvulsant drugs includes a review of their clinical pharmacology,[263] the use of dual-wavelength LC analysis,[264] an analysis of carbamazepine and its epoxide metabolite,[265] an analysis of dyphylline,[266] and the use of reverse-phase LC to determine less common anticonvulsants.[266 a]

In the category of *antiarrhythmics,* an ultramicro method for lidocaine and procainamide appeared,[267] along with procedures for quinidine,[268,269] quinidine and dihydroquinidine,[270,271] quinidine and 3-hydroxyquinidine,[272,272a] and tocainide.[272b] The latter was determined as its dansyl derivative. *Theophylline* by LC has continued to attract much interest[273–278] LC has been used for the physiological analysis of various *antibiotics*[279–281] including sulfisoxazole,[282] gentamicin,[283–284a] tetracyclines,[284b] and trisulfapyrimidines.[284c] *Acetominophen* was analyzed on a reverse-phase column involving detection at 225 nm.[285]

Daunomycin and daunomycinol in rabbit plasma above 10 ng/ml were quantitated at 490 nm after separation on a silica column.[286] Digitalis glycoside standards were separated on a reverse-phase column and quantitated with a post-column fluorescence detector involving reaction with hot hydrochloric acid.[287] The detection limit was 0.5 ng for a nonretained glycoside. This limits the direct applicability of this procedure to physiological analysis of digoxin (therapeutic range 0.5–2.0 ng/ml in serum). *Methotrexate* in plasma was quantitated by oxidizing it to a fluorescent product prior to reverse-phase LC.[288] The *tricyclic antidepressants,* which seem to be headed for considerable therapeutic monitoring (e.g., reference 289), were found to undergo analysis more effectively on silica rather than on reverse-phase LC,[290] and two other analyses on silica also have been reported.[291,292] *Miscellaneous physiological drug analyses* include azaribine,[293] cefuroxime,[294] flunitrazepam,[295] nitrofurantoin,[296] penicillamine,[297] phentolamine,[298] tolbutamide and chlorpropamide,[299] salicyclic acid and salicyluric acid,[300] and salicylic acid.[301]

In regard to *steroid hormones*: androgens and estrogens were chromatographed on silica with UV detection allowing measurements as low as 5 ng[302]; hydrocortisone was measured in patients dosed with methylprednisolone[303]; conjugated and free hormonal steroids were separated by reverse phase LC[304]; aldosterone was quantitated in urine[305]; cortisol in plasma was analyzed using a *p*-nitroaniline-modified-silica packing[306]; and the general subject has been reviewed.[307]

Lipid analysis has been extended by LC procedures for higher glycolipids as their *O*-acetyl-*N*-*p*-nitrobenzoyl derivatives,[308] phospholipids on a silica column with detection at 206 nm,[309] and fecal bile acid metabolites as their *p*-nitrobenzyl ester derivatives.[310]

Both fecal and urinary *porphyrins* were quantitated as their methyl esters on a silica column with detection at 400 nm.[311] The fecal porphyrins were esterified directly with boron trifluoride in methanol; the urinary porphyrins were absorbed on talcum before esterification in this manner.

Vitamin D and its important metabolites have continued to be a popular pursuit by LC.[312–315] Tocopherol analyses have been reported in red blood cells[316] and in plasma.[317] Vitamin A from both serum and liver was analyzed on a reverse-phase column with fluorescence detection.[318] Finally, LC has been applied to the analysis of methylated purines in urine,[319] to adenosine and other adenine compounds in patients with immunodeficiency diseases,[320] to purine bases in urine,[321] and to nucleotide pools in platelets.[322] The use of LC in research on purine nucleoside analogues has been reviewed.[323]

Acknowledgments

The authors thank Dr. Charles E. Pippenger for his helpful comments regarding the section on drugs. Dr. B. L. Karger acknowledges the support of NIH under grant GM15847. This paper is contribution number 38 from the Institute of Chemical Analysis.

5. References

1a. L. R. Snyder and J. J. Kirkland, *Introduction to Modern Liquid Chromatography,* 2nd ed., Wiley-Interscience New York (1978).
1b. N. A. Parris, *Instrumental Liquid Chromatography. A Practical Manual,* Elsevier, New York (1975).
1c. C. F. Simpson, ed., *Practical High Performance Liquid Chromatography,* Heyden Press, New York (1976).
1d. P. A. Bristow, *LC in Practice,* HETP Publ., 10 Langley Dr., Handforth, Wilmslow, Cheshire, UK (1976).
2. B. L. Karger, Modern liquid chromatography in clinical chemistry, ACS Symposium Series, No. 36, *Clinical Chemistry* (D. T. Forman and R. W. Mattoon, eds.), p. 226, American Chemical Society (1976).
3. P. F. Dixon, M. S. Stoll, and C. K. Lim, *Ann. Clin. Biochem., 13,* 409 (1976).
4. P. F. Dixon, C. H. Gray, C. K. Lim, and M. S. Stoll, eds., *High Pressure Liquid Chromatography in Clinical Chemistry,* Academic Press, New York (1976).
5. J. Butler, V. Fantl, and C. K. Lim, *High Pressure Liquid Chromatography in Clinical Chemistry,* p. 59, Academic Press, New York (1976).
6. T. J. X. Mee and J. A. Smith, *High Pressure Liquid Chromatography in Clinical Chemistry,* Academic Press, New York (1976).
7. E. Grushka, ed., *Bonded Stationary Phases in Chromatography,* Ann Arbor Science Publishers, Ann Arbor, Michigan (1974).
8. L. R. Snyder, *Principles of Adsorption Chromatography,* Marcel Dekker, New York (1968).
9. J. H. Knox and G. R. Laird, *J. Chromatogr., 122,* 17 (1976).
10. J. H. Knox and G. Vasvari, *J. Chromatogr., 83,* 181 (1973).
11. K. K. Unger, N. Becker, and P. Roumeliotis, *J. Chromatogr., 125,* 115 (1976).
12. J. J. Kirkland, *Chromatographia, 8,* 661 (1975).
13. K. Karch, I. Sebestian, and I. Halasz, *J. Chromatogr., 122,* 3 (1976).
14. I. E. Bush, *Chromatography of Steroids,* Pergamon Press, London (1961).
15. R. B. Sleight, *J. Chromatogr., 83,* 31 (1973).
16. C. Horvath, W. Melander, and I. Molnar, *J. Chromatogr., 83,* 31 (1973).
17. B. L. Karger, J. R. Gant, A. Hartkopf, and P. H. Weiner, *J. Chromatogr., 128,* 65 (1976).
18. S. R. Bakalyar, R. McIlwrich and E. Roggendorf, *J. Chromatogr., 142,* 353 (1977).
19. N. Tanaka, H. Goodell, and B. L. Karger, submitted to *J. Chromatogr.*
20. G. Schill, in: *Ion Exchange and Solvent Extraction* (J. A. Marinsky and Y. Marcus, eds.), Vol. 6, Marcel Dekker, New York (1974).
21. D. P. Wittmer, N. O. Nuessle, and W. G. Haney, *Anal. Chem., 47,* 1422 (1975).
22. J. H. Knox and J. Jurand, *J. Chromatogr., 125,* 89 (1976).
23. H. Colin, C. Eon, and G. Guiochon, *J. Chromatogr., 122,* 223 (1976).
24. H. Colin and G. Guiochon, *J. Chromatogr., 126,* 32 (1976).
25. H. Colin and G. Guiochon, *J. Chromatogr., 137,* 19 (1977).
26. G. Vigh and J. Inczédy, *J. Chromatogr., 129,* 81 (1976).
27. C. D. Scott, *Science 186,* 226 (1974).
28. I. Halasz, H. Englehardt, J. Asshauer, and B. L. Karger, *Anal. Chem., 42,* 1460 (1970).

29. J. F. K. Huber, D. A. M. Meijers, and J. S. R. J. Hulsman, *Anal. Chem., 44,* 111 (1972).
30. B. Fransson, K. G. Wahlund, I. M. Johansson, and G. Schill, *J. Chromatogr., 125,* 327 (1976).
31. J. H. Knox and M. J. Saleem, *J. Chromatogr. Sci., 7,* 614 (1969).
32. G. J. Kennedy and J. H. Knox, *J. Chromatogr. Sci., 10,* 550 (1972).
33. I. Halasz, H. Schmidt, and P. Vogtel, *J. Chromatogr., 126,* 19 (1976).
34. M. Martin, C. Eon, and G. Guiochon, *J. Chromatogr., 110,* 213 (1975).
35. J. H. Knox and A. Pryde, *J. Chromatogr., 112,* 171 (1975).
36. J. C. Kraak, H. Poppe, and F. Smedes, *J. Chromatogr., 122,* 147 (1976).
37. L. R. Snyder, *J. Chromatogr. Sci., 15,* 441 (1977).
38. R. P. W. Scott, *Contemporary Liquid Chromatography,* Interscience, New York (1976).
39. F. W. Karasek, *Res. Dev., 28*(6), 28 (1977).
40. R. P. W. Scott and C. E. Reese, *J. Chromatogr., 138,* 283 (1977).
41. S. R. Bakalyar and R. A. Henry, *J. Chromatogr., 126,* 327 (1976).
42. G. Brooker, *Anal. Chem., 43,* 1085 (1971).
43. Product literature, Waters Associates, Milford, Massachusetts.
44. D. R. Gere, *Res. Dev., 27*(10), 22 (1976).
45. M. Martin, G. Glu, C. Eon, and G. Guiochon, *J. Chromatogr., 112,* 399 (1975).
46. S. R. Abbott, J. R. Berg, P. Achener, and R. L. Stevenson, *J. Chromatogr., 126,* 421 (1976).
47. F. W. Karasek, *Res. Dev., 28,* 38 (1977).
48. J. J. Kirkland, *Analyst, 99,* 859 (1974).
49. B. L. Karger, M. Martin, and G. Guiochon, *Anal. Chem., 46,* 1640 (1974).
49a. P. Schauwecker, R. W. Frei, and F. Erni, *J. Chromatogr., 136,* 63 (1977).
50. J. Dolan, N. Tanaka, J. R. Gant, R. Giese, and B. L. Karger, in preparation.
51. L. R. Snyder, *J. Chromatogr. Sci., 8,* 692 (1970).
52. J. F. K. Huber and R. van der Linden, *J. Chromatogr., 83,* 267 (1973).
53. A. Wehrli, V. Hermann, and J. F. K. Huber, *J. Chromatogr., 125,* 59 (1976).
54. A. P. Graffeo and B. L. Karger, *Clin. Chem., 22,* 184, (1976).
55. J. F. Lawrence and R. W. Frei, *Chemical Derivatization in Liquid Chromatography,* Elsevier, New York (1976).
56. R. W. Frei, L. Michel, and W. Santi, *J. Chromatogr., 126,* 665 (1976).
57. R. S. Deelder, M. G. F. Kroll, and J. H. M. Van Den Berg, *J. Chromatogr., 125,* 307 (1976).
58. R. R. Schroeder, P. J. Kudirka, and E. C. Torren, Jr., *J. Chromatogr., 134,* 83 (1976).
59. S. H. Chang, K. M. Gooding, and F. E. Regnier, *J. Chromatogr., 125,* 103 (1976).
60. R. S. Deelder and P. J. H. Hendricks, *J. Chromatogr., 83,* 343 (1973).
61. L. R. Snyder, *J. Chromatogr., 125,* 287 (1976).
62. K. Asai, Y. Kanno, A. Nakamoto, and T. Hara, *J. Chromatogr., 126,* 369 (1976).
63. R. M. Rocco, D. C. Abbott, R. W. Giese, and B. L. Karger, *Clin. Chem., 23,* 705 (1977).
64. L. R. Shukur, J. L. Powers, R. A. Margues, M. E. Winter, and W. Sadee, *Clin. Chem., 23,* 636 (1977).
65. J. Novak, *Quantitative Analysis by Gas Chromatography,* Marcel Dekker, New York (1975).
66. L. R. Snyder, *J. Chromatogr. Sci., 8,* 692 (1970).
67. W. B. Furman, *Continuous Flow Analysis. Theory and Practice,* Marcel Dekker, New York (1976).
68. L. R. Snyder, J. Levine, R. Stoy, and A. Conetta, *Anal. Chem., 48,* 942A (1976).
69. D. A. Burns, *Res. Dev.,* 22 (April 1977).

69a. J. C. MacDonald, *Am. Lab.*, 69 (August 1977).
70. H. Adler, D. A. Burns, and L. R. Snyder, *Advances in Automated Analysis. 1972 Technicon Intern. Cong.*, Vol. 6, p. 23, Mediad, Tarrytown, New York (1973).
71. S. A. Margolis, B. F. Howell, and R. Schaffer, *Clin Chem.*, *22*, 1322 (1976).
72. J. R. Miksic and P. R. Brown, *J. Chromatogr.*, *142*, 641 (1977).
73. C. K. Lim, D. P. Robinson, and S. S. Brown, *J. Chromatogr.*, *145*, 41 (1978).
74. W. D. Slaunwhite, D. C. Wenke, M. E. Chilcote, P. T. Kissinger, and L. A. Pachla, *Clin. Chem.*, *21*, 1015 (1975).
75. H. K. Berry, *Fed. Proc.*, *34*, 2134 (1975).
76. I. Molnar and C. Horvath, *J. Chromatogr.*, *143*, 391–400 (1977).
77. M. Iwamori and H. W. Moses, *Clin. Chem.*, *21*, 725 (1975).
78. F. B. Jungalwala and A. Milunsky, High performance liquid chromatography for the detection of homozygotes and heterozygotes of Niemann–Pick disease, personal communication.
79. S. Natelson, in: *Clinical Chemistry* (D. T. Forman and R. W. Mattoon, eds.), p. 95, ACS Symposium Series 36, American Chemical Society, Washington, D.C. (1976).
80. D. S. Young and J. M. Hicks, *The Neonate*, Wiley, New York (1976).
81. I. Molnar and C. Horvath, *Clin. Chem.*, *22*, 1497 (1976).
82. R. W. Stout, R. J. Michelot, I. Molnar, C. Horvath, and J. K. Coward, *Anal. Biochem.*, *76*, 330 (1976).
83. J. Knox and J. Jurand, *J. Chromatogr.*, *125*, 89 (1976).
84. P. T. Kissinger, R. M. Riggin, R. L. Alcorn, and L. O. Rau, *Biochem. Med.*, *13*, 299 (1975).
85. R. M. Riggin and P. T. Kissinger, *Anal. Chem.*, *49*, 2109 (1977).
86. L. J. Felice and P. T. Kissinger, *Clin. Chim. Acta*, *76*, 317 (1977).
87. L. J. Felice and P. T. Kissinger, *Anal. Chem.*, *48*, 794 (1976).
88. L. J. Felice, C. S. Bruntlett, and P. T. Kissinger, *J. Chromatogr.*, *143*, 407–410 (1977).
89. A. Yoshida, M. Toshioka, T. Yamazaki, T. Sakai, and Z. Tamura, *Clin. Chim. Acta*, *73*, 315 (1976).
90. A. Yoshida, M. Yoshida, T. Tanimura, and Z. Tanuro, *J. Chromatogr.*, *116*, 240 (1976).
91. L. D. Mell and A. B. Gustafson, *Clin. Chem.*, *23*, 473 (1977).
92. Thomas Moyer, personal communication.
93. R. E. Shoup and P. T. Kissinger, *Clin. Chem.*, *23*, 1268 (1977).
94. K. Imai, *J. Chromatogr.*, *105*, 135 (1975).
95. G. Schwedt and H. H. Bussemas, *Chromatographia*, *9*, 17 (1976).
96. G. Schwedt and H. H. Bussimas, *Z. Anal. Chem.*, *283*, 23 (1977).
97. P. G. Passon and J. D. Peules, *Anal. Biochem.*, *51*, 618 (1973).
98. N. Seiler, *J. Chromatogr.*, *143*, 221 (1977).
99. H. H. Brown, M. C. Rhindress, and R. E. Groswold, *Clin. Chem.*, *17*, 92 (1971).
100. A. P. Graffeo and B. L. Karger, *Clin. Chem.*, *22*, 184 (1976).
100a. A. Yoshida, T. Yamazaki, and T. Sakai, *Clin. Chim. Acta*, *77*, 95 (1977).
100b. O. Beck, G. Palmskog, and E. Hultman, *Clin. Chim. Acta*, *79*, 149 (1977).
100c. G. M. Anderson and W. C. Purdy, *Anal. Lett.*, *10*, 493 (1977).
101. J. Savory and J. R. Shite, *Ann. Clin. Lab. Sci.*, *5*, 110 (1975).
102. D. H. Russell, *Nature*, *233*, 144 (1971).
103. D. H. Russell, ed., *Polyamines in Normal and Neoplastic Growth*, Raven Press, New York (1973).
104. D. H. Russell and S. D. Russell, *Clin. Chem.*, *21*, 860 (1975).
105. R. Dreyfuss, R. Chayen, G. Dreyfuss, *Isr. J. Med. Sci.*, *11*, 785 (1975).
106. L. J. Marton and P. L. Y. Yee, *Clin. Chem.*, *21*, 1721 (1975).

107. C. W. Gehrke, K. C. Kuo, R. W. Zumwalt, and T. P. Waalkes, *J. Chromatogr., 89,* 231 (1974).
108. H. Veening, W. W. Pitt, Jr., and G. Jones, Jr., *J. Chromatogr., 90,* 129 (1974).
109. C. W. Gehrke, K. C. Kuo, and R. C. Ellis, *J. Chromatogr., 143,* 345–361 (1977).
110. H. Adler, M. Margoshes, L. R. Snyder, and C. Spitzer, *J. Chromatogr. Biomed. Appl., 143,* 125 (1977).
110a. N. Seiler, *Clin. Chem., 23,* 1519 (1977).
111. N. D. Brown, H. C. Sing, G. E. Demaree, and W. E. Neeley, *Clin. Chem., 23,* 1281 (1977).
112. "How to deal with poisoning emergencies," *Medical World News,* p. 77 (March 22, 1976).
113. "Diagnoses of overdoses may not jibe with lab's," *Medical World News,* p. 38 (September 22, 1975).
114. M. M. McCarron, C. B. Walberg, and G. D. Lundberg, *Clin. Chem., 20,* 118 (1974).
115. E. Berman, *Analysis of Drugs of Abuse,* Heyden and Sons, London (1977).
116. K. K. Kaistha, *J. Chromatogr., 141,* 145 (1977).
117. J. R. Boyd, T. R. Covington, W. F. Stanaszek, and R. T. Coussons, *Am. J. Hosp. Pharm., 31,* 362 (1974).
117a. S. J. Soldin and J. G. Hill, Liquid Chromatography Symposium I, Oct. 13–14, Waters Associates, Boston, Massachusetts (1977).
118. C. E. Costello, H. S. Hertz, T. Sakai, and K. Biemann, *Clin. Chem., 20,* 255 (1974).
119. R. H. Greeley, *Clin. Chem., 20,* 192 (1974).
120. J. J. Mule, M. L. Bastos, and D. Jukofsky, *Clin. Chem., 20,* 243 (1974).
121. D. P. Kehane, *Am. Lab.,* 25–32 (May 1977).
122. C. T. Viswanathan, H. E. Booker, and P. G. Welling, *Clin. Chem., 23,* 873 (1977).
123. E. S. Vesell and G. T. Passananti, *Clin. Chem., 17,* 851 (1971).
124. S. H. Atwell, V. Green, and W. G. Haney, *J. Pharm. Sci., 64,* 806 (1975).
124a. G. Morrell and H. C. Pribor, *Lab. Manag.* 40 (June 1977).
125. M. G. Horning, L. Brown, J. Nowlin, K. Lertratanangkoon, P. Kellaway, and T. E. Zion, *Clin. Chem., 23,* 157 (1977).
125a. H. E. Booker and B. Darcey, *Epillepsia (Amst.) 14,* 177 (1973).
126. S. A. Killman and J. H. Thaysen, *Scand. J. Clin. Lab. Invest. 76,* 86 (1955).
127. J. J. McAuliffe, A. L. Sherwin, I. E. Leppik, S. A. Fayle, and C. E. Pippenger, *Neurology, 27,* 409 (1977).
128. R. F. Adams, and F. L. Vandemark, in: *High Pressure Liquid Chromatography in Clinical Chemistry* (P. F. Dixon, C. H. Gray, C. K. Lim, and M. S. Stoll, eds.), p. 143, Academic Press, New York (1976).
129. J. W. Nelson, A. L. Cordry, C. G. Aron, and R. A. Bartell, *Clin. Chem., 23,* 124 (1977).
130. S. J. Soldin and J. G. Hill, *Clin. Chem., 22,* 856 (1976); *23,* 782 (1977).
130a. J. J. Kirkland, W. W. Yau, H. J. Stoklosa, and C. H. Dilks, Jr., *J. Chromatogr., 15,* 303 (1977).
131. L. C. Franconi, G. L. Hawk, B. J. Sandmann, and W. G. Haney, *Anal. Chem., 48,* 372 (1976).
132. H. G. M. Westenberg and R. A. DeZeeuw, *J. Chromatogr., 118,* 217 (1976).
133. J. Kiffin, L. D. Smith, and B. G. White, Blood Level Determinations of Anti-epileptic Drugs. Clinical Value and Methods, DHEW Publ. No. (NIH) 73-396, Washington, D.C. (1972). A more up-to-date source is reference 262.
134. R. F. Adams, in: *Advances in Chromatography* (J. C. Giddings, J. Cazes, E. Grushka, and P. R. Brown, eds.), Vol. 15, p. 131, Marcel Dekker, New York (1976).
135. P. M. Kabra, B. E. Stafford, and L. J. Marton, *Clin. Chem., 23,* 1284 (1977).

136. U. R. Tjaden, J. C. Kraak, and J. F. K. Huber, *J. Chromatogr. Biomed. Appl., 143,* 183 (1977).
137. M. Eichelbaum and L. Bertilsson, *J. Chromatogr., 103,* 135 (1975).
138. P. M. Kabra and L. J. Marton, *Clin. Chem., 22,* 1672 (1976).
139. J. E. Evans, *Anal. Chem., 45,* 2428 (1973).
140. S. Kitazawa and T. Komuro, *Clin. Chem. Acta 73,* 31 (1976).
141. P. M. Kabra, G. Gotelli, R. Stanfill, and L. J. Marton, *Clin. Chem., 22,* 824 (1976).
141a. R. T. Chamberlain, D. T. Stafford, A. G. Majub, and B. C. McNatt, *Clin. Chim., 23,* 1764 (1977).
142. I. M. House and D. J. Berry, in: *High Pressure Liquid Chromatography in Clinical Chemistry* (P. F. Dixon, C. M. Gray, C. K. Lim, and M. S. Stoll, eds.), p. 155, Academic Press, New York (1976).
143. G. Gauchel, F. D. Gauchel, and L. Birkofer, *A. Klin. Chem. Klin. Biochem., 11,* 459 (1973).
143a. H. G. Westerberg, R. A. DeZeeuw, E. Van Der Kleijn, and T. T. Oei, *Clin. Chim. Acta, 79,* 155 (1977).
144. K. Harzer and R. Barchet, *J. Chromatogr., 132,* 83 (1977).
145. R. F. Adams, F. L. Vandemark, and G. Schidt, *Clin. Chim. Acta, 69,* 515 (1976).
146. K. Carr, R. L. Woosley, and J. A. Oates, *J. Chromatogr., 129,* 363 (1976).
147. R. M. Rocco, D. C. Abbott, R. W. Giese, and B. L. Karger, *Clin. Chem., 23,* 705 (1977).
148. L. R. Shukur, J. L. Powers, R. A. Marques, M. E. Winter, and W. Sadee, *Clin. Chem., 23,* 636 (1977).
149. J. S. Dutcher and J. M. Strong, *Clin. Chem., 23,* 1318 (1977).
149a. O. H. Weddle and W. Mason, *J. Pharm. Sci., 66,* 874 (1977).
150. B. A. Persson and P. O. Lagerstrom, *J. Chromatogr., 122,* 305 (1976).
150a. W. D. Mason, E. N. Amick, and O. H. Weddle, *Anal. Lett., 10,* 515 (1977).
150b. G. J. Schmidt and F. L. Vandemark, *Chromatogr. Newsl. 5,* 42 (1977).
151. P. J. Meffin, S. R. Harapat, and D. C. Harrison, *J. Chromatogr., 132,* 503 (1977).
151a. P. J. Meffin, S. R. Harapat, Y. G. Yee, and D. C. Harrison, *J. Chromatogr., 138,* 183 (1977).
152. D. S. Sitar, K. M. Piafsky, R. E. Rangno, and R. I. Ogilvie, *Clin. Chem., 21,* 1774 (1975).
153. G. D. Bates, *Clin. Chem., 23,* 1167 (1976).
154. C. V. Manion, D. W. Shoeman, and D. L. Azarnoff, *J. Chromatogr., 101,* 169 (1974).
155. R. F. Adams, F. L. Vandmark, and G. J. Schmidt, *Clin. Chem., 22,* 1903 (1976).
156. M. A. Evanson and B. L. Warren, *Clin. Chem., 22,* 851 (1976).
157. M. Weingerger and C. Chidsey, *Clin. Chem., 21,* 834 (1975).
158. J. J. Orcutt, P. P. Kozak, Jr., S. A. Gillman, and L. H. Cummins, *Clin. Chem., 23,* 599 (1977).
159. M. J. Cooper, A. R. Sinaiko, M. W. Anders, and B. L. Mirkin, *Anal. Chem., 48,* 1110 (1976).
160. A. Christopherson, K. E. Rasmussen, and B. Salveson, *J. Chromatogr., 132,* 91 (1977).
161. M. J. Cooper, M. W. Anders, A. R. Sinaïko and B. L. Mirkin, in: *High Pressure Liquid Chromatography in Clinical Chemistry* (P. F. Dixon, C. H. Gray, C. K. Lim, and M. S. Stoll, eds.), p. 175, Academic Press, New York (1976).
162. R. M. Riggin, R. L. Alcorn, and P. T. Kissinger, *Clin. Chem., 22,* 782 (1976).
163. U. RL Tjaden, J. Lankelma, and H. Poppe, *J. Chromatogr., 125,* 275 (1976).
164. D. N. Bailey and P. I. Jatlow, *Clin. Chem., 22,* 777 (1976).
164a. J. H. M. van Den Berg, H. J. J. M. De Ruwe, R. S. Deelder, and Th. A. Plomp, *J. Chromatogr., 138,* 431 (1977).

164b. R. R. Brodie, L. F. Chasseaud, and D. R. Hawkins, *J. Chromatogr., 143,* 535 (1977).
164c. J. C. Kraak and P. Bijster, *J. Chromatogr., 143,* 499 (1977).
165. B. Mellstrom and S. Eksborg, *J. Chromatogr., 116,* 475 (1976).
165a. J. M. Grindel, P. F. Tilton, and R. D. Shaffer, *J. Pharm. Sci., 66,* 834 (1977).
165b. T. B. Vree, B. Lenselink, and E. Van Der Kleijn, *J. Chromatogr., 143,* 530 (1977).
165c. R. J. Koenng, C. M. Peterson, R. L. Jones, C. Saudek, M. Lehrman, and A. Cerami, *N. Engl. J. Med. 295,* 417 (1976).
165d. K. Gabbay, *N. Engl. J. Med. 295,* 443 (1976).
165e. R. A. Cole, J. S. Soeldner, P. J. Dunn, and H. F. Bunn, Liquid Chromatography Symposium I, Oct. 13–14, Waters Associates, Boston, Massachusetts (1977).
166. N. D. Brown and R. T. Lefberg, *J. Chromatogr., 99,* 635 (1974).
167. N. D. Brown and E. J. Michalski, *J. Chromatogr., 121,* 76 (1976).
167a. T. I. Bjornsson, T. F. Blaschke, and P. J. Meffin, *J. Chromatogr., 137,* 145 (1977).
168. N. J. Pound, I. J. McGilveray and R. W. Sears, *J. Chromatogr., 89,* 23 (1974).
169. R. Endele and G. Lettenbauer, *J. Chromatogr., 115,* 228 (1975).
170. W. F. Bayne, G. Rogers, and N. Crisologo, *J. Pharm. Sci., 64,* 402 (1975).
171. J.-B. Lecaillon and G. Souppart, *J. Chromatogr., 121,* 227 (1976).
172. M. C. Cosums-Duyck, A. P. De Leenheer, and P. M. Van Baerenbergh, *Clin. Chem., 21,* abstract 438 (1975).
173. S. Roggia, G. Grossoni, G. Pelizza, B. Ratti, and G. G. Gallo, *J. Chromatogr., 124,* 169 (1976).
173a. C. V. Puglisi, J. C. Meyer, and J. A. F. DeSilva, *J. Chromatogr., 136,* 391 (1977).
174. G. G. Duggin, *J. Chromatogr., 121,* 156 (1976).
174a. G. R. Gotelli, P. M. Kabra, and L. J. Marton, *Clin. Chem., 23,* 957 (1977).
175. D. Blair and B. H. Rumack, *Clin. Chem., 23,* 743 (1977).
175a. R. A. Horvitz and P. I. Jatlow, *Clin. Chem., 23,* 1596 (1977).
176. L. S. Goodman and A. Gilman, *Pharmacological Basis of Therapeutics,* 5th ed., Macmillan, New York (1975).
176a. A. Broughton, *Lab. Manag.,* 10 (June 1977).
176b. J. P. Anhalt, *Antimicrob. Agents Chemother., 11,* 651 (1977).
176c. C. J. Little, A. D. Dale, D. A. Ord, and T. R. Marton, *Anal. Chem., 49,* 1311 (1977).
176d. J. S. Wold and S. A. Turnipseed, *Clin. Chim. Acta, 78,* 203 (1977).
177. N. R. Bachur, C. E. Riggs, M. R. Green, J. J. Langone, H. Van Vunakis, and L. Levine, *Clin. Pharmacol. Ther., 21,* 70 (1977).
177a. M. Israel, W. J. Pegg, P. M. Wilkinson, and M. B. Garnick, Liquid Chromatography Symposium I, Oct. 13–14, Waters Associates, Boston, Massachusetts (1977).
177b. J. J. Langone, H. Van Vunakis, and N. R. Bachur, *Biochem. Med., 12,* 283 (1975).
178. R. F. Adams, G. J. Schmidt, and F. L. Vandemark, *Clin. Chem., 23,* 1226 (1977).
179. W. C. Butts, V. A. Raisys, M. A. Keeny, and C. W. Bierman, *J. Lab. Clin. Med., 84,* 451 (1974).
180. Z. K. Shihabi and R. P. Dave, *Clin. Chem., 23,* 942 (1977).
181. T. J. Butler, *Clin. Chem., 20,* 868 (1974).
182. E. B. Solow, J. M. Metaxas, and T. R. Summers, *J. Chromatogr. Sci., 12,* 256 (1974).
183. D. M. Woodbury, J. K. Penry, and R. P. Schmidt, eds., *Antiepileptic Drugs,* Raven Press, New York (1972).
184. T. J. X. Mee and J. A. Smith, in: *High Pressure Liquid Chromatography in Clinical Chemistry* (P. F. Dixon, C. H. Gray, C. K. Lim, and M. S. Stoll, eds.), p. 119, Academic Press, New York (1976).
185. F. K. Trefz, D. J. Byrd, and W. Kochen, *J. Chromatogr., 107,* 181 (1975).
185a. J. H. M. Van Den Berg, Ch. R. Mol, R. S. Deelder, and J. H. H. Thijssen, *Clin. Chim. Acta, 78,* 165 (1977).

186. R. Menard, personal communication.
187. Sj. van der Wal and J. F. K. Huber, *J. Chromatogr., 135,* 305 (1977).
188. Sj. van der Wal and J. F. K. Huber, private communication.
189. M. Lafosse, G. Keravis, and M. H. Durand, *J. Chromatogr., 118,* 283 (1976).
190. R. J. Dolphin and P. J. Pergande, *J. Chromatogr., 143,* 267 (1977).
191. V. Fantyl, in: *High Pressure Liquid Chromatography in Clinical Chemistry* (C. K. Lim and C. H. Gray, eds.), p. 51, Academic Press, New York (1976).
192. G. Schwedt, H. H. Bussemas, and C. L. Lippman, *J. Chromatogr., 143,* 259 (1977).
193. K. Horn, I. Marschner, and P. C. Scriba, *J. Clin. Chem. Clin. Biochem., 14,* 353 (1976).
194. B. L. Karger, S. C. Su, S. Marchese, and B. A. Persson, *J. Chromatogr. Sci., 12,* 678 (1974).
195. J. F. Kachmar and D. W. Moss, in: *Fundamentals of Clinical Chemistry,* (N. W. Tietz, ed.), Chapter 12, W. B. Saunders, Philadelphia (1976).
196. D. W. Mercer, S. P. Peters, R. H. Glew, R. E. Lee, and D. M. Wenger, *Clin. Chem., 23,* 631 (1977).
197. M. E. Legaz and M. A. Kenny, *Clin. Chem., 22,* 57 (1976).
198. O. Hetland, R. R. Andersson, and T. Gerney, *Clin. Chim. Acta, 62,* 425 (1975).
199. D. W. Mercer, *Clin. Chem., 23,* 611 (1977).
200. R. S. Galen, *Clin. Chem., 22,* 120 (1976).
201. E. Jockers-Wretou, *Clin. Chim. Acta, 73,* 183 (1976).
202. W. G. Yasmineh, R. B. Pyle, N. Q. Hanson, and B. K. Hultman, *Clin. Chem., 22,* 63 (1976).
203. W. H. Lederer and H. L. Gerstbrein, *Clin. Chem., 22,* 1748 (1976).
204. D. W. Mercer and M. A. Varat, *Clin. Chem., 21,* 1088 (1975).
205. J. Griffiths and G. Handschuh, *Clin. Chem., 23,* 567 (1977).
206. S. H. Chang, K. M. Gooding, and F. E. Regnier, *J. Chromatogr., 125,* 103 (1976).
207. F. E. Regnier, K. M. Gooding, and S. H. Chang, in: *Contemporary Topics in Analytical and Clinical Chemistry* (D. Hercules, G. Hieftje, L. Snyder, and M. Evenson, eds.), Vol. 1, p. 1, Plenum Press, New York (1977).
208. R. R. Schroeder, P. J. Kudirka, and E. C. Toren, Jr., *J. Chromatogr., 134,* 83 (1977).
209. T. D. Schlabach, S. H. Chang, K. M. Gooding, and F. E. Regnier, *J. Chromatogr., 134,* 91 (1977).
210. I. Duncan and P. Culbreth, *Clin. Chem., 23,* 1138 (1977).
211. W. H. Coch, G. Kessler, and J. S. Meyer, *Clin. Chem., 20,* 1368 (1974).
212. F. B. Jungawala, J. E. Evans, and R. H. McCluer, *Biochem. J., 155,* 55 (1976).
213. L. Gluck, M. Kulovich, R. C. Borer, Jr., P. H. Brenner, G. A. Anderson, and W. Spellacy, *Am. J. Obstet. Gynecol., 109,* 440 (1971).
214. L. Gluck, *Hosp. Pract.,* 45 (November 1971).
215. L. M. Demers and G. Hepner, *Clin. Chem., 22,* 602 (1976).
216. I. Makino, S. Nakagawa, and K. Maskino, *Gastroenterology, 56,* 1033 (1969).
217. N. A. Parris, *J. Chromatogr., 133,* 273–279 (1977).
218. R. Shaw and W. H. Elliott, *Anal. Biochem., 74,* 273–281 (1976).
219. C. A. Bloch and J. B. Watkins, personal communication.
220. R. H. Mcluer and F. B. Jungawala, in: *Current Trends in Sphingolipidoses and Allied Disorders* (B. W. Volk and L. Schneck, eds.), p. 533, Plenum Press, New York (1976).
221. K. Aitzetmuller, *J. Chromatogr., 113,* 231 (1975).
221a. E. G. Perkins, J. C. Means, and M. F. Picciano, *Rev. Fr. Corps Gras., 24,* 73 (1977); *CA 86,* 167226r (1977).
222. A. R. Battersby, D. G. Buckley, G. L. Hodgson, R. E. Markwell, and E. McDonald, in: *High Pressure Liquid Chromatography in Clinical Chemistry* (P. F. Dixon, C. H. Gray, C. K. Lim, and M. S. Stoll, eds.), p. 63, Academic Press, New York.

223. N. Evans, A. H. Jackson, S. A. Matlin, and R. Towill, *J. Chromatogr., 125,* 343 (1976).
224. R. F. Adams, W. Slavin, and A. R. Williams, *Chromatogr. Newsl. 4,* 24 (1976).
225. W. Slavin, A. T. R. Williams, and R. F. Adams, *J. Chromatogr., 134* 121 (1977).
226. C. H. Gray, C. K. Lim, and D. C. Nicholson, *Clin. Chim. Acta 77,* 167 (1977).
226a. Z. J. Petryka and C. A. Pierach, Liquid Chromatography Symposium I, Oct. 13–14, Waters Associates, Boston, Massachusetts (1977).
227. M. G. M. DeRuyter and A. P. DeLeenheer, *Clin. Chem., 22,* 1593 (1976).
227a. K. Abe, K. Ishibashi, M. Ohmae, K. Kawabe, and G. Katsui, *Vitamins, 51,* 275 (1977).
228. L. S. Reed and M. C. Archer, *J. Chromatogr., 121,* 100 (1976).
229. R. C. Williams, D. R. Baker, and J. A. Schmit, *J. Chromatogr. Sci., 11,* 618 (1973).
230. Perkin-Elmer, *Separation Abstracts,* Number 4 (April 1977).
231. H. F. Deluca, Vitamin D endocrine system, in: *Advances in Clinical Chemistry* (O. Bodansky and A. L. Catner, eds.), Vol. 19, p. 163, Academic Press, New York (1977).
231a. J. A. Eisman, R. M. Shepard, and H. F. Deluca, *Anal. Biochem. 80,* 298 (1977).
232. T. J. Gilbertson and R. P. Stryd, *Clin. Chem., 23,* 1139 (1977).
232a. K. T. Koshy and A. L. Van Der Silk, *Anal. Lett., 10,* 523 (1977).
232b. P. W. Lambert, E. A. Lindmark, and T. C. Spelsberg, Liquid Chromatography Symposium I, Oct. 13–14, Waters Associates, Boston, Massachusetts (1977).
233. K. Abe and G. Katsui, *Vitamins, 49,* 259 (1975) (in Japanese).
234. L. W. Bond and R. B. McComb, *Clin. Chem., 21,* 1016 (1976).
235. J. C. Liao, N. E. Hoffman, J. J. Barboriak, and D. A. Roth, *Clin. Chem., 23,* 802 (1977).
235a. K. Gossler, K. Schaller, and W. Zschiesche, *Clin. Chim. Acta, 78,* 91 (1977).
236. A. Kouichi, I. Kyoko, O. Masahiko, K. Kiyoshi, and K. Goichiro, *Vitamins, 51,* 111, 119 (1977) (in Japanese); *Chem. Abstr. 86,* 239 (1977).
237. T. P. Waalkes, C. W. Gehrke, and W. A. Bleyer, *Cancer Chemother. Rep., 59,* 721 (1975).
238. T. P. Waalkes, C. W. Gehrke, R. W. Zumwalt, S. Y. Chang, D. B. Lakings, D. C. Tormey, D. L. Ahmann, and C. G. Moertel, *Cancer, 36,* 392 (1975).
239. G. E. Davis, R. D. Suits, K. C. Kuo, G. W. Gehrke, T. P. Waalkes, and E. Borek, *Clin. Chem., 23,* 1427 (1977); Liquid Chromatography Symposium I, Oct. 13–14, Waters Associates, Boston, Massachusetts (1977).
240. P. R. Brown, R. A. Hartwick, and A. M. Krstulovic, Liquid Chromatography Symposium I, Oct. 13–14, Waters Associates, Boston, Massachusetts (1977).
241. R. A. Hartwick, and P. R. Brown, *J. Chromatogr., 126,* 679 (1976).
242. G. E. Davis, R. D. Suits, K. C. Kuo, C. W. Gehrke, T. P. Waalkes, and E. Borek, *Clin. Chem., 23,* 1427 (1976).
243. S. Moulton, *Med. Lab. World 1,* 27 (1977).
244. P. H. Culbreth, I. W. Duncan, and C. A. Burtis, *Clin. Chem., 23,* 2288 (1977).
245. F. Geeraerts, L. Schimpfessel, and R. Crokaert, *J. Chromatogr., 145,* 63 (1978).
246. P. W. Banda, A. E. Shemy, and M. S. Biois, *Clin. Chem., 23,* 1397 (1977).
247. G. Schwedt and H. H. Bussemas, *Z. Anal. Chem., 285,* 381 (1977).
248. G. Schwedt, *Anal. Chim. Acta, 92,* 337 (1977).
249. B. M. Eriksson, I. Andersson, K. L. Borg, and B. A. Persson, *Acta Pharm. Suec., 14,* 451 (1977).
250. S. Thelma, C. N. Corder, R. H. McDonald, Jr., and J. A. Feldman, *J. Lab. Clin. Med., 90,* 604 (1977).
251. F. Engbaek and I. Magnussen, *Clin. Chem., 24,* 376 (1978).
252. K. Mori, *Sangyo Igaku, 18,* 528 (1976).

252a. E. Grushka, E. J. Kikta, Jr., and E. N. Naylor, *J. Chromatog., 143,* 51 (1977).
252b. A. M. Krstulovic, P. R. Brown, D. M. Rosie, and P. B. Champlin, *Clin. Chem., 23,* 1984 (1977).
252c. R. E. Majors, *Liquid Chromatography at Work,* No. 52, Varian, Palo Alto, California (1977).
252d. R. W. A. Oliver, *Perkin-Elmer Anal. News, 7,* 8 (1977).
253. T. Hayashi, T. Sugiura, S. Kawai, and T. Ohno, *J. Chromatogr., 145,* 141 (1978).
254. Y. Soejima and S. Otsuji, *Igaku No. Ayumi 103,* 31 (1977); *CA 88,* 2717e (1978).
255. E. Roeder, I. Pigulla, and J. Troschutz, *Fresenius Z. Anal. Chem., 288,* 56 (1977).
256. N. Seiler, B. Knodgen, and F. Eisenbeiss, *J. Chromatogr., 145,* 29 (1978).
257. E. I. Johnson, *Liquid Chromatography at Work,* No. 54, Varian, Palo Alto, California, Nov. 21 (1977).
258. M. M. Abdel-Monem, in: *Advances in Chromatography,* Vol. 16 (J. C. Giddings, J. Cazes, E. Grushka, and P. R. Brown, eds.), p. 249, Marcel Dekker, New York (1978).
258a. N. Seiler, *Clin. Chem., 23,* 1519 (1977).
259. W. L. Chiou, M. A. F. Gadalla, and G. W. Peng, *J. Pharm. Sci., 67,* 182 (1978).
260. W. L. Chiou, G. W. Peng, M. A. F. Gadalla, and S. T. Nerenberg, *J. Pharm. Sci., 67,* 292 (1978).
261. B. B. Wheals and I. Jane, *The Analyst, 102,* 625 (1977).
262. C. E. Pippenger, J. K. Perry, and H. Kutt, *Antiepileptic Drugs: Quantitative Analysis and Interpretation,* Raven Press, New York (1978).
263. K. W. Leal and A. S. Troupin, *Clin. Chem., 23,* 1964 (1977).
264. S. J. Soldin and J. G. Hill, *Clin. Chem., 23,* 2352 (1977).
265. G. W. Mihaly, J. Phillips, W. J. Louis, and F. J. Vajada, *Clin. Chem., 23,* 2283 (1977).
266. D. C. Drummond, *Clin. Chem., 23,* 2172 (1977).
266a. R. F. Adams and G. Schmidt, *Chromatogr. Newsl., 4,* 8 (1976).
267. G. Schmidt, F. Vandemark, and R. Adams, *Chromatogr. Newsl., 4,* 32 (1976).
268. E. Gaetani, C. F. Laureri, and G. Vaona, *Liquid Chromatography at Work,* No. 51, Varian, Palo Alto, California (1977).
269. J. L. Powers and W. Sadee, *Clin. Chem., 24,* 299 (1978).
270. R. E. Kates, D. W. McKennon, and T. J. Comstock, *J. Pharm. Sci., 67,* 269 (1978).
271. W. G. Crouthamel, B. Kowarski, and P. K. Narang, *Clin. Chem., 23,* 2030 (1977).
272. D. E. Drayer, K. Restivo, and M. M. Reidenberg, *J. Lab. Clin. Med., 90,* 816 (1977).
272a. D. E. Drayer, K. Restivo, and M. M. Reidenberg, *J. Lab. Clin. Med., 90,* 816 (1977).
272b. P. J. Meffin, S. R. Harapat, and D. C. Harrison, *J. Pharm. Sci., 66,* 583 (1977).
273. R. K. Desiraju, R. L. Mayock, and E. T. Sugita, *J. Chromatogr. Sci., 15,* 563 (1977).
274. W. J. Jusko and A. Poliszczuk, *Am. J. Hosp. Pharm., 33,* 1193 (1976).
275. G. Peng, M. A. F. Gadalla, and W. L. Chiou, *Clin. Chem., 24,* 357 (1978).
276. J. V. Aranda, D. S. Sitar, W. D. Parsons, P. M. Loughnam, and A. H. Neims, *N. Engl. J. Med., 295,* 413 (1976).
277. S. J. Soldin and J. G. Hill, *Clin. Biochem., 10,* 74 (1977).
278. K. K. Midha, S. Sved, R. D. Hossie and I. J. McGilveray, *Biomed. Mass. Spectrom., 4,* 172 (1977).
279. I. Nilsson-Ehle, *Acta Pathol. Microbiol. Scand. Suppl., 259,* 61 (1977).
280. M. A. Carrol, E. R. White, Z. Jancsik, and J. E. Zarembo, *J. Antibiot., 30,* 397 (1977).
281. J. S. Wold and S. A. Turnipseed, *Clin. Chim. Acta, 78,* 203 (1977).
282. G. W. Peng, M. A. F. Gadalla, and W. L. Chiou, *Res. Commun. Chem. Pathol. Pharmacol., 18,* 233 (1977).
283. G. W. Peng, M. A. F. Gadalla, A. Peng, V. Smith, and W. L. Chiou, *Clin. Chem., 23,* 1838 (1977).

284. S. K. Maitra, T. T. Yoshikawa, J. L. Hansen, I. Nilsson-Ehle, W. J. Palin, M. C. Schotz, and L. B. Guze, *Clin. Chem., 23,* 2275 (1977).
284a. J. P. Anhalt, *Antimicrob. Agents Chemother. 11,* 651 (1977).
284b. J. P. Sharma, G. D. Koritz, E. G. Perkins, and R. F. Bevill, *J. Pharm. Sci., 66,* 1319 (1977).
284c. T. J. Goehl, L. K. Mathur, J. D. Strum, J. M. Jafte, W. H. Pitlick, V. P. Shah, R. I. Poust, and J. L. Colaizzi, *J. Pharm. Sci., 67,* 404 (1978).
285. T. G. Rosano, C. A. Brito, and J. M. Meola, *Chromatogr. Newsl., 6,* 1 (1978).
286. R. Hulboven, J. P. Desager, G. Sokal, and C. Harvengt, *Arch. Int. Pharmacodyn. Ther., 226,* 344 (1977).
287. J. C. Gfeller, G. Frey, and R. W. Frei, *J. Chromatogr., 142,* 271 (1977).
288. J. A. Nelson, B. A. Harris, W. J. Decker, and D. Farquhar, *Cancer Res., 37,* 3970 (1977).
289. S. H. Preskorn and J. T. Biggs, *N. Engl. J. Med., 298,* 166 (1978).
290. F. L. Vandemark, R. F. Adams, and G. J. Schmidt, *Clin. Chem., 24,* 87 (1978).
291. J. H. M. van den Berg, *Improvement of Selectivity in Column Liquid Chromatography,* p. 48, J. H. M. van den Berg, The Netherlands (1978).
292. N. A. Klitgaard, *Arch. Pharm. Chemi, Sci. Ed., 5,* 61 (1977).
293. C. H. Doolittle, C. J. McDonald, and P. Calabresi, *J. Lab. Clin. Med., 90,* 773 (1977).
294. I. Nilsson-Ehle and P. Nilsson-Ehle, *Clin. Chem., 24,* 365 (1978).
295. T. B. Vree, B. Lenselink, E. van der Kleijn, and G. M. Nijhuis, *J. Chromatogr., 143,* 530 (1977).
296. M. B. Aufrere, B. A. Hoener, and M. E. Vore, *Clin. Chem., 23,* 2207 (1977).
297. R. Saetre and D. L. Rabenstein, *Anal. Chem., 50,* 276 (1977).
298. F. deBros and E. M. Wolshin, *Anal. Chem., 50,* 521 (1978).
299. R. E. Hill and J. Crechiolo, *J. Chromatogr., 145,* 165 (1978).
300. I. Bekersky, H. G. Boxenbaum, M. H. Whitson, C. V. Puglisi, R. Pocelinko, and S. A. Kaplan, *Anal. Lett., 10,* 539 (1977).
301. C. P. Terweij-Groen, T. Vahlkamp, and J. C. Kraak, *J. Chromatogr., 145,* 115 (1978).
302. P. G. Satyaswaroop, E. Lopez de la Osa, and E. Gurpide, *Steroids, 30,* 139 (1977).
303. D. C. Garg, J. W. Ayres, and J. G. Wagner, *Res. Commun. Chem. Pathol. Pharmacol., 18,* 137 (1977).
304. G. Keravis, M. Lafosse, and M. H. Durand, *Chromatographia, 10,* 678 (1977).
305. C. P. DeVries, C. Popp-Snijders, W. DeKieviet, and A. C. Akkerman-Faber, *J. Chromatogr., 143,* 624 (1977).
306. J. H. M. van den Berg, *Improvement of Selectivity in Column Liquid Chromatography,* p. 29, J. H. M. van den Berg, The Netherlands (1978).
307. F. A. Fitzpatrick, in: *Advances in Chromatography,* Vol. 16 (J. C. Giddings, J. Cazes, E. Grushka, and P. R. Brown, eds.), p. 37, Marcel Dekker, New York (1978).
308. A. Suzuki, S. Handa, and T. Yamakawa, *J. Biochem., 82,* 1185 (1977).
309. W. M. A. Hax and W. S. M. Geurts van Kessel, *J. Chromatogr., 142,* 735 (1977).
310. B. Shaikh, N. J. Pontzer, J. E. Molina and M. I. Kelsey, *Anal. Biochem., 85,* 47 (1978).
311. Z. J. Petryka and C. J. Watson, *Anal. Biochem., 84,* 173 (1978).
312. P. W. Lambert, B. J. Svverson, C. D. Arnaud, and T. C. Spelsberg, *J. Steroid Biochem., 8,* 929 (1977).
313. G. Jones, *Clin. Chem., 24,* 287 (1978).
314. R. A. Wiggins, *Chem. Ind., 20,* 841 (1977).
315. P. C. Schaefer and R. S. Goldsmith, *J. Lab. Clin. Med., 91,* 104 (1978).
316. K. Ishibashi, K. Abe, M. Ohmae, K. Kawabe, and G. Katsui, *Vitamins, 51,* 415 (1977); *CA 87*, 196608t (1977).

317. B. Nilsson, B. Johansson, L. Jansson and L. Holmberg, *J. Chromatogr., 145,* 169 (1978).
318. K. Abe, K. Ishibashi, M. Ohmae, K. Kawabe and G. Katsui, *Vitamins, 51,* 275 (1977); *CA 87,* 98289v (1977).
319. B. Shaikh, S.-K.S. Huang, N. J. Pontzer, and W. L. Zielinski, Jr., *J. Liq. Chrom., 1,* 75 (1977).
320. J. F. Kuttesch, F. C. Schmalstreg, and J. A. Nelson, *J. Liq. Chrom., 1,* 97 (1977).
321. A. Bye and M. E. Brown, *J. Chromatogr. Sci., 15,* 365 (1977).
322. L. D'Souza and H. I. Glueck, *Thromb. Haemostasis., 38,* 990 (1977).
323. W. Plunkett, in: *Advances in Chromatography,* Vol. 16 (J. C. Giddings, J. Cazes, E. Grushka, and P. R. Brown, eds.), p. 211, Marcel Dekker, New York (1978).

Index